A Writer's Handbook
for ENGINEERS

David A. McMurrey

Joanne Buckley

Australia   Canada   Mexico   Singapore   Spain   United Kingdom   United States

THOMSON

A *Writer's Handbook for Engineers*
by David A. McMurrey and Joanne Buckley

Publisher: Chris Carson	**Proofreader:** Martha McMaster	**Cover Design:** Andrew Adams
Developmental Editor: Hilda Gowans	**Indexer:** Shelly Gerger-Knechtl	**Compositor:** Integra
Permissions Coordinator: Melody Tolson	**Production Manager:** Renate McCloy	**Printer:** Transcontinental
Production Services: RPK Editorial Services, Inc.	**Creative Director:** Angela Cluer	**Cover Image Credit:** Vladimir Mucibabic Shutterstock; Lisa F. Young/Shutterstock; Franc Podgorsek/ Shutterstock
Copy Editor: Shelly Gerger-Knechtl	**Interior Design:** Katherine Strain	

North America
Thomson Learning
1120 Birchmount Road
Toronto, Ontario MIK 5G4
Canada

Asia
Thomson Learning
5 Shenton Way #01-01
UIC Building
Singapore 068808

Australia/New Zealand
Thomson Learning
102 Dodds Street
Southbank, Victoria
Australia 3006

Europe/Middle East/Africa
Thomson Learning
High Holborn House
50/51 Bedford Row
London WCIR 4LR
United Kingdom

Latin America
Thomson Learning
Seneca, 53
Colonia Polanco
11560 Mexico D.F.
Mexico

Spain
Paraninfo
Calle/Magallanes, 25
28015 Madrid, Spain

BRIEF CONTENTS

CONTENTS

CONTENTS

CONTENTS

CONTENTS

CONTENTS

PREFACE

This handbook is for people in engineering programs in colleges and universities as well as for practicing engineers who must develop writing projects such as those listed in the following sections of this preface. You'll find a wide range of writing projects and writing-project development tools in this handbook. You'll also find examples from a wide range of engineering fields and disciplines.

How Writing and Engineering Relate and Connect

In their article "Beginnings and Endings: Keys to Better Engineering Technical Writing," Marcia Martens Pierson and Bion L. Pierson[1] cite a survey that found that practicing aerospace engineers and scientists spend on average between 19.6 and 23.3 hours per week writing documents. Even so, they point out that "engineers as a group dread writing." That's unfortunate because, as the examples included in this handbook show, the field of engineering possesses a wonderful, fascinating array of things to write about. And to the question posed in the title of her article "Can Engineers Write?" Joan Knapp answers with a resounding "yes."[2]

What This Handbook Covers: Writing Projects

This handbook covers numerous types of common writing projects that you are likely to confront in your career as an engineering student and as a practicing engineer. Your writing projects are likely to go by a confusing variety of names. Skim Chapter 11, "Engineering Documents," to get a sense of the purpose and contents of the documents described there. Then make sure you understand your audience's needs and requirements; you should be able to find the type

---

[1] Marcia Martens Pierson and Bion L. Pierson, "Beginnings and Endings: Keys to Better Engineering Technical Writing," *IEEE Transactions in Professional Communications*, vol. PC-40, no. 4, pp. 299–304, December 1997.
[2] Joan Knapp, "Can Engineers Write?" *IEEE Transactions in Professional Communications*, vol. PC-33, no. 1, pp. 12–18, March 1990.

of writing project you need in Chapter 11. Here's a list of the documents covered in this handbook:

accident reports	adjustment messages	application letters	claim messages
complaint messages	consumer instructions	cover letters	design specifications
engineering instructions	evaluation reports	feasibility reports	field reports
handbooks	inspection reports	investigation reports	lab reports
literature reviews	manufacturing specifications	memoranda	operating specifications
oral presentations	policies and procedures	progress reports	proposals
recommendation reports	résumés	survey reports	trip reports

What This Handbook Covers: Project-Development Tools

To support you in these document-development efforts, you'll also find a useful variety of tools to plan, develop, format, and finalize your engineering writing projects:

alert notices	APA documentation	audience analysis	brainstorming
bulleted, numbered lists	business-letter format	cause–effect analysis	classification
coherence revision	comparison	contrast	copyedit-level revision
CSE documentation	description	division	document-design revision
document-level revision	e-mail strategies	equations	format revision

graphs and charts	headings	highlighting	IEEE documentation
illustrations	information gathering	library research	memo format
MLA documentation	narration	nominalization revision	online research
organization revision	outlining	passive-voice revision	process analysis
proofreading	tables	technical material	topic narrowing
web pages, sites	wordiness problems	translation	

This handbook also covers basic mechanics topics such as the following:

abbreviations	agreement errors	apostrophes	capitalization
clauses, phrases	colons	comma splices	commas
fragments	hyphens	modifier problems	numbers vs. words
parallelism	parts of sentences	parts of speech	pronoun errors
semicolons	spelling	tense, pronoun shifts	

How to Use This Handbook

Obviously, this handbook covers a lot of ground. It's not something you want to read from cover to cover some rainy weekend. Here are some ideas on how to use this handbook:

- *Writing-process overview.* If you're not sure how to go about a writing project, read Chapter 1, "Writing Process." That will give you a game plan.

- *Specific writing projects.* If you have a specific writing project to do—for example, a proposal, a progress report, a set of instructions—find it in Chapter 11, "Engineering Documents." If you don't find the exact name of the writing project, skim Chapter 11; you're likely to find it under a different name. (Names for these documents are not standard.)
- *Research: gathering information.* For your writing project, you probably need to gather information. Read Chapter 10, "Research Process," for strategies on finding information in the library and on the Internet.
- *Special formatting.* As you draft an engineering document, you're likely to need headings, bulleted and numbered lists, and other such formatting. See Chapter 2, "Professional Document Design."
- *Special-formatting overview.* If you've not used headings, lists, notices, highlighting, tables, graphs, charts, and illustrations in your writing projects before, it's time to start! Read Chapter 2, "Professional Document Design" carefully.
- *Documenting sources of borrowed information.* Either as you gather information for a writing project or as you write the rough draft, you may need to know how to document the sources of your borrowed information. See Chapter 12, "Documentation."
- *Checking up on punctuation rules.* As you write the rough draft, maybe you can't remember some of the rules for commas, semicolons, colons, apostrophes, or hyphens. See specific sections in Chapter 5, "Punctuation."
- *Revision and proofreading strategies:* You've completed the rough draft, and now it's time to finalize it. See the final two sections of Chapter 1, "Writing Process."
- *Responding to review comments.* You send your draft engineering document out for review to colleagues, editors, and maybe an instructor. The reviews come back with comments about "comma splices," "dangling modifiers," "pronoun–antecedent agreement," and other things you're not sure about. See Chapter 4, "Grammatical Sentences."
- *Interpreting reviewer comments.* Some of the reviewer comments on your draft document may use such terms as "subordination," "antecedent," "modifier," "conjunction," "gerund," "subject complement," "linking verb," and "objective case." Look these up in Chapter 3, "Basic Grammar."

What This Handbook Covers

As you can see from the previous sections, this handbook covers a wide range of writing-related topics. Here are some examples; for full coverage, see the Index.

YOU NEED TO:	SEE THIS CHAPTER:
Analyze an audience	Chapter 1
Create an outline	Chapter 1
Create navigation tools for a web site	Chapter 2
Design a web page	Chapter 2
Find information on the Internet	Chapter 10
Find out about the different alert notices	Chapter 2
Find out how to punctuate a subordinate clause	Chapter 5
Find out how to search for journal articles	Chapter 10
Find out what an appositive is	Chapter 3
Find out what a comma splice is and how to fix it	Chapter 4
Fix a parallelism problem	Chapter 6
Format graphs and charts correctly	Chapter 2
Format headings and lists correctly	Chapter 2
Format tables correctly	Chapter 2
Get help on using articles	Chapter 8
How to design a résumé	Chapter 13
How to write a progress report	Chapter 11
How to write an application letter for a job	Chapter 13
Prepare and deliver an oral report	Chapter 13
Review how to avoid sexist language	Chapter 7
Review the rules for colons and semicolons	Chapter 5
Review the rules on apostrophes	Chapter 5

What Examples Are Used

This handbook contains plenty of examples from engineering fields such as the following:

aerospace engineering	agricultural engineering	architectural engineering	biochemical engineering
bioengineering	biomedical engineering	chemical engineering	civil engineering
computer engineering	construction engineering	electrical engineering	environmental engineering
industrial engineering	materials engineering	mechanical engineering	mechanical engineering
nuclear engineering	petroleum engineering	software engineering	structural engineering

The world of engineering is loaded with fascinating subject matter to write about. True, there are also topics like corrosion and prestressed concrete, but consider these topics, all of which are used in the examples in this handbook:

Albert Einstein	Alessandro Volta	Apollo 13	architectural daylighting
artificial intelligence	Babbage's calculating machines	Bhopal nuclear accident	bridge design
Charles Kuen Kao	Charles Townes	cleanrooms	coal gasification
computer face-recognition systems	corrosion	DARPA Grand Challenge	Deep Blue
earthquake damage	ENIAC computer	fracture mechanics	fuzzy logic
Gamma Quadrant	Global Positioning System satellites	Grace Hopper	green roofs
Heinrich Hertz	high-occupancy vehicle lanes	Hubbert peak oil theory	hydrogen-powered vehicles
Isaac Newton	Jack Kilby	James Clerk Maxwell	Kyoto Protocol
lasers	long-term sequestration of CO_2	Lotfi A. Zadeh	low-cost carbon fibers
maglevs	Marie and Pierre Curie	Mark I and II computers	Mars Global Surveyor
Mars Rover Vehicle	Mars surface	Michael Faraday	nanotechnology
Nikola Tesla	offshore wind turbines	outsourcing	photovoltaic cells
Pluto	refrigeration technology	Robert Royce	RoboCup
rotating architectural design	semiconductors	Seymour Cray	solar heating system

space elevators	SQUID	Theodore Maiman	Thomas Alva Edison
torque	transistors	ultra-wideband	vacuum tube
Vannevar Bush	William Shockley	Willis Haviland Carrier	wind-turbine design
wireless communications	World Trade Center		

Acknowledgments

A special thanks must also go to Michael Laine and Nyein Aung of the Liftport Group for going the extra mile for this book. They provided us with an illustration that reflects the vision and enthusiasm they both feel for their compelling project.

Many thanks to Chris Carson for getting me involved in this project; Hilda Gowans, for shepherding me through it; Gerald E. Wheeler, PE, for his insightful technical reviews of these chapters; Rose Kernan, for her expert handling of the production phase; and finally to Phoebe, Patrick, and Jane for giving me, shall we say, incentives to write this handbook.

1

WRITING PROCESS

1

Writing Process

1   Writing Process

Writing an engineering document is a project like any other project you may have been involved in. Having a reliable process, one that you follow systematically, is the key. The writing process begins with planning; planning before you do any research or any writing is essential. The documentation specialist, JoAnne Hackos, claims that planning should take 60 percent of the entire amount of time spent developing a document.[1] The writing process also includes writing (obviously), review, testing (in some cases), revision, and proofreading.

1.1 PLANNING

When you plan a writing project, spend the time necessary on each of these phases:

- Analyze the writing project—what is needed, what is expected.
- Analyze the audience—its needs, expectations, limitations. Interview members of your audience if possible.
- Be able to state the specific topic and purpose of the document.
- Identify the type of document that you must write (for example, proposal, progress report, or recommendation report), determine specific requirements, and find some documents that are generally accepted as good models.

---

[1] JoAnne Hackos, *Managing Your Documentation Projects*. New York: Wiley, 1994, p. 278

- Brainstorm the writing project, identify useful writing structures (for example, description, process, definition) to use in the document, and develop a tentative outline.
- Find information resources for your writing project.

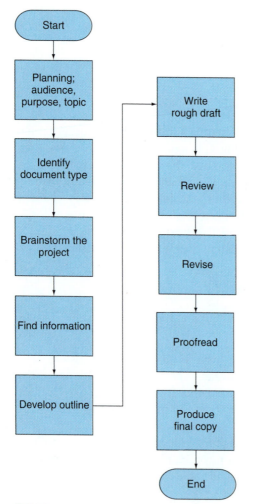

FIGURE 1.1
Overview of the writing process. Many of these phases overlap and return to previous phases, as this chapter will explain.

Some of these activities overlap or must be repeated. As you research your project, your outline is likely to change. As you proceed through the project, your sense of its purpose and topic may evolve.

1.1.1 ANALYZING THE PROJECT, AUDIENCE, TOPIC, AND PURPOSE

A writing project is made up of a situation that requires a document to be written, a reader or readers who need to read the document, the topic or focus of the document, the purpose of that document in terms of both its readers and its writer. You must analyze these elements to be sure that you understand them thoroughly and that you are not making any bad assumptions.

SITUATION

The **situation** of a writing project is that combination of people, organizations, and requirements that bring about the need for the document. Here are some examples:

- Your engineering firm sees an opportunity to land a lucrative development project but must first write a proposal and submit it to the potential client.
- Your software development company has produced a new application; now someone must write the user guide and other documents to support that project.
- Your engineering firm has been awarded a contract by the city to study the feasibility of wireless communications in major city buildings. The results of that feasibility study must be summarized in a report, submitted to the city.
- As an engineer for a state agency, you've been asked to investigate damage to several bridges after recent area flooding. You will summarize your findings in a report.

AUDIENCE (CLIENTS)

If you've taken writing courses or seminars in the past, you've probably heard the standard wisdom, "write for your audience," more than you care to remember. It is such an obvious, simplistic statement that it is hard to take it seriously. However, not attending to the audience's needs is the primary reason documents fail—especially technical ones.

To gain a fresh perspective on this important issue, think of your audience as your client, yourself as a supplier, and your writing project as a deliverable. Clients can be both external and internal.

An external client pays you for your expertise in recommending a technical solution to a problem. Your deliverables may be twofold: the solution and, most likely, the document that describes that solution. This scenario applies to organizations internally as well. If your job is to test a product before it is released to the marketplace, your deliverables are likely to be the test results and the report that summarizes those results.

When you think of the audience of your document as a client, things change: external clients supply you with business and income; satisfied clients return and help keep you in business. Satisfied internal clients enable you to keep your job and earn promotions and raises—not to mention supporting and advancing the organization as a whole.

To satisfy the clients of your engineering documents, be mindful of these issues and write your documents accordingly:

- Why do they need this document: recommendations, instructions, background, specifications?
- What response or action do you hope to generate in your clients: correctly performing a procedure, hiring you as the consultant, following your recommendations, taking some important matter seriously?
- How will your clients use the document: following your recommendations, using your instructions, considering hiring you as a consultant?
- What are their primary interests, their areas of expertise: technical, theoretical, financial, legal, administrative?
- What are your clients' limitations? What areas of your engineering document are they less likely to understand?

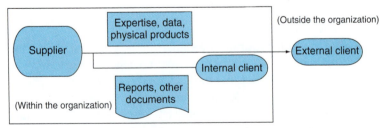

FIGURE 1.2
Suppliers, deliverables, and internal and external clients. Engineering documents are deliverables too.

If you write in an academic setting, obviously your instructors are your clients. Check with them to find out how they want to address the client–supplier–deliverable connections.

TOPIC NARROWING

Your engineering writing project focuses on a specific topic—or should. Be careful to identify your exact, specific topic and to narrow it as much as you can; otherwise, you may be committing yourself to writing a book. Consider the example topic, `wireless communications`; that in itself is a topic for not just a book but a multivolume work. The topic `wireless technology` doesn't work any better. After all, these two topics promise everything under the sun: the history of the development of this technology, the major milestones, the major researchers and developers, and much more. What we're really interested in is `wireless computer networking`. How about `wireless devices: design and operation`? Better, but we're still missing the target: specific clients and specific situation. How about `components of a wireless-communication system`? Good, but our clients need more than just that. `Equipment and costs`—yes, that too. `Design for the designated city area and buildings`—yes. `Range and effectiveness`—yes. This narrowing process is diagrammed in Figure 1.3.

As you can see, the **narrowing** process zeroes in on an increasingly specific topic, tossing out unnecessary topics and collecting necessary topics along the way. And all of it is driven by your understanding of the project—its audience, purpose, and situation.

PURPOSE

Engineering documents seek to fulfill a single purpose or a combination of two major purposes: to inform or persuade:

- *Inform*. The informative purpose is deceptively simple—in fact, simplistic. To describe the purpose of a report as "to inform anyone interested in hydrogen vehicles" is to say nearly nothing at all. Reporting performance data on three top-ranking hydrogen-powered vehicles—without any effort to summarize, compare, conclude, explain, evaluate, recommend, or persuade—would indeed be an instance of the simple informative purpose. But your clients are not likely to be satisfied with just a simple rendition of

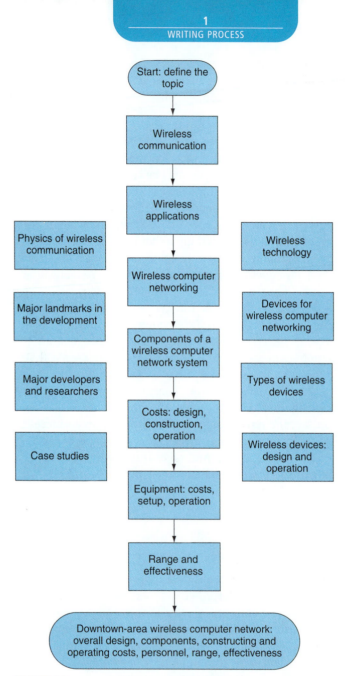

FIGURE 1.3
Narrowing process. The unattached side topics are peripheral to the focus of the project.

data; they want you to apply your expertise to that data using purposes such as the following.

- *Explain, conclude, interpret.* When you go beyond the simple presentation of data, you explain or interpret that data. Conclusions work the same way.
- *Summarize.* Summaries are condensed versions of larger presentations of data. For example, a literature review is a summary of research on a topic. Your expertise is useful in this case because you can sift through the data and see the major developments and trends.
- *Instruct.* When you provide instructions on how to do something, you are writing informatively. However, because instructions seek to get people to do things a certain way, because there is something "pushy" about them, they seem not quite to fit in the category of informative purposes.
- *Evaluate.* When you evaluate, you make statements as to the value or usefulness of something, such as an ongoing program or project. You compare that program or project to its expectations and determine whether it meets those expectations.
- *Recommend.* When you recommend something, you are writing persuasively in a limited way. Recommendations use data and logic to point to one option as being the best choice. Recommendation reports are not fully persuasive, however, because they do not use the full arsenal of persuasive strategies.
- *Persuade or convince.* Plenty of situations require that you, as a practicing engineer, must write persuasively. Good examples are employment-seeking documents and proposals. Résumés and application letters show readers that you are well qualified for the job. Proposals seek to persuade potential clients that you or your firm is well-qualified to undertake the project. If you recall your studies of persuasive rhetoric, documents like these use not only logical appeals but personal and emotional appeals as well.

1.1.2 IDENTIFYING REQUIREMENTS, DOCUMENT TYPE, MODELS

Engineering documents—and professional, workplace documents in general—commonly fall into a particular category or, in some cases, a combination. These types are covered in Chapter 11, "Engineering Documents." One of your jobs early in the planning of your engineering document is to identify which of the following types best meet your requirements:

- Proposal
- Progress report
- Instructions
- User guide
- Recommendation, evaluation, or feasibility report
- Primary research report (lab report or field study)
- Literature review
- Background report
- Short report—accident, trip, investigation, and so on
- Handbook
- Specifications

Of course, there are numerous other names for different types of engineering and workplace documents, such as reconnaissance reports and design reports. When your clients call for a specific type of document, ask them specific questions about the document's purpose, content, and organization; ask them to show you some examples. Chances are good that the type of document they want is similar to one or a combination of those described in Chapter 11.

When you develop an engineering document based on a model, remember that you probably must customize that model to meet your requirements, as Figure 1.4 illustrates. Don't include a section in your engineering document just because it's there in the model. For example, the technical background may not be needed in a recommendation report, because the target audience knows that background all too well. At the same time, try to imagine information that is not included in the model but that is vitally needed in your own engineering document.

1.1.3 BRAINSTORMING AND IDENTIFYING WRITING STRUCTURES

In past writing courses, you may have experienced strategies for generating and exploring ideas and, in some cases, shaping and connecting thoughts to create new ideas. Don't think that the engineering context makes those strategies obsolete. You may need

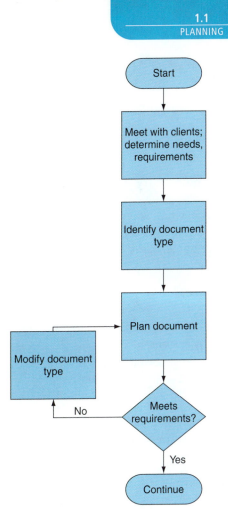

FIGURE 1.4
Initial document planning: understand clients' requirements, identify document type, and customize that document type accordingly.

some brainstorming methods to think of useful, essential material to include in your engineering documents. These idea-generating strategies include:

- Asking questions
- Brainstorming and listing

- Clustering and branching
- Using the writing structures
- Free-writing
- Writing journal entries

Try these strategies and find one that works best for you.

ASKING QUESTIONS

At the initial stage of the writing process, try thinking like a journalist. Reporters are known for probing subjects by asking the 5 Ws and H, which is short for who, what, where, when, why, and how. Suppose you're writing an engineering report on a new, energy-efficient, circular, rotating home. Some questions you might ask are

Why is it circular? Why does it rotate?
Statistically, how energy efficient is it?
Where is it more energy efficient—in which geographical areas?
What design details make it energy efficient?
How does it achieve its energy efficiency operationally?
How much does it cost to construct? Can the general public afford such a residence?

Your first set of questions can easily lead to other questions that will help you identify even more ideas. For instance, another W question might produce this:

What materials were used in the design?
What problems are there; what improvements are needed?

BRAINSTORMING AND LISTING

Use brainstorming as a method of generating ideas for writing. **Brainstorming** means listing words and phrases related to the topic of your engineering document. Don't worry about whether these words and phrases relate precisely to material you must cover in your document; list all the ideas relating to your topic that come into your mind. Don't censor or criticize; your goal is quantity, not quality.

You may have heard of space elevators, that outlandish notion of using some form of physical ladder or cable to get vehicles into space rather than rocket propulsion. What if, for some professional or academic reason, you were tasked with writing about space elevators? Here's a brainstorming list you might come up with:

SPACE ELEVATORS

- How exactly would space elevators work? Are they even theoretically possible?
- Why is there interest in such a crazy idea? Are there organizations and institutions currently working on this idea?
- How long would the tether (the mechanism from earth's surface to a point beyond geosynchronous orbit) have to be?
- What are the chief components of a space elevator?
- What materials and construction would the tether require?
- How would vehicles traverse the tether?

FIGURE 1.5
Design for a seagoing anchor station for a geosynchronous orbital tether (beanstalk).

(Source: Wikipedia http://en.wikipedia.org/wiki/Tether_propulsion 1 Jan. 2007. Accessed 2 Jan. 2007.)

- What technical problems stand in the way?
- Currently, have any designs been proposed?
- What are some of the projected costs?

When you have given yourself plenty of brainstorming time, take a critical look at your list of words and phrases. Reject weak or irrelevant ideas, add any new ideas that occur to you, and start looking for idea patterns and linkages. This kind of shaping and organizing is a preliminary step to the writing outline.

CLUSTERING AND BRANCHING

The goal of **clustering** and **branching** is to go beyond the brainstorming stage and to think about possible relationships between the words and phrases in your list. To start clustering, put a key subject word at the top or center of a blank page. Look for ideas related to the key word and connect them to it, and to other ideas, using circles and lines. Let your mind explore all possible associations freely. Figure 1.6 shows a sample cluster to generate ideas for an engineering report on the components and materials items in the brainstorming list shown within the figure.

Branching is another strategy for generating ideas and showing how they are linked. To create a branch diagram, write a topic at the top of a blank page, think of related subtopics, and write them branching downward. Next, extend the branching options as far as you think necessary. Finally, write in any supporting ideas and details on branches below each main idea.

USING A LIST OF WRITING STRUCTURES

Any brainstorming activities (such as those previously discussed) you have already done will supply you with raw materials for an outline. But another major source is writing structures. **Writing structures** are combinations of content and organization that make up major sections of an engineering document. You can use these writing structures to further brainstorm and plan the outline for your engineering document.

FREE-WRITING AND JOURNAL WRITING

You may have experienced this common writing phenomenon: you didn't actually know what you already knew about a topic, you didn't know what you did not know about the topic, nor did you know how

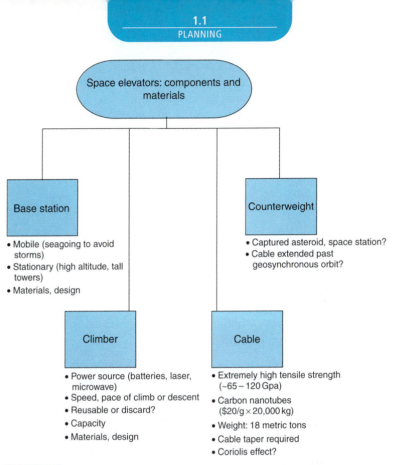

FIGURE 1.6
Branching (clustering) diagram for brainstorming. List ideas for each of the nodes.

you could write about that topic until you actually started writing about the topic? In this respect, writing certainly is a process of discovery.

Free-writing and journal writing are two methods that writers use to discover what they know and don't know about a topic and how they might write about that topic.

In **free-writing**, you designate a period of time—five to ten minutes—or a given amount of space—part of a page or a whole page—and write about a more general subject area related to the topic of your engineering document. You make a game of writing as fast as you can without ever lifting up your pen or pencil off the paper or fingers off the keyboard. In free-writing, you don't worry if at times you are writing gibberish—the point is to "let it all hang out." After you have recovered—perhaps after a full day has passed—go back and see what you wrote; look for patterns, questions, and new ideas. If this strategy works for you, it will be the source of material for your engineering documents.

With **journal writing**, the process is not so frantic. However, its goal is also to open you up to explore your topic, find out what you know and don't know, and practice writing about that topic. The main difference is that you make regular journal entries on areas such as the following:

- Your observations and questions about the topic
- Interesting ideas and facts that you've learned from research on the topic
- Your thoughts about whether you can achieve the purpose you have stated
- Your thoughts about whether your audience will understand your engineering document and find it useful

Plenty of scientists and engineers regularly keep journals about their professional work. Your engineering documents may be a good a place to start your own professional journal.

1.1.4 DEVELOPING AN OUTLINE

We all know the old joke about writing the outline after writing the paper—merely to satisfy an instructor's requirements. Indeed, for short papers of 350 words or less, the outline seems unnecessary. You know what you want to cover; you don't need an outline for planning or to remind you.

Larger writing projects, however, usually require a lot of work and time and are subject to many interruptions. A formal outline enables you to establish a plan and to "segment" your work. Instead of working on the entire document, which may be as many as 35 pages, you work on section 3, for example, which is a less-daunting 6 pages. Outlines enable you to put your writing project on a schedule and to make steady progress.

To generate an outline, you first need lots of raw material—lots of "stuff"—to work with and select from. The brainstorming strategies covered in the preceding text are great ways to generate those raw materials. And as discussed in the preceding section,

writing structures are particularly useful tools for getting ideas for the sections or subsections you will need in your engineering documents. For example, process: by skimming through Table 1.1, you realize you need to explain a mechanical process (how a system works) in detail in your report and that discussion may take up one whole section in the outline and in the body of the document.

Description	What are the physical details of the objects, mechanism, places, or events you discuss? These include height, weight, length, width, color, texture, ingredients, parts, methods of attachment, and so on.
Process	What are the events in the natural, biological, and mechanical processes you discuss?
Causes and effects	What are the known and hypothesized causes of things you discuss in your engineering document? What are the known and potential effects of things you discuss?
Benefits, advantages	What are the good things that result from a topic or topics you discuss in your engineering document?
Applications	What are practical uses of things you discuss in your report?
Categories	Which category does something you discuss belong to? What categories can the topic you discuss be divided into?
Examples	What are some good illustrative examples of the topic of your engineering document?
Definition	Which terms in your report need extra discussion to enable readers to understand them?
Comparison, contrasts, analogies	What similar or familiar things can your topic (or some aspect of it) be compared to? What are important differences? What is a useful analogy (extended comparison to something familiar) that can be used to discuss your topic?

TABLE 1.1
Writing Structures: Methods for Brainstorming a Topic

Figure 1.7 shows a formal outline for a report on the current status of research and development being done on space elevators. You would not actually see the subentries under "Introduction" as headings in the introduction portion of the report. These four subentries are here strictly as reminders that an introduction must contain elements such as these. Indeed, the background material in the introduction would be brief—no more than a paragraph and possibly just a sentence or two.

Thesis. Although there is much research and development still to do, designs for the chief components of a space elevator have become quite detailed.

I.   Introduction
   A.   Brief background
   B.   Thesis statement
   C.   Audience and situation
   D.   Overview of topics covered
II. Space elevator: conceptual overview
   A.   Problems with rocket propulsion
   B.   Space elevators: theory and models
   C.   Physics of space elevator
III. Chief components of space elevators
   A.   Tether
      1.   Length, weight, other dimensions
      2.   Materials: tensile strength
      3.   Taper requirements
      4.   Launch and construction
      5.   Environmental hazards
   B.   Climber
      1.   Power source and requirements
      2.   Ascent and descent mechanisms
      3.   Discard or reuse options
      4.   Capacity
      5.   Dimensions
      6.   Materials
      7.   Radiation shielding
   C.   Base station
      1.   Location: land or sea; mobile or stationary
      2.   Geographical location: equator or other

    D. Counterweight
        1. Captured asteroid
        2. Space station
        3. None
IV. Summary and conclusions
    A. Estimated ascent time
    B. Projected costs
    C. Projected first implementation date
    D. Remaining obstacles
V. References

TRADITIONAL OUTLINE	DECIMAL OUTLINE
I.	1
A.	1.1
B.	1.2
1.	1.2.1
2.	1.2.2
a.	1.2.2.1
b.	1.2.2.2
(1)	2
(2)	2.1
(a)	2.1.1
(b)	2.1.1
II.	2.1.2

FIGURE 1.7
Example outline for a report on current status of space elevator design.

The traditional outline and decimal outline follow the patterns shown in Figure 1.7. Here are the essential guidelines:

- Each outline level should have more than one entry.
- Entries within the same section at the same level should be grammatically parallel. In other words, items 2.1.1 and 2.1.2 must

be parallel to each but need not be parallel to anything else in the outline.

- Subordinate items should be indented to the *text* of the next-higher-level item. (See Figure 1.7 for examples of this format.)
- Avoid vague, meaningless outline entries such as "Technical Background" or "Basic Information."

1.1.5 GATHERING INFORMATION

You may have read the preceding section on outlining and wondered how you can develop an outline if you haven't researched the topic yet. Good point. Actually, you probably know plenty about the topic already, as the various brainstorming strategies covered earlier in this chapter may have demonstrated. Also, you can still predict much of the outline from just common sense. Engineering topics typically

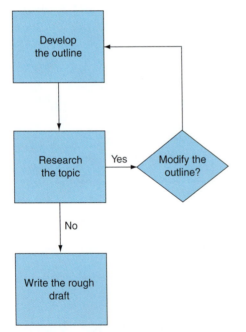

FIGURE 1.8
Information-gathering phase: researching a topic for a writing project typically necessitates changes and improvements to the outline.

have mechanisms, mechanisms have components, and components have descriptive detail. Engineering topics typically involve processes, and processes have steps and outcomes. Even so, researching your topic will enable you to add detail and fine-tune your outline.

See Chapter 10 for information-gathering and strategies. In that chapter, you will review how to:

- Develop an information-gathering strategy.
- Find information resources pertinent to your writing project, both online and in print.
- Choose the best information resources for your specific topic.
- Evaluate the information resources that you find in terms of their currency and reliability.

1.2 DRAFTING THE ENGINEERING DOCUMENT

If you've done a thorough, systematic job of planning an engineering writing project, the drafting stage should go far more smoothly than for relatively unplanned projects. Writing at this early state is called "drafting" or "rough drafting."

1.2.1 LIMITATIONS OF ROUGH DRAFTING

When you write a rough draft, write rapidly—almost to the point of carelessness. You can tie up the loose ends later. If you can't think of a word, put xxxxx in its place. If you know you need to elaborate on something but can't at that moment, just type [more on this later!]. If you know you need to plug in statistical information, type in [add data here] or some such. You need not construct perfect source citations in this phase, but be sure to include the author names, page numbers, or both, of the sources you use at the point where you use them in the rough draft.

When you write a rough draft, it would be nice to be in full control of such things as organization, topic sentences, transitions, technical detail, supporting information, sentence clarity and economy, and even logical development of ideas. But it's difficult to have that much control over the rough-drafting process; in fact, if you attempt to write a perfect draft the first time, you may prevent yourself from writing much at all.

Think of the rough-drafting process as the design phase of a project rather than the manufacturing phase. In the manufacturing phase of a physical product, the process is straightforward, predictable, and logical. In the manufacturing phase, you worry about defects and the overall efficiency of the process. The product itself has already been designed. The design phase, however, is rather unpredictable, uncontrollable, and illogical. If you are attempting to design a physical product, you may have to throw any number of initial efforts in the trash; you can't predict when that great idea will hit you—maybe in the middle of the night. Also, the process turns back upon itself; you may have to go back and modify or overhaul things you thought you had finished. Such is the case with the writing process as well.

1.2.2 TOOLS FOR ROUGH DRAFTING

However, you do have some tools and controls at your disposal during the rough-drafting phase:

AUDIENCE (CLIENTS), PURPOSE, SITUATION
Your sense of the project as a whole and your clients' requirements should guide you as to which topic to cover, which topics to avoid, and how to write about the topics you do cover.

OUTLINE
Having a detailed outline is a great help, but remember that rough drafting typically uncovers problems in the outline—topics left out, unnecessary topics, outline detail missing, and topics out of order.

WRITING STRUCTURES
In the preceding pages, you have read about using writing structures to brainstorm and outline a writing project. They also function as tools or resources as you create a rough draft. For example, in the thick of your rough-drafting, you may realize that you need to explain a mechanical process in detail or that you have already started such an explanation and just need to be more thorough about it. At other points, you may realize that an extended description is needed, comparison to something familiar would help, or a definition of essential terminology is critical to readers' comprehension.

LEVELS OF DETAIL

One final control available to you in the rough-drafting stage involves your awareness of levels of detail. After you have started a document, section, or paragraph, you always have the option to go into further detail about what you've just written or to stay at that same level of detail. Your sense of the project—your clients, the document's purpose, and the topic—govern these decisions. Look at Figure 1.9 to get a sense of how text has levels of detail. The opening sentence promises to talk about two benefits of green roofs. It provides a single sentence of detail about how roof longevity is increased but chooses to go into several sentences of detail about how green roofs help reduce problems involving storm-water runoff. The paragraph could start with a more ambitious sentence such as this:

> Establishing plant material on rooftops provides numerous ecological and economic benefits, including stormwater management, energy conservation, mitigation of the urban heat-island effect, and increased longevity of roofing membranes, as well as providing a more aesthetically pleasing environment in which to work and live.

Establishing plant material on rooftops provides numerous ecological and economic *benefits*, one of the most important of which includes increased longevity of roofing membranes and stormwater management.	Level 1
The mitigation of stormwater runoff, however, is considered by many to be the primary *benefit* because of the prevalence of impervious surfaces in urban areas.	Level 2
The rapid runoff from ordinary roof surfaces can exacerbate flooding, increase erosion, and may result in raw sewage being discharged directly into our rivers. This large amount of runoff also results in a greater quantity of water that must be treated before it is potable. Instead, green roofs can absorb stormwater and release it slowly over a period of several hours. Green roof systems have been shown to retain 60–100% of the stormwater they receive.	Level 3

FIGURE 1.9
Levels of detail. The writer of this paragraph chooses to zero in on how green roofs mitigate problems involving stormwater runoff rather than staying at a higher level by overviewing the other benefits of green roofs. (Italics added.)

The paragraph could then provide sentences detailing each of these benefits. The resulting paragraph would have covered more ground but at the same level of detail.

1.2.3 DOCUMENT EVOLUTION

The rough-drafting phase is a serious test of your document-planning skills. Don't be disheartened if things go scarcely as you had expected—that's normal. That's why there is a revision stage. Lots of things can change during the rough-drafting stage, just as they can in the information-gathering stage, including:

- Your understanding of your clients' requirements changes.
- Your tentative outline changes.
- You discover that you need to find additional information resources.
- Even your sense of the purpose of the document may subtly change.

As you can see, the rough-drafting phase can result in a rather sobering overhaul of your planning decisions. Scholars of rhetoric and composition describe these phenomena as "iterative" or "recursive" or "nonlinear." Just as in any design phase, you cannot expect the production of an engineering document to be a linear, lock-step matter, as Figure 1.10 illustrates.

1.3 SELECTING AND USING THE WRITING STRUCTURES

As you plan and write an engineering document, use the writing structures strategically. As the preceding and following sections of this chapter explain, writing structures can help you at every phase of a writing project: brainstorming, outlining, rough drafting, and revising. Here are some of the common writing structures:

- Narration
- Description
- Definition
- Classification and division
- Process analysis
- Comparison and contrast
- Cause and effect
- Examples and illustrations
- Analogy

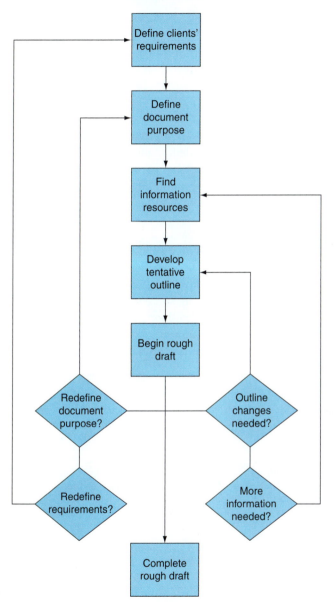

FIGURE 1.10
Recursive nature of the writing process—in this case, the rough-draft phase. As you write the rough draft, you will likely find that you need to find additional information resources, change your outline, and even make small changes to your project definition.

1.3.1 NARRATION

Narration is storytelling—either fictional or historical. In a narrative paragraph, the writer tells all or part of a story, usually in chronological order. The text in Figure 1.11 narrates efforts to monitor and ensure the safety of U.S. bridges.

1.3.2 DESCRIPTION

In **description**, you use words to create a picture of an object, mechanism, place, person, or event. Details include materials, methods of attachment, orientations, textures, colors, parts, contents, ingredients, dimension, and so on. In the example in Figure 1.12, NASA describes the construction of Mars Exploration Rover vehicles.

1.3.3 DEFINITION

In your engineering document you may need **definition**—explanations of important terms that may be unfamiliar to your readers. Don't assume that your readers know everything you know, and don't assume they know the topic the same way you know that topic. Unless your introduction warns them, nonspecialists are likely to read your engineering documents. Often, defining terms can be done in a few sentences, but occasionally it will require an

---

HISTORY OF BRIDGE-MAINTENANCE LEGISLATION

The collapse of the Silver Bridge in Ohio in 1967 and the loss of 47 lives due to the instantaneous fracture of an eyebar made obvious the need to carefully monitor the condition of bridges. At the time, no systematic monitoring and maintenance program was in place. To address this problem, the Federal Highway Act of 1968 created the National Bridge Inspection Program (NBIP), which ordered states to track the condition of bridges on principal highways. The Federal Highway Act (FHWA) of 1970 provided federal funding to states in order to replace bridges that were in the most danger of failure. Also, FHWA published the Recording and Coding Guide for the Structure Inventory and Appraisal of the Nation's Bridges. In 1978, the Surface Transportation Assistance Act and, later, the Highway Bridge Replacement and Rehabilitation Program (HBRRP), changed the basis for eligibility of bridges for federal funding so that bridges could be repaired before they deteriorated into a critical state. Further, the Intermodal Surface Transportation Efficiency Act of 1991 sought to establish cost-effective maintenance schedules for a network of bridges.

FIGURE 1.11
Narration: This example narrates the history of the monitoring of bridge conditions and bridge repair.

On each Mars Exploration Rover, the core structure is made of composite honeycomb material insulated with a high tech material called aerogel. This core body, called the warm electronics box, is topped with a triangular surface called the rover equipment deck. The deck is populated with three antennas, a camera mast and a panel of solar cells. Additional solar panels are connected by hinges to the edges of the triangle. The solar panels fold up to fit inside the lander for the trip to Mars, and deploy to form a total area of 1.3 square meters (14 square feet) of three layer photovoltaic cells. Each layer is made of different materials: gallium indium phosphorus, gallium arsenide and germanium. The array can produce nearly 900 watt hours of energy per Martian day, or sol. However, by the end of the 90-sol mission, the energy-generating capability is reduced to about 600 watt hours per sol because of accumulating dust and the change in season. The solar array repeatedly recharges two lithium ion batteries inside the warm electronics box.

— Mars Exploration Rover Landings (January 2004)
http://mars.jpl.nasa.gov/newsroom/pressreleases/pressreleases 2004.html

FIGURE 1.12
Description: This descriptive example provides details on the Mars Exploration Rover.

entire paragraph. The example in Figure 1.13 provides an extended definition of a relatively new technology. Notice that it begins with a general overview definition. The extended definition continues with an explanation of just how weak detectable signals can be, what makes up a SQUID, and how it can be used.

A superconducting quantum interference device (SQUID) is a mechanism used to measure extremely weak signals, such as subtle changes in the human body's electromagnetic energy field. Using a device called a Josephson junction, a SQUID can detect a change of energy as much as 100 billion times weaker than the electromagnetic energy that moves a compass needle. A Josephson junction is made up of two superconductors, separated by an insulating layer so thin that electrons can pass through. A SQUID consists of tiny loops of superconductors employing Josephson junctions to achieve superposition: each electron moves simultaneously in both directions. Because the current is moving in two opposite directions, the electrons have the ability to perform as qubits (that theoretically could be used to enable quantum computing). SQUIDs have been used for a variety of testing purposes that demand extreme sensitivity, including engineering, medical, and geological equipment. Because they measure changes in a magnetic field with such sensitivity, they do not have to come in contact with a system that they are testing.

— Source: http://whatis.techtarget.com/definition/0,,sid9_gci816722,00.html

FIGURE 1.13
Extended definition: This example spends extra time explaining what a SQUID is.

1.3.4 CLASSIFICATION AND DIVISION

Classification involves grouping items such as objects, mechanisms, phenomena, ideas, and people according to some system of categorization, known as the **basis of classification**.

Figure 1.15 shows an example of **"true" classification**, in which the effort is to explain why an item fits into a particular category or why it does not.

1.3.5 PROCESS ANALYSIS

In **process analysis**, the writer analyses and explains how something works or how to do something. The paragraph pattern closely follows the chronological pattern in the process being described. Put one

---

3.3.1 CONCENTRATION CELL CORROSION.

Concentration cell corrosion occurs when two or more areas of a metal surface are in contact with different concentrations of the same solution. There are three general types of concentration cell corrosion: (1) metal ion concentration cells, (2) oxygen concentration cells, and (3) active-passive cells.

3.3.1.1 Metal Ion Concentration Cells. In the presence of water, a high concentration of metal ions exists under faying surfaces and a low concentration of metal ions exists adjacent to the crevice created by the faying surfaces. An electrical potential will exist between the two points. The area of the metal in contact with the low concentration of metal ions is cathodic and is protected, and the area of metal in contact with the high metal ion concentration is anodic and corroded. This condition can be eliminated by sealing the faying surfaces to exclude moisture. Proper protective coating application with inorganic zinc primers is also effective in reducing faying surface corrosion.

3.3.1.2 Oxygen Concentration Cells. A water solution in contact with the metal surface normally contains dissolved oxygen. An oxygen cell can develop at any point where the oxygen in the air is not allowed to diffuse uniformly into the solution, thereby creating a difference in oxygen concentration between two points. Typical locations of oxygen concentration cells are under either metallic or nonmetallic deposits (dirt) on the metal surface and under faying surfaces such as riveted lap joints. Oxygen cells can also develop under gaskets, wood, rubber, plastic tape, and other materials in contact with the metal surface. Corrosion will occur at the area of low-oxygen concentration (anode). The severity of corrosion due to these conditions can be minimized by sealing, maintaining surfaces clean, and avoiding the use of material that permits wicking of moisture between faying surfaces.

FIGURE 1.14
Categorization-style classification: This example discusses several categories of corrosion.

Pluto was officially proclaimed a planet in 1930 by Clyde Tombaugh at the Lowell Observatory. However, the discovery in 1992 that the Kuiper belt, an immediate neighbor to Pluto, held other Pluto-like objects raised the question as to whether Pluto could be considered a true planet.

Figure 3. Pluto size as compared to Earth.

In 2006, the International Astronomical Union approved and published a set of criteria for an astronomical object to be classified as a planet: (a) it must orbit the sun, (b) it must not be a satellite, (c) it must be massive enough for its own gravity to keep it round, and (d) it must be big enough to dominate its own orbit. Pluto does not meet this final requirement and thus cannot be considered a true planet according to the International Astronomical Union's definition. Instead, astronomers prefer the loose classification of Pluto as a "dwarf planet."

FIGURE 1.15
True classification: This example explains why Pluto does not fit in a certain category.

essential step out of sequence, or miss a step, and the reader can be in big trouble. In Figure 1.16, the writer explains how a photovoltaic cell works.

1.3.6 COMPARISON AND CONTRAST

When you make **comparisons**, you examine similarities; with **contrasts**, you examine differences. Two major approaches are available for organizing comparisons and contrasts. In the first approach, sometimes called the block or **subject-by-subject method**, you discuss one of the topics first and then the other. Although this approach is the one that you might grab first, it is the least effective in accomplishing its purpose—to compare or contrast. Instead, the **point-by-point approach** provides a much more effective way of discussing similarities and differences. Each "point" represents a different way of comparing the topics. In Figure 1.17, speed is one way that two early computers, the Mark I and the Mark II, can be compared. You can imagine other paragraphs comparing the

In this diagram of a photovoltaic (PV) cell, the N-type ("n" for negative) semiconductor, because it has been doped with phosphorus, is loaded with free electrons. The P-type semiconductor ("p" for positive), instead of having free electrons, has free holes, which are simply the absence of electrons. They can carry the opposite (positive) charge and move around the way electrons do. When the N-type and P-type silicon are placed in contact, an electric field is formed: the free electrons on the N side move to fill in the free holes on the P side. The antireflective coating applied to the top of the cell reduces reflection losses to less than 5 percent. The glass cover plate protects the photovoltaic cell from the elements. PV modules consist of several PV cells (usually 36) in series and parallel to produce useful levels of voltage and current.

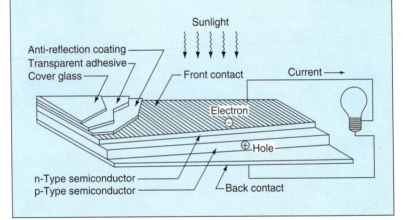

FIGURE 1.16
Process analysis: this example explains how a PV cell works.

In terms of speed, the Mark II, completed in 1947, made huge advances over the Mark I, which was completed in 1944. The Mark I could do an addition in 0.3 seconds and a multiplication in 6 seconds. By contrast, the Mark II's addition time was 0.125 seconds; its multiplication time was 0.75 seconds. Using built-in hardware, the Mark II could do square root, logarithmic, and some trigonometric functions between 5 and 12 seconds. In contrast, the Mark I required over 60 seconds to complete a logarithmic or a trigonometric function.

FIGURE 1.17
Comparison–contrast: This example uses the recommended point-by-point approach to compare two early computers on one comparative point—processing speed.

machines' sizes, their methods of computing, their reliability, and so on.

1.3.7 CAUSE AND EFFECT

Using **cause–effect**, you show the causes of situations or events or the consequences (effects) of situations or events. Figure 1.18 shows the results (benefits, essentially consequences or effects) of architecturally designing for greater daylighting. The conclusion sections of research reports, of which the following is an example, are primarily cause–effect discussions.

1.3.8 EXAMPLES AND ILLUSTRATIONS

An **example** helps readers understand general concepts by giving them something practical and concrete. Examples also keep writers from having to cover everything. A phrase such as *for example* implies that coverage will not be complete. In Figure 1.19, the writer explores just one of apparently many examples concerning problems with the idea of space elevators.

1.3.9 ANALOGY

Writers can use a type of comparison called an **analogy** to help readers understand a difficult concept by relating that concept to something with which they are familiar. In Figure 1.20, the writer uses an analogy of a string attached to a stone to explain why a space elevator in geosynchronous orbit would not fall back to Earth.

CONCLUSIONS

The most obvious conclusion is that daylighting, even excluding all of the productivity and health benefits, makes sense from a financial standpoint. The daylit schools in the study indicated energy-cost reductions of between 22% to 64% over typical schools. With break-even on all the new daylit schools occurring in less than three years, the long-term benefits to a school system are enormous. In North Carolina, a 125,000 square foot middle school that incorporates a well-integrated daylighting scheme is likely to save $40,000 per year over typical constructions. And, if energy costs go up by 5% per year, the savings on just this one school, over the next ten years, would exceed $500,000.

FIGURE 1.18
Cause–effect discussion: This example discusses the effects (results, benefits) of architectural daylighting.

The concept of the space elevator has many technical problems to overcome before it can ever become a practical working reality. For example, along the estimated 100,000 km that the space elevator would have to travel from Earth's surface to a point just beyond geosynchronous orbit, a significant portion of its journey would occur within the Van Allen radiation belts. This treacherous distance is maximized by the current designs which call for space elevators to be anchored to an ocean platform near the equator. At the equator, the radiation belts extend from about 1000 to 20,000 km in altitude. This region did not hurt the Apollo astronauts in the 1960s and 1970s because their rockets delivered them swiftly through it. However, a space elevator traveling at the current proposed speed of 200 km/h might spend half a week in the belts. During that time, passengers would receive 200 times the radiation experienced by the Apollo astronauts.

FIGURE 1.19
Examples: This example discusses one example of the technological hurdles facing the idea of space elevators.

1.4 REVIEWING AND REVISING

In some ways, the **revision process** is a repeat of the planning process. Although the issues are the same, think of it as an inspection process in which you search for problems—problems in the following specific areas.

To the ordinary person, the notion of a space elevator—specifically, the tether (or ribbon as it has also been called)—is fraught with impossibilities. For example, why won't the whole length of the tether, some 60,000 to 100,000 km long, just fall back to Earth? Why won't it just wrap around the surface of Earth as our planet spins? To answer these alarmists, consider a simple analogy. Take a little piece of string and attach to it a stone. Begin to rotate this primitive sling. Under the influence of centrifugal force, neither the string nor the stone will fall back to you, but instead the stone will try to pull itself away from you and tightly stretch the string. The same principles in this simple example apply to the space elevator!

Adapted from Yuri Artsutanov, "To the Cosmos by Electric Train," *Young Person's Pravda* (July 31, 1960) Trans. Joan Barth Urban and Roger G. Gilbertson. <http://www.liftport.com/files/Artsutanov_Pravda_SE.pdf>

FIGURE 1.20
Analogy: This example is based on an extended comparison between a string attached to a stone and a space elevator in geosynchronous orbit.

1.4.1 REVIEWING FOR PROJECT-LEVEL PROBLEMS

Think of **project-level problems** as involving audience, purpose, and situation:

- Does the draft report meet the requirements of your clients, the audience?
- Can your clients read and understand the draft you have written?
- Does the draft fulfill its purpose?
- Does the draft meet the requirements of the situation in which it is needed?

How can you know whether your clients can understand your engineering document and whether it meets their needs? Unless you are good at assuming the perspective of your clients, you may need to find readers similar to your clients and have them review your work. Why not get your actual clients involved? Send them the draft, telling them it is strictly preliminary and asking for a review.

In whatever way you discover problems in these areas, you may find remedies for those problems in the revision strategies discussed in the following subsections.

1.4.2 REVISING CONTENT PROBLEMS

One of the first potential problems to search for involves the content that you have included in the rough draft. Writing teachers call content "development"—in other words, how fully and adequately you have developed the topic. When you review the content—the materials, the information—you have included in a rough draft, ask yourself these questions:

- Does the draft include all the information these clients need, according to their requirements and uses for this document?
- Does this rough draft leave out essential information that the clients need?
- Does this draft include information that the clients do not need?
- Is the discussion in this draft at a level that the clients can understand?

If you find major problems in this area, you need to go back to that outline, revise it with the missing pieces or yank the unnecessary pieces, and rough-draft the sections you left out. If your clients cannot or are not likely to understand what you have written, see the suggestions in "Translating Technical Discussions" later in this chapter.

Content problems can occur at any level of a document: in its outline, in its sections and subsections, and in its individual paragraphs and sentences.

Consider the outline of the space-elevator report in Figure 1.7. If section II, "Space elevator: conceptual overview," were omitted from the outline and from the draft, the report would have serious development (or content) problems. Clients may need first to understand what the space-elevator idea is before they can intelligently read about the individual components.

Development (content) problems can also occur at lower levels of a rough draft, specifically, within paragraphs. Figure 1.21 shows an example. The problem with the under-developed version is that it never defines "self-replication" or "self-replicating components." That content is italicized and added to the beginning of the revised paragraph.

Lacking content (development)	The idea of self-replication was originally put forth by John von Neumann in the 1940s. In *Engines of Creation*, Drexler foresees replicators whose components will include molecular "tape" to supply instructions; a reader to translate those instructions into arm motions; and several assembler arms to hold and move workpieces. Drexler calculates that such a replicator will add up to one billion atoms or so and, working at one million atoms per second, will copy itself in one thousand seconds (about 15 minutes). Working singly, a replicator would need a century to stack up enough copies to make a "respectable speck." But with replicators making copies of themselves, the process would produce a ton of copies in less than a day.
Improved content (development)	*For objects to be manufactured inexpensively and efficiently, self-replicating components are also needed. Such a component would make*

FIGURE 1.21 (*Continued*)

	copies of itself and manufacture the desired objects. The idea of self-replication was originally put forth by John von Neumann in the 1940s. In *Engines of Creation*, Drexler foresees replicators whose components will include molecular "tape" to supply instructions; a reader to translate those instructions into arm motions; and several assembler arms to hold and move workpieces. Drexler calculates that such a replicator will add up to one billion atoms or so and, working at one million atoms per second, will copy itself in one thousand seconds (about 15 minutes). Working singly, a replicator would need a century to stack up enough copies to make a "respectable speck." But with replicators making copies of themselves, the process would produce a ton of copies in less than a day.

FIGURE 1.21
Improved content: The initial version is unclear because there is no definition of self-replication. (Italics added.)

1.4.3 TRANSLATING TECHNICAL DISCUSSIONS

If you have nonspecialist, nontechnical readers for an engineering document, you will need to "translate" that document so that they can read and understand it. Typical nontechnical readers for engineering documents are administrative, financial, or legal people. The old joke about managers who come from engineering positions is that, as soon as they become managers, they forget everything technical they ever knew. In any case, nontechnical people involved in your project must make executive decisions about such things as whether to implement the project or to continue the project. They may need to assess the financial aspects of the project or assess its legal ramifications.

To translate your engineering document for these people, you engage in a specialized form of revision for development:

- Define terms that your readers may not understand, or avoid those terms altogether.
- Explain the reasons and importance of things in the draft.
- Provide plenty of examples.

- Use analogies to common, familiar things.
- Provide conceptual graphics that show simplified versions of the mechanism, process, or phenomenon.
- Carefully explain processes step by step.
- Use the "in-other-words" technique; restate things in simpler terms.
- Add the human perspective; explain how something can be used, what effects, good or bad, it may have.
- Move seriously technical material that is "untranslatable" to appendixes.

1.4.4 TROUBLESHOOTING ORGANIZATIONAL PROBLEMS

If your draft has the right content, you still must consider how that content is organized. Once again, this is an issue that involves all levels of a document: in the outline, in the sections and subsections, and in the paragraphs and sentences.

Consider the outline of the space-elevator report in Figure 1.7. If section II, "Space elevator: conceptual overview," were moved toward the end of the outline, the report would have serious organizational problems. The report must first establish what the space-elevator idea is before it can get into the individual components.

Organizational problems can also occur at lower levels of a rough draft—specifically, with paragraphs. Figure 1.22 shows an example. In the problem version, the definition of self-replication and self-replicating devices (italicized) occurs too late: readers need that essential bit of information before they can read and understand the components and operation of such a device.

1.4.5 REVISING FOR COHERENCE

You can have great content and logical organization, but there may still be problems. Readers may find it hand to follow the discussion, despite perfectly adequate content and organization. The next problem to look for involves **coherence**—the way the discussion flows logically and naturally from one sentence to the next, from one point to the next, from one paragraph to the next, and from one section to the next. As with development and organization, coherence can be an issue at all levels of a document. You can use a number of tools to strengthen the coherence of a document, including:

- Topic sentences
- Key words and previous-topic summarizers

Organizational problems	The idea of self-replication was originally put forth by John von Neumann in the 1940s. In *Engines of Creation*, Drexler foresees replicators whose components will include molecular "tape" to supply instructions; a reader to translate those instructions into arm motions; and several assembler arms to hold and move workpieces. Drexler calculates that such a replicator will add up to one billion atoms or so and, working at one million atoms per second, will copy itself in one thousand seconds (about 15 minutes). Working singly, a replicator would need a century to stack up enough copies to make a "respectable speck." *However, for objects to be manufactured inexpensively and efficiently, self-replicating components are needed. Such a component would make copies of itself and manufacture the desired objects.* With replicators making copies of themselves, the process would produce a ton of copies in less than a day.
Revised organization	*For objects to be manufactured inexpensively and efficiently, self-replicating components are also needed. Such a component would make copies of itself and manufacture the desired objects.* The idea of self-replication was originally put forth by John von Neumann in the 1940s. In *Engines of Creation*, Drexler foresees replicators whose components will include molecular "tape" to supply instructions; a reader to translate those instructions into arm motions; and several assembler arms to hold and move workpieces. Drexler calculates that such a replicator will add up to one billion atoms or so and, working at one million atoms per second, will copy itself in one thousand seconds (about 15 minutes). Working singly, a replicator would need a century to stack up enough copies to make a "respectable speck." But with replicators making copies of themselves, the process would produce a ton of copies in less than a day.

FIGURE 1.22
Organizational problems. In this example, the definition is out of place; it occurs too late to help the reader readily understand the discussion. (Italics added.)

> The *surface of Mars* is thought to be primarily composed of basalt, based upon the Martian meteorite collection and orbital observations. There is some evidence that a portion of the Martian surface might be more silica-rich than typical basalt, perhaps similar to andesitic stones on Earth, though these observations may also be explained by silica glass. Much of the surface is deeply covered by iron oxide dust as fine as talcum powder. There is conclusive evidence that on the surface of Mars liquid water existed at one time. Key discoveries leading to this conclusion include the detection of various minerals such as hematite and goethite which usually only form in the presence of water.

FIGURE 1.23
Topic-reference topic sentence: The focus of the paragraph is indicated in the first sentence, but nothing else.

- Parallel phrasing
- Transitional words and phrases

CREATING OR STRENGTHENING TOPIC SENTENCES

A **topic sentence** orients the reader to the discussion in the rest of the paragraph in several ways, as discussed in the following paragraphs.

Topic reference: simple occurrence of the main topic word or phrase in the first sentence of a paragraph. In Figure 1.23, the topic of the paragraph "surface of Mars," occurs in the first sentence, but the sentence as a whole does nothing else to orient readers to the paragraph as a whole.

Key phrase: a phrase that projects what the paragraph will cover. In Figure 1.24, the key phrase, "chemical characteristics," alerts readers to the fact that these characteristics will be the focus of the rest of the paragraph.

> The *chemical characteristics* of methyl isocyanate gas (MIC) made it particularly deadly to the residents of Bhopal. MIC is twice as dense as air. Because of MIC's high density, it does not diffuse readily into the atmosphere, and it stays close to the ground instead. MIC is flammable, reactive and toxic. . . .

FIGURE 1.24
Key-phrase topic sentence: The focus of the paragraph is more specifically indicated in the word *characteristics*, than in the example in Figure 1.23. (Italics added.)

> Most brains exhibit a visible distinction between *gray matter* and *white matter*. The gray matter makes up the outer layer of the brain, called cerebral cortex. Gray matter consists of the cell bodies of the neurons, while white matter consists of the fibers (axons) that connect neurons. The axons are surrounded by a fatty insulating sheath called myelin, giving the white matter its distinctive color. Deep in the brain, compartments of white matter (fasciculi, fiber tracts), gray matter (nuclei) and spaces filled with cerebrospinal fluid (ventricles) are found. . . .

FIGURE 1.25
Overview topic sentence: This type provides a list of subtopics to be covered in the paragraph (or following paragraphs). (Italics added.)

Overview: a series of words or phrases that specify what the paragraph will cover. In Figure 1.25, the topic sentence forecasts that the paragraph will discuss "gray matter" and "white matter."

Thesis-like statement: a sentence that states the main point or argument of the rest of the paragraph. In Figure 1.26, the topic sentence makes a thesis-like assertion about Babbage's calculating machines; the rest of the paragraph must support that assertion.

> Although Babbage's machines were mechanical monsters, *their basic architecture was astonishingly similar to a modern computer*. The data and program memory were separated, operation was instruction based, control unit could make conditional jumps, and the machines had separate I/O units.

FIGURE 1.26
Thesis-statement topic sentence: This type makes an assertion that the rest of the paragraph or following paragraphs must support. (Italics added.)

As mentioned throughout this chapter, the writing process is a messy affair, a process of discovery. When you begin writing a paragraph, you may not be able to construct a topic sentence for it or to decide which type of topic sentence is best. However, it may occur to you as you write the paragraph; in fact, you may even write that topic sentence toward the end of the paragraph. When you come back and revise, move it to the top.

REPEATING KEYWORDS AND USING PREVIOUS-TOPIC SUMMARIZERS
You may be surprised to know that carefully repeating keywords in your document can have a positive effect on the coherence of that

document. A sequence of repeated keywords in a paragraph (or at higher levels in the document) is called a **topic string**. There are two fundamental types of topic strings:

- *Static topic strings.* The same keyword or closely related synonyms occur toward the beginning of the second and following sentences of that paragraph. Figure 1.27 shows an example. Although `solar` cell seems like the topic, the conversion process is the real focus. The topic string is `converts` → `conversion` → `conversion`. Beginning with the second sentence, these keywords occur toward the beginning of each sentence.
- *Dynamic topic strings.* In what is also known as the "old-to-new" pattern, the new topic in a sentence becomes the old (or familiar) topic in the next sentence. Sentences begin with an old, familiar topic and end with new topics, new information. Here's an example (Figure 1.28):

The old-to-new pattern looks roughly like this:

> Solar heating system → solar collector
> Solar collector → types
> Types → flat-plate collector
> Flat-plate collector → absorber; layers
> Layers → heat-trapping effect
> Water; heated → absorber

A perfect old-to-new pattern would be a→b | b→c | c→d, and so on. However, in this solar heating system example, that pattern is not "perfect," but still quite evident. You will rarely find perfect

A solar cell (or a "photovoltaic" cell) is a semiconductor device that *converts* photons from the sun (solar light) into electricity. To achieve this *conversion*, the device needs to fulfill only two functions: photogeneration of charge carriers (electrons and holes) in a light-absorbing material, and separation of the charge carriers to a conductive contact that will transmit the electricity. This *conversion* is called the photovoltaic effect, and the field of research related to solar cells is known as photovoltaics.

FIGURE 1.27
Static topic string: Not `solar cell` but variants of `conversion` make up the topic string here. (Italics added.)

The most important part of a *solar heating system* is the *solar collector* whose main function is to heat water to be used in space heating. There are various *types* of *collectors*. However, the *flat-plate collector* is the most common and the focus of the following discussion. A *flat-plate collector* consists of a box-shaped black plate *absorber* covered by one or more transparent *layers* of glass or plastic with the sides and the bottom of the box insulated. These *layers* of glass or plastic have an intervening air space that produces the *heat-trapping effect*. *Water* is *heated* as it circulates through or below the *absorber* component, which is heated by solar radiation.

FIGURE 1.28
Dynamic topic string: In this pattern, what is a new topic in one sentence becomes the old topic in the next sentence. (Italics added.)

static or dynamic topic strings in real-world writing. Still, knowing and loosely applying those patterns to your paragraphs can have a strong, positive effect on the coherence of your writing.

You may have some reservations about repeating words in a document. You may have been encouraged to vary your word choice in the past. While "elegant variation" may be good in light, journalistic, entertainment-oriented writing, it is not good in business, scientific, technical, and engineering writing. Varying words that are the essential focus can only cause readers to become confused. In the earlier days of computer documentation, hard drives were also called fixed disks, fixed-disk drives, non-volatile storage devices, DASD, and other such. Similarly, floppy disks were also called removable disks and diskettes. Resetting the system was also referred to as "rebooting" the system. Variations between these essential terms drove people crazy! When it comes to keywords, use the same words for the same objects and events—every time!

Related to keywords is the previous-topic summarizer, a particularly powerful device for strengthening coherence. The **previous-topic summarizer** is a word or phrase, often beginning with "this," that captures the topic of several preceding sentences. It essentially says, "Okay, here's what we've been talking about up to this point in a nutshell; now let's move on." Although they can occur within paragraphs, these special phrases typically occur between paragraphs (at the beginning of a following paragraph). In Figure 1.29, the entire discussion in the first paragraph—the process of

Outsourcing is nothing new. In the early 1800s, Britain imported cotton from the U.S., spun and weaved the fabric in England, and exported the textiles abroad, primarily to India, then its colony. This arrangement kept about 80 percent of the monetary equivalent of these exports in England, thus contributing to the national wealth. Soon, the mill owners realized that they could grow the cotton in India, move the textile machinery there, and save money on wages and shipping. However, this new arrangement meant that only 15 to 20 percent of the total value of the textiles was being returned to England. Instead, 80 percent of the created wealth remained in India, a great loss to the British economy resulting in unemployment and social unrest.

We here in the United States face *similar problems*. By outsourcing, we lose not only skilled labor but monetary wealth.

http://www.memagazine.org/contents/current/webonly/wex30905.html

FIGURE 1.29
Previous-topic summarizer: The phrase *similar problems* summarizes the focus of the discussion in the preceding paragraph. (Italics added.)

outsourcing to India and the loss of wealth in Britain—is summarized in the phrase "similar problems" at the beginning of the next paragraph.

USING PARALLEL PHRASING

Using parallel phrasing in sentences with similar content can also help to achieve coherence. Sentences using **parallel phrasing** use the same style of phrasing. In the first example in Figure 1.30, What if is repeated; in the second example, It can.

Reverse engineering lies at the very heart of the engineering profession. But *what if* the very act of opening a device destroys it? *What if* its details are too fine to discern with the naked eye? *What if* you want to probe the secrets of an integrated circuit or a microelectromechanical system?

With its 3-axis sensing, the ADXL330, a tiny accelerometer from Analog Devices Inc., is the first step toward cheap, low-power gyroscopes. *It can* provide motion-sensitive flip-wrist scrolling in mobile phones or image stabilization in digital cameras. *It can* secure a hard drive so it survives the drop of a notebook computer or media player. *It can* make video games more interactive and intuitive.

FIGURE 1.30
Parallel phrasing for increased continuity. The words *what if* help emphasize the flow of ideas, as do the words *It can*, in the second example. (Italics added.)

Three traits were essential in a semiconductor laser if it were to be used for telecommunications. *It would have to* generate a continuous beam rather than pulses. *It would need* to function at room temperature and operate for hundreds of thousands of hours without failure. Finally, the laser's output *would have to be* in the infrared range, optimal for transmission down a fiber of silica glass. (http://www.greatachievements.org/?id=3713)

Lasers have found many applications since their development in the 1950s and 1960. *In manufacturing*, infrared carbon dioxide lasers cut and heat-treat metal, trim computer chips, drill tiny holes in tough ceramics, silently slice through textiles, and pierce the openings in baby bottle nipples. *In construction*, the narrow, straight beams of lasers guide the laying of pipelines, drilling of tunnels, grading of land, and alignment of buildings. *In medicine*, detached retinas are spot-welded back in place with an argon laser's green light, which passes harmlessly through the central part of the eye but is absorbed by the blood-rich tissue at the back. (http://www.greatachievements.org/?id=3711)

FIGURE 1.31
More parallel phrasing. (Italics added.)

The parallel phrasing tells readers to treat these sentences in the same way, as adding similar kinds of detail. Of course, the effect can be rather dramatic—if not melodramatic. Perhaps this style of parallel phrasing should be saved for situations where lots of emphasis for persuasive purposes is needed. Figure 1.31 shows two final examples that are not quite so histrionic.

USING TRANSITIONAL WORDS AND PHRASES

If you have managed the location and repetition of keywords well, you may find that transitional words and phrases are not so critical in achieving or improving coherence. Transitional words and phrases are most critical at those points in a discussion where you cannot begin a sentence with a familiar keyword. Consider the examples shown in Figure 1.32.

Transitional words and phrases are about logic. They indicate the logical connection between two ideas (clauses, sentences, paragraphs, groups of paragraphs). In the following, "For example" indicates that what follows is an example of the preceding statement:

Rather than giving names to the features of Mars that they mapped, Beer and Mädler simply designated them with letters. *For example*, Meridian Bay was feature "a."

Transition needed	Lasers are at the core of many everyday devices. A CD or DVD player reads the digital contents of a rapidly spinning disc by bouncing laser light off minuscule irregularities stamped onto the disc's surface. Barcode scanners in supermarkets play a laser beam over a printed pattern of lines and spaces to extract price information and keep track of inventory.
Revision with transitions	Lasers are at the core of many everyday devices. A CD or DVD player, *for example*, reads the digital contents of a rapidly spinning disc by bouncing laser light off minuscule irregularities stamped onto the disc's surface. Barcode scanners in supermarkets play a laser beam over a printed pattern of lines and spaces to extract price information and keep track of inventory.
Transition needed for emphasis	At the time lasers emerged, the ability of flexible strands of glass to act as a conduit for light was a familiar phenomenon, useful for remote viewing and a few other purposes. Such fibers were considered unsuitable for communications because any data encoded in the light were quickly blurred by chaotic internal reflections as the waves traveled along the channel.
Revision with transitions	At the time lasers emerged, the ability of flexible strands of glass to act as a conduit for light was a familiar phenomenon, useful for remote viewing and a few other purposes. Such fibers were considered unsuitable for communications, *however*, because any data encoded in the light were quickly blurred by chaotic internal reflections as the waves traveled along the channel. http://www.greatachievements.org/?id=3713

FIGURE 1.32

Transitional words and phrases: The italicized words are transitional words and phrases that guide readers between sentences where repeated keywords are unavailable. (Italics added.)

In the following example, the transitional word "Today" indicates chronology:

Rather than giving names to the features of Mars that they mapped, Beer and Mädler simply designated them with letters. *Today*, features on Mars are named from a number of sources. Large albedo features retain many of the older names, but are often updated to reflect new knowledge of the nature of the features.

In the following example, the transitional word "However" indicates a contrast or contradiction between the previous and following sentences:

Features on Mars are named from a number of sources. Large albedo features retain many of the older names. *However*, these names are often updated to reflect new knowledge of the nature of the features.

In this final example, "As a result" indicates a cause–effect relationship between the two sentences:

In 2004, NASA's Mars Exploration Rover Opportunity detected various minerals such as hematite and goethite which usually only form in the presence of water. *As a result*, scientists believe that liquid water existed at one time on the surface of Mars.

The following table lists the common transitional words and phrases and the logic that they indicate:

To show time	after, as, before, next, during, eventually, later, finally, meanwhile, then, when, while, immediately, soon, subsequently, next, today, tomorrow, yesterday
To show direction or place	above, around, below, beyond, beside, farther on, nearby, opposite to, close, to the right, elsewhere, here, there
To show addition	additionally, and, again, also, too, at the same time, besides, equally important, finally, further, furthermore, in addition, lastly, moreover, next, first, second

FIGURE 1.33
Transitional words and phrases grouped according to logical function.

FIGURE 1.33 (Continued)

To compare	also, similarly, likewise, compare, by way of comparison, in the same way
To contrast	but, however, at the same time, on the contrary, in contrast, yet, on the other hand, nevertheless, in spite of, conversely, still, although, even though, instead, though, despite
To give examples	for instance, for example, specifically, to illustrate, in fact, indeed, that is, in particular, namely, thus
To show logical relationship	consequently, thus, as a result, if, so, therefore, hence, accordingly, because, otherwise, then, to this end
To concede	of course, naturally, granted, although, certainly, even though, with the exception of
To conclude or summarize	altogether, in brief, in conclusion, in other words, in short, in summary, to summarize, to sum up, therefore, that is, in general, finally

Notice how the transitions in Figure 1.34 guide the reader. In the early 1800s is a transitional phrase indicating time as is Soon, occurring later in the paragraph. The transitional phrase As a result shows a logical (in this case, causal) relationship as

Outsourcing is nothing new. *In the early 1800s*, Britain imported cotton from the U.S., spun and weaved the fabric in England, and exported the textiles abroad, primarily to India, then its colony. *As a result*, this arrangement kept about 80 percent of the monetary equivalent of these exports in England, *thus* contributing to the national wealth. *Soon*, the mill owners realized that they could grow the cotton in India, move the textile machinery there, and save money on wages and shipping. *However*, this new arrangement meant that only 15 to 20 percent of the total value of the textiles was being returned to England. *Instead*, 80 percent of the created wealth remained in India, a great loss to the British economy resulting in unemployment and social unrest.

http://www.memagazine.org/contents/current/webonly/wex30905.html

FIGURE 1.34
Transitional words and phrases at work. (Italics added.)

does `thus` later in the paragraph. The transitional words, `However` and `Instead`, are contrastive transitional words. Overall, the paragraph discusses changes occurring over time and the contrastive effects of those changes. Thus transitions indicating temporal, causal, and contrastive relationships are appropriate.

1.4.6 REVIEWING FOR FORMATTING PROBLEMS

Chapter 2, "Professional Document Design," covers the important issues involving format and style. Observing well-established formatting and stylistic practices can increase not only the professional look of your engineering documents but their readability and reader comprehension. The following list contains a review of the important things to look for in a rough draft:

- Does the draft include headings that are meaningful, adequate, properly hierarchical and subordinated, and parallel in phrasing? Are there any lone headings, stacked headings, or pronouns referring to headings?
- Does the draft use the different types of lists properly? Are numbered and bulleted lists used correctly, are list items parallel in phrasing, and are sublist items formatted correctly?
- Is highlighting (italics, bold, color, and alternate fonts) used functionally, consistently, and not overused? For extended emphasis (one or more sentences), are notices used instead of highlighting?
- Are tables used to present large amounts of data? Are those tables formatted in a standard way? Does preceding text introduce and explain those tables?
- Are figures (illustrations, drawings, diagrams) used to enhance the discussion? Are those figures formatted properly? Does preceding text refer to and introduce those figures?
- Are appropriate front-matter components included (title page, transmittal letter, table of contents, list of figures, abstract)? Are those components formatted in a standard way?
- Are appropriate back-matter components included (list of information sources, appendixes)? Are those components formatted in a standard way?
- Is a system of documentation correctly used to indicate sources of borrowed information?
- Are equations and formulas properly formatted?

1.4.7 REVIEWING FOR SENTENCE-STYLE PROBLEMS

Chapters 6 and 7 cover sentence-style problems. A **sentence-style problem** makes a sentence wordy, indirect, unemphatic, and unclear, even though that same sentence may have no grammar, usage, or punctuation errors. Review your draft for these sentence-style problems:

- Weak, inappropriate passive voice. (But remember that passive voice is a useful writing tool; see Chapters 6 and 7.)
- Excessive nominalization, over-reliance on the *be* verb. Put active verbs in those sentences.
- Unnecessary, wordy expletives (variations of *it is, there is*).
- Wordiness and redundancy.
- Lack of subordination, leading to weak use of *and* or to short, choppy sentences.
- Excessive subordination, leading to long, dense sentences.
- Overly formal, stuffy writing style or overly informal, unprofessional writing style.

1.5 PROOFREADING THE FINAL MANUSCRIPT

Proofreading is the final stage of the writing process.

1.5.1 PROOFREADING OVERVIEW

Proofreading involves looking for correcting errors involving the following:

- Grammar
- Usage
- Punctuation
- Spelling
- Capitalization
- Typography
- Missing words or letters
- Any other mechanical or writing convention problems

  Proofreading your own work is difficult because you probably read quickly, skipping over words and phrases because you assume what they will be—or ought to be. This problem is exacerbated if you have not had time away from your draft so that you can see it with "fresh" eyes. Proofreading is painstaking. It demands slow, careful, and methodical reading so you can identify any errors in the final draft.

1.5.2 PROOFREADING STRATEGIES

Because proofreading requires checking and rechecking your work, you might consider combining strategies, or experimenting with a series of strategies to reach your goal—an error-free manuscript. Here are some proofreading strategies to choose from:

- Read slowly, examining each word separately; consider proof-reading using a ruler so that you focus on one line at a time.
- Make a list of your most common grammar, punctuation, spelling, and mechanical errors and then check your draft thoroughly for occurrences of these errors.
- Use or adapt a proofreading checklist to check for errors systematically. Look for each error type one a time.
- Proofread aloud, emphasizing each part of a word as you read.
- Proofread your sentences in reverse order (this will take your attention off meaning so that you can focus on words, letters, and punctuation).
- Read "against copy"; this means comparing your final draft one sentence at a time against the edited draft to ensure that all editorial changes have been implemented.
- Use computer spell checkers and grammar checkers, but never rely on them exclusively. All such tools have limitations and should be employed as only part of your extensive proofreading repertoire.
- Watch out for similar-sounding words like *affect* and *effect, principal* and *principle*, and *its* and *it's*. (These are listed in Chapter 7.)
- Work with someone else to proofread your work.

Submitting a manuscript riddled with errors can undermine your credibility and professionalism. It can make readers think, perhaps unfairly, that you didn't bother to check supporting facts or even think rigorously about large ideas. In the working world, employers have been known to reject applications containing a single spelling or punctuation error that should have been caught by careful proofreading.

2

PROFESSIONAL DOCUMENT DESIGN

2

Professional Document Design

2   Professional Document Design

The documents you write in your engineering studies and in your engineering career, of course, must be well written. But in these professional contexts, you have some additional expectations—to produce documents that use design, format, and style that make those documents readable, accessible, and professional-looking. In plenty of instances, both academically and professionally, these expectations include a standard framework and structure for certain documents such as proposals, progress reports, and other specialized reports.

These expectations have to do with **document design**. They include decisions about:

- Fonts

- Margins, line spacing, and alignment

- Headings and lists

- Highlighting and notices

- Graphics and tables

- Equations

- Online media

All of these issues are covered here. A well-designed document is readable, accessible, and professional-looking.

Note: The guidelines presented in the following text are not hard-and-fast rules. Requirements for fonts, margins, line spacing, alignment, headings, lists, notices, highlighting, tables, and graphics vary from industry to industry and from organization to organization.

2.1 FONTS, MARGINS, LINE SPACING, ALIGNMENT

When you begin a document project, make sure you have made the appropriate settings:

- Font—including typeface style and size
- Margins
- Line spacing
- Alignment

Please note that for the instructional purposes of this text, sample document margins and page sizes throughout are not necessarily to scale.

2.1.1 FONTS

Choose a standard font size for your engineering documents. Normal font sizes for the body text of professional documents range from 10 points to 12 points:

10-point font
11-point font
12-point font

Choose a standard font style. Fonts are categorized as serif or sans serif. **Serif fonts** have little curlicues at the end of the main strokes of each letter which, according to printing-industry tradition, make text easier to read. In the body text of professional documents, use a serif font, such as Times New Roman or Garamond. **Sans serif fonts** (without serif) do not have the little curlicues. Use a sans serif font, such as Arial or Helvetica, for headings and figure and table titles. (See the sections on headings, graphics, and tables later in this chapter.)

Avoid unusual font styles such as Impact, Benguiat, Lithograph, Haettenschweiler, or Futura. These fonts are designed to be

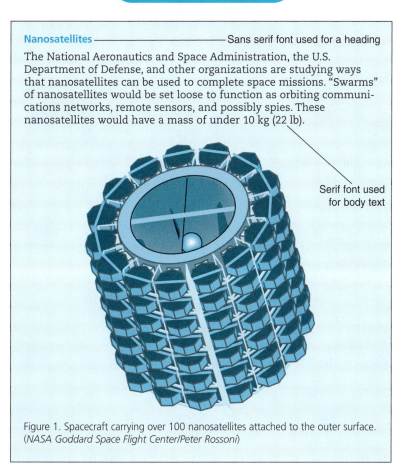

Nanosatellites ─────────────────── Sans serif font used for a heading

The National Aeronautics and Space Administration, the U.S. Department of Defense, and other organizations are studying ways that nanosatellites can be used to complete space missions. "Swarms" of nanosatellites would be set loose to function as orbiting communications networks, remote sensors, and possibly spies. These nanosatellites would have a mass of under 10 kg (22 lb).

Serif font used for body text

Figure 1. Spacecraft carrying over 100 nanosatellites attached to the outer surface. (*NASA Goddard Space Flight Center/Peter Rossoni*)

FIGURE 2.1
Serif fonts for body text and sans serif fonts headings.

used on small amounts of text and are difficult to read for any long period.

Impact font	Benguiat font	LITHOGRAPH FONT
Haettenschweiler font	**FuturaBlack BT font**	Allegro BT

Don't use all italics, bold, or all capitals in the main text of your essay. These effects should be used only for emphasis, of a word or phrase, or on headings. If you believe that you must emphasize text (such as cautions or warnings), use the notice format discussed later in this chapter. (Apply this to your résumé design as well.)

2.1.2 MARGINS

Standard margins are one-inch (about 2.5 cm) on the top, bottom, and both sides of a page. Some commonly used word-processing software uses 1.25-inch left and right margins. Check with your organization or instructor to see if there are stringent requirements on margins. If not, you can adjust these settings to make text fit properly on pages.

2.1.3 LINE SPACING

You may have double-spaced your academic papers. This require-ment is likely to change in your engineering coursework and in your engineering career. Ask questions; look around; find out.

You can adjust line spacing in single-spaced text for greater read-ability. In commonly used word-processing software, the setting for line spacing in Times New Roman 10-point font is 12 points. For Times New Roman 12, it is 14 points. If you find that these settings for line spacing create text that seems too "tight," add one or two points to the line-spacing value.

2.1.4 ALIGNMENT

For the normal text of standard professional documents, you can choose between three types of alignment:

- *Justified.* **Full justification** means that both left and right margins are aligned, such as you see in professionally printed materials.
- *Left-aligned.* When a document is left-aligned (often called "**ragged right**"), only the left margin is aligned, while the right margin is "ragged." In other words, the line endings are uneven.
- *Ragged-right.* Use ragged-right style in professional and academic documents. Unless you have access to a professional document designer and sophisticated publishing equipment, your justified text will be difficult to read. Poor justification can create "rivers" of white space on the page and odd spacing between words and necessitates frequent use of hyphenation—all of which makes text difficult to read.

2.2 HEADINGS

The titles and subtitles of sections within a document are called **headings**. They function as if the writer had spliced in items from the document's outline right into the text where the corresponding sections begin. Aside from the fact that headings give your professional documents a polished professional look, headings have some important functions; they

- Indicate the topic, purpose, or both of the corresponding section.
- Focus readers' attention on the topic or purpose of the corresponding section.
- Enable readers to skip sections they are not interested in.
- Provide an overview of the document, a sort of ongoing outline right in the text.
- Add white space to a document, thus increasing its readability.

Headings also have the indirect advantage of focusing writers' attention on the current topic and enabling writers to stay organized. When you first start using headings, you will probably write the draft and then go back and insert the headings. As you get used to using headings in your professional documents, you'll begin to add headings as you write the draft—a sort of outlining on the fly. Such practice is a good thing: when you add a heading during the rough-drafting stage, you are saying to yourself, "okay, now I'm going to focus on the applications of this technology."

As you can see in Figure 2.2, headings correspond directly to the outline of a document.

When you create headings, keep the following guidelines in mind.

2.2.1 USE ADEQUATELY SELF-DESCRIPTIVE HEADINGS

For example, headings like "Technical Background," "Overview," or "Technical Discussion" don't tell readers anything about the topic of the upcoming section. Instead, add the specifics—for example, "Technical Background: Computer Vision Systems" or "Overview: Cryogenic Applications."

Sometimes, a section will cover multiple subtopics. For example, if in a section on biomedical engineering safety, you discuss both

SQUID APPLICATIONS

Applications of superconducting quantum interference devices (SQUIDs) include magnetic-field measurement, electrical measurement, noise thermometry, as well as geophysical and biological applications.

Magnetic Field Measurement using SQUIDs

A SQUID is capable of detecting a change in magnetic field without any extra equipment. However, in practice obtained by linking the SQUID to a flux tr of transducers. Currently, SQUID measure are accomplished primarily with (a) magn ters, and (c) scanning SQUID microscopes

Magnetometers. One measurement devi which measures the magnitude of an appl case, the flux transformer is a simple two-(Figure 10).

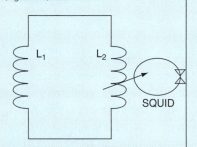

Figure 10: Basic SQUID magnetometer

Gradiometers. Another measurement de which uses an arrangement similar to the two or more pick-up coils. The two pick-u site senses. If a uniform field is applied . .

Scanning SQUID Microscopy. One last to be considered is the scanning SQUID n extremely sensitive instrument for imagin scans the SQUID relative to the sample to fields with extraordinary sensitivity. As with most scanning probe microscopes, the probe is moved about using piezoelectric scanning elements . . .

—Adapted from the Superconducting Quantum Interference Devices (SQUIDs) Research Group, Bristol University Physics Department. http://homepages.nildram.co.uk/~phekda/richdawe/squid/

FIGURE 2.2

Headings in a document and document outlines.

safety and regulatory issues, it would be insufficient to use a heading like "Biomedical Engineering Safety"; instead, use "Biomedical Engineering: Safety and Regulatory Issues."

2.2.2 USE HEADINGS AT A REASONABLE FREQUENCY

There is no set rule as to how many headings and subheadings to create per page. However, if you have a standard page of single-spaced text with only paragraph breaks, look carefully for opportunities to add headings and subheadings.

At the same time, avoid having too many headings. If most subheadings in a section are followed by only a sentence of body text, you need to use some other format such as a table.

2.2.3 SUBORDINATE HEADINGS PROPERLY

When you "subordinate" one topic to another, you indicate that the subordinated topic is a subset of the other topic. The same thing occurs in outlining: when you have II. Vegetables and you make onions A, potatoes B, squash C, and so on, you are subordinating the items that belong to the vegetable category. The same process applies to headings. You must use typographical and format techniques to indicate the subordination levels of headings. Most word-processing software provides styles, which include formats for title, heading 1, heading 2, and heading 3. All you have to do is put your cursor on the text you want to set as a heading 2 and click Heading 2 in the style box. In Figure 2.3, the style called `squid_heading1` has been applied to SQUID APPLICATIONS; `squid_heading2`, to Magnetic Field Measurement using SQUIDs; `squid_heading3`, to Magnetometers; and `squid_parag0`, to the regular body paragraphs.

When you designate headings to indicate levels, notice what happens: font, font size, bold, italics, above and below spacing, and in some cases horizontal lines indicate the level of subordination of each heading. You can tell just by looking at the heading (or you should be able to) what level in the hierarchy it is. The three levels of headings in Figure 2.3 use Arial 14 bold, Arial 12 bold, and Arial 10 bold, respectively.

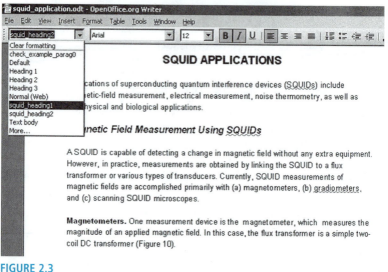

Using software "styles" to create headings.

In addition to taking care of the phrasing, frequency, and subordination of headings, be aware of the following guidelines.

2.2.4 AVOID LONE HEADINGS

A **lone heading** is a certain level of heading all by itself within a section without another heading at its same level either directly before or directly after it. For example, picture a document with a heading 1, followed by some text, followed by a heading 2, again followed by some text, and then followed by a heading 1. The heading 2 in between is the lone heading. Lone headings are like a 1 without a 2 or an A without a B in outlines. To fix lone headings, look first at the text to see if you can create a companion heading. If you can't, just delete the lone heading. Before you do, see if you need to incorporate some of its phrasing into the preceding heading.

2.2.5 AVOID STACKED HEADINGS

When two or more consecutive headings occur without intervening text, you've got **stacked headings**. While you don't want to jam meaningless text between stacked headings, there are some useful things you can do to unstack headings:

- Provide some transition from the preceding section.
- Provide an overview of the upcoming section.
- Give some essential background information, such as a definition.

2.2.6 MAKE HEADINGS PARALLEL IN PHRASING

Parallelism, as Chapter 6 explains, refers to the use of a similar style of phrasing or items in a list. Headings at the same level within the same section of a document must also use parallel phrasing. Look at Figure 2.4 to see what this means.

2.2.7 AVOID PRONOUN REFERENCE TO HEADINGS

Avoid using pronouns like "this" or "it" to refer to a topic stated in the preceding heading. For example, if you have a heading such as "Computer-Vision Systems: Recent Developments," don't begin the

Biological Applications

Stacked headings

Biomagnetism. The invention of SQUIDs has enabled the measurement of magnetic fields, ranging from the susceptibility of tissue to applied magnetic fields to ionic healing currents and those associated with neural or muscle activity. With fields of such a s

Biological Applications

The extreme sensitivity of SQUIDs make them ideal for studies in biology. While a typical refrigerator magnet is ten thousand microteslas, some processes in animals produce very small magnetic fields, typically between a nanotesla and a microtesla.

Biomagnetism. The invention of SQUIDs has enabled the measurement of magnetic fields, ranging from the susceptibility of tissue to applied magnetic fields to ionic healing currents and those associated with neural or muscle activity. With fields of such a small magnitude, ambient magnetic noise is of comparable magnitude to the signal being measured . . . .

Brain Imaging. By far the largest area of study within biomagnetism is brain imaging. Most existing non-invasive brain imaging methods such as Computerised Tomography (CT), Magnetic Resonance Imaging (MRI) and Positron Emission Tomography (PET), measure the distribution of some kind of matter and are therefore primarily a measure of structure. . . .

FIGURE 2.4
Stacked headings. Try to include regular text between any two headings.

Brain Imaging. This area is by far the largest area of study within biomagnetism is brain imaging. Most existing non-invasive brain imaging methods such as Computerised Tomography (CT), Magnetic Resonance Imaging (MRI) and Positron Emission Tomography (PET), measure the distribution of some kind of matter and are therefore primarily a measure of structure. . . .

Pronoun referring to heading

Brain Imaging. By far the largest area of study within biomagnetism is brain imaging. Most existing non-invasive brain imaging methods such as Computerised Tomography (CT), Magnetic Resonance Imaging (MRI) and Positron Emission Tomography (PET), measure the distribution of some kind of matter and are therefore primarily a measure of structure. . . .

FIGURE 2.5
Pronoun reference to a heading. Write text as if the headings were not there.

following sentence like this: "In this area . . . ." Instead, repeat the topic, varying the words if you can or pushing the repetitious words toward the end of the sentence: "Recent developments in the field of computer vision have . . . ."

2.2.8 DO NOT MAKE A HEADING A GRAMMATICAL PART OF THE FOLLOWING SENTENCE

A heading simply cannot act as part of the grammar of the body sentence that follows it. Just repeat the key words in the heading topic in the following body text, perhaps altering the wording a bit or shifting those words down toward the end of that first body sentence.

2.3 LISTS

Lists are those numbered, bulleted or otherwise itemized chunks of text you see in business and technical documents. Types of lists include:

- Numbered lists
- Bulleted lists
- Simple lists
- In-sentence lists

The first three can all be categorized as "vertical" because, as you can see in Figure 2.6a, they are listed vertically on the page. In-sentence lists can be categorized as "horizontal" because, as you can also see in Figure 2.6a, they occur within regular paragraph format.

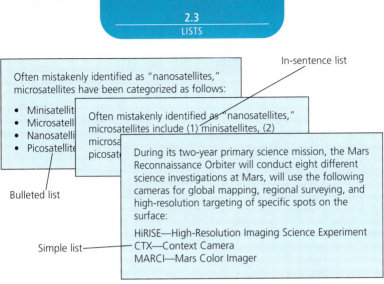

FIGURE 2.6A
Common types of lists.

Lists are useful because they make text easier to follow sequentially, enable text to be scanned more readily, emphasize text, and make certain kinds of text easier to find. Some guidelines on using lists follow.

Mars Reconnaissance Orbiter launched in August, 2005 from Space Launch Complex 41 at Cape Canaveral Air Force Station and followed these steps through the launch phase:

1. After countdown and systems checks, lift-off through Atlas Stage 1 booster accelerates the spacecraft to supersonic speeds of about 4,500 meters per second (10,000 miles per hour).
2. The booster engine for the Atlas Stage I booster cuts off, separates from Centaur Stage II, and falls back to Earth into the Atlantic Ocean.
3. Centaur Stage II main engine starts and first burn phase lasts 9.5 minutes, boosting the spacecraft into a "parking orbit" in which the spacecraft and Centaur coast for about 33 minutes in between the first and second Centaur burns.
4. Payload fairing is jettisoned and falls back to Earth into the Atlantic Ocean, and main engines cut off for the Centaur first burn.
5. Centaur second burn phase lasts about 10 minutes, accelerating the spacecraft out of Earth orbit and setting it on its way to Mars.
6. The main engine separates from the Centaur Stage II, and Centaur is maneuvered to avoid collision with the spacecraft.

FIGURE 2.6B
Numbered lists: used for items in a required order.

2.3.1 USE NUMBERED AND BULLETED LISTS CORRECTLY

Don't send readers the wrong signal by numbering list items that are not in any required order. For example, if you want to emphasize four important points, use bullets, not numbers. Similarly, if you are providing step-by-step instructions, use numbered lists to emphasize the order and sequence of the procedure. The numbered-list format helps readers track their actions to the instructions much more readily.

2.3.2 USE IN-SENTENCE LISTS FOR SECTION OVERVIEWS

In-sentence lists are a less emphatic format for list items. A common use for in-sentence lists is to provide an overview of the sections or subsections with a document.

2.3.3 INTRODUCE ALL LISTS WITH A LEAD-IN

Readers scan engineering, technical, and business documents. They are likely to slow down and pay attention to lists. They need list lead-ins to help them understand the meaning of lists. Always punctuate list lead-ins with a colon. The colon says "here it comes!"

2.3.4 MAKE THE PHRASING OF LIST ITEMS PARALLEL

Parallelism, as explained in Chapter 6, refers to the use of the same style of phrasing for elements in a series. When items in a list are not parallel in phrasing, the effect is distracting.

2.3.5 MAKE SURE THAT EACH LIST ITEM READS GRAMMATICALLY WITH THE LEAD-IN

In some cases, list items actually complete a sentence started by the list lead-in. When that happens, make sure *every* list item reads grammatically with the lead-in.

Magnetic Field Measurement using SQUIDs

A SQUID is capable of detecting a change in magnetic field without any extra equipment. However, in practice, measurements are obtained by linking the SQUID to a flux transformer or various types of transducers. Currently, SQUID measurements of magnetic fields are accomplished primarily with (a) magnetometers, (b) gradiometers, and (c) scanning SQUID microscopes.

FIGURE 2.7
In-sentence list format used to indicate section overview.

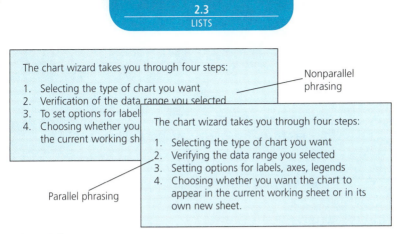

FIGURE 2.8
Parallel phrasing for list format items.

2.3.6 BE CONSISTENT WITH PUNCTUATION AND CAPITALIZATION

If you work for state or federal government, you may be required to end list items with a semicolon and next-to-last list items with a semicolon followed by *and* or *or*. If not, use a period on list items that are complete sentences or that contain a dependent clause. (See Chapter 3 for dependent clauses.)

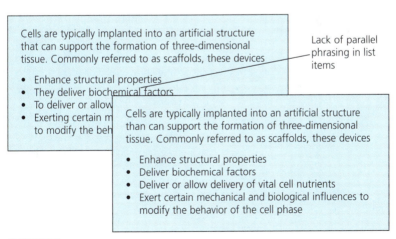

FIGURE 2.9
Grammatical connections between lead-ins and list items

If you selected several templates, you must first superimpose them by doing one of the following:

- Click **Fit>Iterative Magic Fit** (the structural alignment will be automatically done).
  or
- Click **Fit>Magic Fit** and **Fit>Generate Structural Alignment**.

FIGURE 2.10
"Or" list. Don't use numbering for alternatives.

2.3.7 USE BULLETS FOR ALTERNATIVES

When you have two choices or options, bullet those choices or options—don't use numbers. Numbering implies a sequence. Numbering two options sends the wrong message. A useful additional touch is to add "or" between the two items.

2.3.8 FORMAT NESTED LISTS CORRECTLY

Some list items are so detailed that you must create sublists. As you can see in Figure 2.11, you can have a bulleted list with sub-bullets, a numbered list with numbered sublist items, a bulleted list with

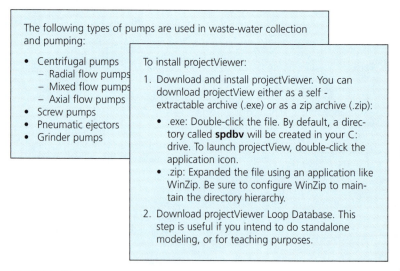

The following types of pumps are used in waste-water collection and pumping:

- Centrifugal pumps
  - Radial flow pumps
  - Mixed flow pumps
  - Axial flow pumps
- Screw pumps
- Pneumatic ejectors
- Grinder pumps

To install projectViewer:

1. Download and install projectViewer. You can download projectView either as a self - extractable archive (.exe) or as a zip archive (.zip):
   - .exe: Double-click the file. By default, a directory called **spdbv** will be created in your C: drive. To launch projectView, double-click the application icon.
   - .zip: Expanded the file using an application like WinZip. Be sure to configure WinZip to maintain the directory hierarchy.
2. Download projectViewer Loop Database. This step is useful if you intend to do standalone modeling, or for teaching purposes.

FIGURE 2.11
Examples of nested lists.

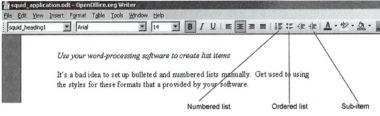

FIGURE 2.12
Using software "style" to create lists.

subnumbered items, and a numbered list with bulleted sublist items—in other words, every combination possible.

Notice how the sublist items indent to the *text* of the parent list items. Notice too that the sub-bullet items use a less noticeable symbol such as a hyphen or a clear circle.

2.3.9 USE WORD-PROCESSING STYLES FOR LISTS

It's a bad idea to set up bulleted and numbered lists manually. Get used to using the styles for these formats that are provided by your software.

2.4 HIGHLIGHTING

Highlighting refers to the use of typographical effects such as bold, italics, underlining, capital letters, alternate fonts, color, or combinations of these to do one of two things:

- Emphasize something important, to ensure that it gets noticed.
- "Cue" readers to expect something, such as a word at its point of definition.

2.4.1 HIGHLIGHTING PROBLEMS

As mentioned in the previous section on fonts, you must avoid unsightly combinations of bold, italics, underlining, capital letters, alternate fonts, and color to create emphasis:

- Avoid using special typographical effects for extended text of a sentence or more. Limit such special typographical effects to a few words.

WARNING: *NEVER* PUT TRANSPARENCIES DESIGNED OR SPECIFIED FOR INKJET PRINTERS IN THE HP4500. THESE TRANSPARENCIES HAVE A ROUGH SURFACE TO ABSORB INK AND WILL *JAM* THE PRINTER AND MELT, POSSIBLY CAUSING *SERIOUS INTERNAL DAMAGE!* USE ONLY TRANSPARENCIES DESIGNED FOR THE HP4500 OR STANDARD LASER PRINTER TRANSPARENCIES.

Highlighting overload

WARNING: Never put transparencies designed or specified for inkjet printers in the HP4500. These transparencies have a rough surface to absorb ink and will jam the printer and melt, possibly causing serious internal damage! Use only transparencies designed for the HP4500 or standard laser printer transparencies.

FIGURE 2.13

Notices and readability. The text overloaded with bold, italics, and caps is uncomfortable for most readers. In the notice format, the label calls attention to the text, but the actual text of the notice remains readable.

- Use italics or bold (but not both) for individual words you want to emphasize (such as *no, not, never, always*).
- If you have a large chunk of information you must emphasize, use a notice format.

2.4.2 HIGHLIGHTING SYSTEMS

Certain types of technical information, such as computer documentation, require a rather complex system of highlighting. Be careful to follow the highlighting standards for your organization or industry as consistently as possible. If you must design your own highlighting system, keep it simple; otherwise, the system will be lost on readers. Here is an example highlighting system commonly used in computer-user guides:

Bold	Interface elements such as on-screen buttons and menu options that caused something to happen when pressed.	Click the **OK** button to save your changes.

FIGURE 2.14

A common highlighting system used in computer documents.

FIGURE 2.14 (*Continued*)

	Command-line commands (such as in Linux).	Type **ls –al** to see file details.
Italic	Words needing emphasis.	Do *not* enter `rm –fr *.*!`
	Variables names: words that users replace with their own text.	Type **rm** *myfile* and press Enter.
`Courier`	Text that users type in verbatim; messages displayed on screen.	Type `rm –f myfile.txt` and press Enter.
Regular	Interface such as field labels and window names that users do not click and that do not cause anything to happen. (But use the same capitalization style.)	Enter your name in the blank next to Your Name.

Even in this system, you can probably see some potential inconsistency: for example, the rm command is bold in one instance and Courier New in another. This system does not overtly indicate how to highlight keyboard key names: Enter or **Enter**? Nor does this system indicate what to do with hardware labels such as on/off labels and the like, especially when they use symbols rather than words.

2.5 NOTICES

Notices, also called "alerts," are those specially formatted cautions and warnings you see in text and product labeling. The notice format spares readers from having to read text that is all bold, all italics, all caps, or combinations thereof as shown in Figure 2.13. Notices are particularly useful in instruction-type text. In fact, notices are a legal issue. Companies have been sued for lack of appropriate notices and for ineffective notices.

2.5.1 TYPES OF NOTICES

Notices use different formatting so that the most urgent notice has the most "noticeability" while less-urgent notices use less-noticeable formatting. Standards for notices vary widely across industries and even organizations within industries. Here is one hierarchy of notices that is well established:

- **Danger**. For situations in which there is the potential for serious or fatal injury to human beings.
- **Warning**. For situations in which there is the potential for minor injury to human beings.
- **Caution**. For situations in which there is the potential for damage to equipment and data, and for situations in which the outcome of a procedure could be ruined.
- **Note**. For all other situations in which something must be emphasized, such as an exception.

Look around at product labeling and user or operator guides: you'll see a lot of variability in notice format. You're likely to see "Important" and "Attention" used instead of "Note."

2.5.2 NOTICE CONTENT

Wording for notices is critical: you must express the essentials in as few words as possible, while maintaining readable, understandable English. In any notice you write, consider including the following components:

Alert	The direct command to do something or not to do something
Conditions	The situation in which the reader should (or should not) heed the alert
Consequences	What will happen if the reader fails to heed the alert
Recovery	What to do if the reader fails to heed the alert

2.5.3 NOTICE FORMAT

Formatting for notices corresponds to the urgency of the notices. Figure 2.15 shows reasonably common formats for the four notices previously described. Notice that the label (Caution, Warning, Danger, Note) has all the typographical effects such as bold, italics, underlining, capital letters, alternate fonts, color, or combinations. The actual text does not; it remains calm, reasonable, and readable.

> *Note:* The odometer cannot distinguish between forward and reverse wheel condition. Operating the LRV in reverse adds to the odometer reading.

> *Caution:*
> If the GCTA staff is not properly locked into place, it can fall on the Lunar Communications Relay Unit and cause severe radiator damage.

> **Warning:**
> When recharging batteries, handle the electrolyte (KOH) with care. It is alkaline and corrosive and can cause serious burns if allowed to contact eyes or skin.

> **Danger**
>
> **The impact of an asteroid of about 0.5 and 5 km diameter, perhaps near 2 km, will cause widespread global mortality and threaten civilization. Because such an object may strike in the next hundred years, chemical rockets, or perhaps mass drivers, should be developed to divert it from its path toward Earth.**

FIGURE 2.15
Four common notice formats. Despite the science-fiction nature of the danger notice, observe that it has all the elements of a notice: alert, conditions, consequences, and recovery.

2.5.4 NOTICE PLACEMENT

Placement of notices is tricky. The standard rule is to place the notice *before* the text where the notice is applicable. However, in practice, this rule is not always followed. Danger and warning notices are often placed at the beginning of the section where they apply—well before the actual steps to which they apply. Less critical notices are placed just after the text to which they apply.

2.6 TABLES

A **table** allows you to summarize large amounts of information, which is usually in statistical form. Since tables are set up in columns, they are also useful for comparing information.

2.6.1 CONVERTING TEXT TO TABLES

Study your draft documents carefully for text that could be turned into tables. Look for clusters of data that occur in pairs or triplets: for example, names of countries, GNP increase or decrease, and corresponding years. But don't look just for numerical data. Textual material can sometimes also be better presented in a table. For example, you might have a paragraph in which you define a number of key terms. Convert that text to a table—terms in the left column, definitions in the right column.

Missions to Mars have surprisingly long history, for the U.S. dating back to 1964. However, the Viking 1 and 2 Orbiter/Lander accomplished the first successful landing in 1975, returning 16,000 images and extensive atmospheric data and soil experiments. In 1992, the Mars Observer was lost prior to Mars arrival. In 1996, the Mars Global Surveyor returned more images than all previous Mars missions. In 1996, the Mars Pathfinder performed highly successful technology experiments and lasted five times longer than warranty. The 1998 Mars Climate Orbiter and the 1999 Mars Polar Lander were lost on arrival. The 1999 Deep Space 2 Probes, carrying the Mars Polar Lander, were also failures, having been lost on arrival. In 2001, the Mars Odyssey sent back high-resolution images of the planet. The mission in 2003 was both a success and failure: the Mars Express Orbiter/Beagle 2 provided images but the lander lost on arrival. In 2003, the two Mars Exploration Rover missions were successful, delivering over 120,000 images and lasting 8 times longer than warranty. The jury is still out on the Mars Reconnaissance Orbiter, launched in 2005.

LAUNCH DATE	MISSION NAME	RESULT	REASON/RESULTS
1975	Viking 1 Orbiter/Lander	Success	Located landing site for Lander and first successful landing on Mars
1975	Viking 2 Orbiter/Lander	Success	Returned 16,000 images and extensive atmospheric data and soil experiments
1992	Mars Observer	Failure	Lost prior to Mars arrival
1996	Mars Global Surveyor	Success	More images than all Mars Missions
1996	Mars Pathfinder	Success	Technology experiment lasting 5 times longer than warranty

FIGURE 2.16
Turning text into tables.

FIGURE 2.16 (*Continued*)

1998	Mars Climate Orbiter	Failure	Lost on arrival
1999	Mars Polar Lander	Failure	Lost on arrival
1999	Deep Space 2 Probes	Failure	Lost on arrival (carried on Mars Polar Lander)
2001	Mars Odyssey	Success	High resolution images of Mars
2003	Mars Express Orbiter/Beagle 2 Lander	Success/ Failure	Orbiter imaging Mars in detail and lander lost on arrival
2003	Mars Exploration Rover (Spirit and Opportunity)	Success	Over 120,000 images lasting 8 times longer than warranty 2003
2005	Mars Reconnaissance Orbiter	TBD	—

2.6.2 DESIGNING TABLES

Here are some essential guidelines to keep in mind when designing tables:

- Create a table title, placed *above* the table, for all but informal tables. **Informal tables** are not cross-referenced from anywhere else in the document and can be introduced with a simple lead-in preceding the table. Include self-descriptive text with the table title. Instead of "Table 3," write "Table 3. Notebook computer sales, 1995–2005."
- Remember that *all* column headings refer downwards. In the example in Figure 2.17, it would be incorrect to insert the column heading "Activity results" in the top left column. The column heading "Apollo mission/activity" correctly provides a general title for the items in that first column.
- Use bold or italics for column headings. Use the same size font for the entire table, including the table title. Don't use any font size in the table larger than the size of the body text.

- Keep measurement indicators out of the table data cells whenever possible. If an entire column or row is psi, put psi in parentheses in the column or row heading. In Figure 2.17, notice that `hr:min` and `Km` are kept in the column headings.
- Left-align textual material, such as words and phrases. If the entire column of left-aligned data cells is too far to the left side of the column, increase the left margin setting for those cells.

As can be seen in Table 1, actual duration and distances of EVAs were reasonably close to the plan and, in some cases, over the plan. During EVA 3, astronauts Young and Duke reported that they set a "new lunar speed record" of 17 Kmph.

Table 1. Apollo 16 and 16 Extra Vehicular Activity (EVA) Comparisons

APOLLO MISSION/ ACTIVITY	DURATION (hr:min)		TRAVERSAL DISTANCE (KM)	
	PLAN	ACTUAL	PLAN	ACTUAL
EVA 1				
Apollo 15	7:00	6:32	8.2	10.3
Apollo 16	7:00	7:11	3.2	4.2
EVA 2				
Apollo 15	7:00	7:12	14.3	12.5
Apollo 16	7:00	7:23	9.5	11.5
EVA 3				
Apollo 15	6:00	4:49	10.5	5.1
Apollo 16	7:00	5:40	12.6	11.4
TOTALS				
Apollo 15	20:00	18:33	33.0	27.9
Apollo 16	21:00	20:14	25.2	27.1
LRV driving time				
Apollo 15		3:08		
Apollo 16		3:17		

(Source: Apollo 17 LRV Technical Information [4].)

FIGURE 2.17
Standard table. Notice how measurement indicators are restricted to column headings. Notice too that the preceding text provides some indication of the table's contents.

- Right-align numeric material; don't center it. If you have a mix of decimal numbers and nondecimal numbers, use decimal alignment. If the entire column of right-aligned data cells is too far to the right side of the column, increase the right margin setting for those cells, which is done in Figure 2.17.
- If you have borrowed information from some other author or source to create the table, indicate that source either in the table title or in a footnote just beneath the table. Check with your instructor or organization for precise formatting details.
- If individual columns, rows, or data cells in the table require explanation, use footnotes.

2.6.3 PLACING TABLES

You can place tables in one of two ways within a document:

- Near the text to which the table relates
- In an appendix, where the table should be labeled—for example, "Appendix C: Notebook Computer Sales, 1995–2005"

  Wherever you place tables, keep these guidelines in mind:

- If you place the table in the text, cross-reference the table in the text before the table: for example, "as shown in Table 3 . . . ."
- If you place the table in the text, briefly explain the gist of the table, particularly if it consists of numerical data. Notice the text preceding the table in Figure 2.17. Do the work for readers: point out the key trend or conclusion to be drawn from the table.
- If you place the table in the appendix, cross-reference the table in the related text: for example, "as shown in the table in Appendix C . . .".

2.7 GRAPHS AND CHARTS

Graphs are the familiar representations of data with lines. **Charts** include all manner of pie charts, bar charts, and flowcharts.

2.7.1 ADVANTAGES OF GRAPHS AND CHARTS

Graphs and charts are simply more dramatic, visual representations of data contained in tables. If you are determined that your readers understand that global warming is for real, use a line graph rather than a table. If you want stockholders to see how much better your

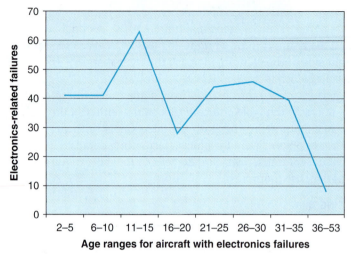

FIGURE 2.18A
Graph depicting failures by aircraft age.
(*Air Force Research Laboratory*)

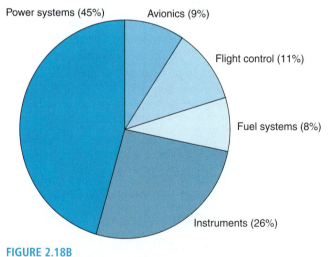

FIGURE 2.18B
Pie chart depicting failure by aircraft system.
(*Air Force Research Laboratory*)

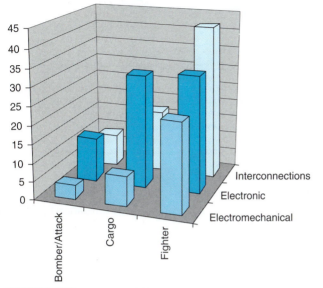

FIGURE 2.18C
3D bar chart depicting failures by aircraft type.
(Air Force Research Laboratory)

company is doing in sales in three different market sectors, a bar chart will be far more visually dramatic than a table. If you want readers to see how little wind-generated power contributes to the national total, use a pie chart as opposed to a table.

Of course, tables provide down-to-the-penny detail. If that's what your purpose and your audience require, then present a table. However, if your purpose leans toward the persuasive, graphs and charts may be preferable.

2.7.2 FLOWCHARTS

FLOWCHARTS AND ORGANIZATION CHARTS
A **flowchart** is a form of diagram, but it is so important that it deserves its own separate focus. A flowchart graphically depicts the movement of information, people, or other elements through a process. Flowcharts can use yes/no branching as the flowcharts in Chapter 1 show. Figure 2.19 shows how aircraft cruising speed affects flight time, fuel cost, and equipment, which in turn affect value, which in turn comes back to affect cruising speed.

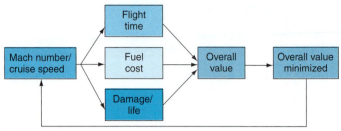

FIGURE 2.19
Flowchart depicting relationships between costs and conditions.
(NASA Glenn Research Center)

The figure shows how flight time, fuel usage, and engine damage are evaluated on the basis of aircraft cruise condition. Optimization is based on minimizing overall costs.

Organization charts illustrate relationships between people and groups within an organization, which in turn define the flow of information and the levels of authority within that organization.

2.7.3 GUIDELINES FOR GRAPHS AND CHARTS

Here are some guidelines to keep in mind for graphs and charts:

- Include a title for every graph and chart in your engineering documents. When you generate a graph and chart from spreadsheet or

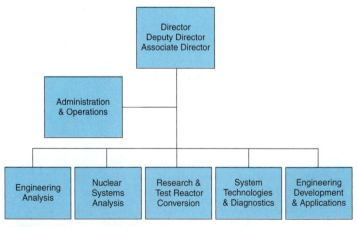

FIGURE 2.20
Organization chart depicting relationships between the different individuals and departments of an engineering firm.

database software, notice the title is centered at the top. Remove that title and create a figure title beneath the graph and chart, as shown in Figures 2.18b and 2.18c.

- Label the axes of graphs and charts. Except for pie charts, both axes of graphs and charts should contain marks indicating years, dollars, tons, or whatever is being graphed or charted. Use your software to turn the vertical axis 90 degrees counterclockwise. Include the measurement in parentheses—for example, "Total Sales (millions U.S. dollars)." Label the horizontal axis similarly unless it is obvious—for example, 5-year increments.

- Either label the lines, bars, or slices in a graph or chart or provide a legend. For the sales graph, you can label each individual graph line with the name of the company whose sales are being plotted, or you can use a legend. In a graph or chart, a *legend* is a key indicating what the different color, textures, or shadings represent—in the case of a sales graph, a different color for each company.

- For precision, you can add numeric detail or percentages to the elements of charts, as is shown in Figure 2.18b. For example, you can include a dollar amount for the bars of different market sectors in a bar chart comparing the sales of three different companies. You could indicate a percentage for each energy source in the pie chart indicating how little wind-generated power contributes to the national total.

- Always discuss the key points in a graph or chart. Refer to the graph and chart in the preceding text and mention its key trend or conclusion.

- If you have borrowed information from some other author or source to create the graph or chart, indicate that source either in the title or in a footnote. Check with your instructor or organization for precise formatting details and see the chapter on documentation.

2.8 ILLUSTRATIONS

Illustrations include the following possibilities:

- Drawings and diagrams
- Flowcharts
- Photographs

2.8.1 DRAWINGS AND DIAGRAMS

For most of your engineering documents requiring illustrations, drawings are likely to be a good choice. To provide a simplified visual representation of objects or ideas, **drawings** use simple lines to indicate the outlines of those objects or ideas along with some shading as necessary. Drawings leave out unnecessary detail and focus on the important objects, tools, actions, and concepts. Use drawings to illustrate relationships and concepts that you discuss in nearby text.

For your engineering documents requiring illustrations, diagrams are also likely to be a good choice. **Diagrams** are even more simplified visual representation of objects or ideas. They strip away even more detail to focus on a few aspects of objects or ideas—to such an extent you may not be able to recognize the object being diagrammed. For example, consider a wiring diagram. Most people might be hard pressed to realize it's the wiring diagram of a radio.

2.8.2 PHOTOGRAPHS

A **photograph** provides much more detail than a drawing or diagram—in fact, too much detail in many situations. If you want to describe the components of the Mars Global Surveyor, a drawing or diagram (Figure 2.21b) is likely to be more usable. If you intend to guide people through some repair procedure on an automobile engine, a drawing or diagram of the engine is likely to be more usable. However, if you have finished designing a solar-power system for an office park, the drawings and diagrams will be essential, but your less technical readers are likely to appreciate the photographs of how such systems look. The same is true for the photograph-like illustration of the Mars Global Surveyor in Figure 2.21a.

Whichever type of illustration you use in your engineering documents, keep these guidelines in mind:

- Place the illustration just after the point in your text where it is relevant. If it won't fit on the page at that point, let it "float" to the top of the next page. Your cross-reference will point readers to it.
- In the text where the illustration is relevant, create a cross-reference. When appropriate, don't just write "See Figure 3"; explain something about Figure 3—for example, what it illustrates.

FIGURE 2.21A
Photograph-like illustration of the Mars Global Surveyor.
(Courtesy NASA/JPL-Caltech)

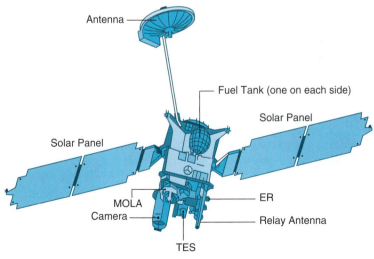

Antenna

Fuel Tank (one on each side)

Solar Panel

Solar Panel

MOLA

Camera

ER

Relay Antenna

TES

FIGURE 2.21B
Drawing of the Mars Global Surveyor.
(Courtesy NASA/JPL-Caltech)

- Create a figure title for each illustration and place it *below* the illustration. Make that figure title descriptive of the illustration: don't just write "Figure 3"; make it descriptive, such as "Figure 3. Visual examples of axial stress, torsional stress, and flexural stress."
- You can leave out the cross-reference and the figure title if the illustration is decorative or if it has only immediate relevance. For example, most instructions use many illustrations, most of which illustrate only a specific step and are introduced by the preceding text. Figure titles in these cases serve no purpose.
- When appropriate, include labels in illustrations, those words and phrases that indicate the names of the various parts shown in the illustration.
- Always indicate the source of illustrations you borrow from other authors. Use the style required by your organization or instructor. See the chapter on documentation in this book.

2.9 EQUATIONS

Presenting equations is essential in many engineering documents. Because word-processing software now handles equations effectively, not much formatting and style detail is needed here. A nice introduction is made available at the University of Waterloo called "Creating Equations with Microsoft Word and PowerPoint": http://ist.uwaterloo.ca/ec/equations/equation.html

Here are some standard guidelines for incorporating equations into your engineering documents:

- Center equations and number them consecutively with parenthetical numbers on the right margin. If you have numbered chapters or sections, use the double-enumeration system (for example, 1.1, 1.2); otherwise, use single numbers. For example: A single-layer neural network can compute a continuous output instead of a step function. A common choice is the so-called logistic function:

$$\frac{1}{1 + e^{-x}} \qquad (2)$$

- When you cross-reference equations, put the equation number in parentheses:

With the logistic function (2), the single-layer network is identical to the logistic regression model, widely used in statistical modeling.

- Italicize regular alphanumeric symbols for quantities and variables. (Some styles direct you not to italicize Greek symbols). Do not use italics on common functions that have their own names (such as log, tanh, sign):

$$f(x) = K(\sum_i (w_i g_i(x))$$

- Use an en dash (–) rather than a hyphen for a minus sign; use parentheses to avoid ambiguities in denominators:

$$y = \text{sign}(x - d)\left(1 - \exp\left(-\left(\frac{x - d}{s}\right)^2\right)\right)$$

- Punctuate equations with commas and periods.
- When you have a multiline equation, do not use punctuation where the lines break. Align the runover lines beginning with a relation sign (for example, an equals sign) to the relation sign above; align runover lines beginning with an operation sign (for example, a plus sign) to the first character to the right of the equals sign on the line above it:

$$I^2 = \int_0^{2\pi} \int_0^\infty e^{-r^2} r \, dr \, d\theta = 2\pi \int_0^\infty e^{-r^2} r \, dr$$

$$= \pi \int_0^\infty e^{-u} \, du = \pi.$$

- Define local symbols and abbreviations in your equation before the equation appears or immediately following:

The double sigmoid is a function similar to the sigmoid function with numerous applications. Its general formula is:

$$y = \text{sign}(x - d)\left(1 - \exp\left(-\left(\frac{x - d}{s}\right)^2\right)\right)$$

where d is its center and s is the steepness factor.

- In regular text, do not use mathematical symbols as shorthand for words:

If \$empty_file = no . . .

If \$empty_file equals no . . .

2.10 WEB PAGES AND WEB SITES

As an engineering student or professional, you may have the opportunity to present your information as a web page or web site. A **web site** is simply a collection of multiple related web pages. While the structure and design of web sites vary enormously, consider using the framework described here.

"Chunk" your information into separate HTML files—separate web pages—so that readers will not have to press the Page Down key more than two or three times. But be sure to make those chunks logical; carve up your document in logical segments such as a section or chapters.

Create an entry page where visitors will start and name it index.html. That way, instead of typing `http://www.io.com/mcmurrey/list_reports.html`, they can just type `http://www.io.com/mcmurrey/`.

Design the entry page to include some sort of table of contents (TOC) that enables visitors to get to every one of the web pages in your web site. A favorite design is to use frames to create a TOC frame on the side of the page and a "reading" frame on the other in which to display whichever item in the TOC the visitor has clicked (Figure 2.22).

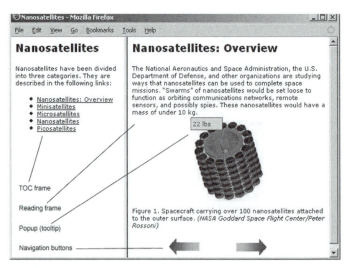

FIGURE 2.22
Entry page to a web site with a TOC frame and a reading frame.

In the individual web pages of your web site, provide links to previous or next web page, when doing so makes sense. If you link to an external web page, open that page in a separate, full browser—not the reading frame. Avoid cluttering up the text of the web page in the reading frame with lots of links. Links are ultimately distracting for visitors; they are not sure whether the link is important enough to interrupt their train of thought and visit the link. If you simply want to provide a definition of a term, pop that definition up in a popup window with JavaScript popup code, a tooltip, or the `<title>` tag.

Obviously, lots of web page design issues are not covered in the preceding. Plenty of advice and tutorials are available on the Web. Web page tools such as Dreamweaver, HomeSite, or FrontPage make creating web pages easy. In any case, the guidelines presented in this book for headings, lists, notices, highlighting, tables, charts, illustrations, and writing style and clarity all still apply.

3

BASIC GRAMMAR

3

Basic Grammar

3 Basic Grammar

In this section, you review the basic concepts of grammar which will give you a common vocabulary when you work on engineering and business documents with other people.

3.1 PARTS OF SPEECH

Every word in a sentence functions as some part of speech. Words can be classified as one or more of the following parts of speech:

- Nouns
- Articles, or determiners
- Pronouns
- Verbs
- Adjectives
- Adverbs
- Prepositions
- Conjunctions
- Interjections

Which part of speech a word plays depends on how it functions in an individual sentence. Notice in these examples how *increase* can be a noun in one and a verb in the other:

Noun
A <u>decrease</u> in the national wealth manifests itself in an increase in foreign debt.

Verb
Between 2002 and 2005, the percentage of U.S. workers employed in industrial jobs <u>decreased</u> from about 14 percent to 10 percent.

Some words are **singular**, which means they denote a single person or thing (*I, calculator*); some are **plural**, designating more than one person or thing (*we, calculators*).

Singular Words
Engineering is the application of scientific and technical knowledge to solve a human problem.

Plural Words
In the second half of the nineteenth century, Willard Gibbs initiated a more scientific understanding of materials by examining how the thermodynamic properties that govern the arrangement of atoms in various phases are related to the properties of a material.

Some words may function as the **subject**, or doer of the action, in a sentence (*Math* is fun) or as the **object**, who or what is affected by the action (*I like math*), or as a **complement** (*Math is fun*).

Subject
Modern materials science is a product of the exploration of space.

Object
Thermodynamics studies the effects of changes in temperature, pressure, and volume on physical systems by analyzing the collective motion of their particles.

Complement
The starting point for most thermodynamics is the laws of thermodynamics, which state that energy can be exchanged between physical systems as heat or work.

A group of words may form a **sentence**, which has two main parts, a **subject** and a **predicate** (which includes the verb, object, and phrases governed by the verb), and can stand alone. A group of words may form an **independent clause**, which is a complete sentence; or it may form a **dependent clause**, which has a subject and predicate, is a part of a larger sentence, and cannot stand alone. Finally, a group of words may form a **phrase**, which does not have a subject or verb and cannot stand alone. A word, phrase, or clause may be a **modifier**, meaning that it describes or limits another word, phrase, or clause within a sentence.

Subject
Fracture <u>mechanics</u> is an important tool in improving the mechanical performance of materials and components.

Predicate
In many cases, failure of engineering structures through fracture <u>can be fatal</u>.

Modifying Clause
Industrial engineering is the engineering discipline that concerns the design, implementation, and evaluation of integrated systems of people, knowledge, equipment, energy, material, and process.

Modifying Phrase
<u>Using higher pressure or temperature</u>, some chemical reactions can be made easier.

The three primary physical laws <u>underlying chemical engineering design</u> are conservation <u>of mass</u>, conservation <u>of momentum</u> and conservation <u>of energy</u>.

Modifying Words
Before the <u>Industrial</u> Revolution, there were <u>military</u> engineers who built such things as fortifications, catapults, and, later, cannons and <u>civil</u> engineers who built bridges, harbors, aqueducts, buildings, and other structures.

To distinguish parts of speech, follow the guidelines that follow.

3.1.1 NOUNS

A **noun** names a person, place, or thing. Whenever an **article** (*a, an, the*) could precede a word without destroying the logic of a sentence, that word is a noun. Nouns can be classified by type:

TYPES OF NOUNS		
NOUN TYPE	**THESE NAME:**	**EXAMPLES:**
Proper	specific people, places, and things (always with an initial capital)	*Buckminster Fuller, Moose Jaw, Honda*
Common	general people, places, or things	*architect, town, car*

Concrete	things you experience through your senses	*gravel, ice cream, storm*
Abstract	things you do not experience through your senses	*knowledge, liberation, fear*
Collective	groups	*jury, police*
Noncount	things that cannot be counted	*snow*
Count	countable things	*snowflakes, cornflakes*
Possessive	indicate ownership	*corporation's, engineer's community's*

ESL NOTE Watch out for certain common suffixes (or endings) on nouns. These endings are often a clue that the word in question is a noun. These endings include *-ance, -ence, -ness, -ion*, and *-ty*.

3.1.2 ARTICLES

Articles, or determiners, come before nouns: *a, an, the.* The **indefinite articles** are *a* and *an* and the **definite** article is *the.* For more advice on how to use articles correctly, see Chapter 8.

3.1.3 PRONOUNS

A **pronoun** can replace a noun. The word (or words) replaced by a pronoun is called the **antecedent**.

Antecedent
Chemical <u>engineers</u> aim for the most economical process.

Pronouns
<u>They</u> plan for and control costs for the entire production chain.

<u>Their</u> work involves the application of chemistry, physics, and mathematics to the process of converting raw materials or chemicals into more useful or valuable forms.

Industries often hire <u>them</u> to design and maintain chemical processes for large-scale manufacture.

TYPES OF PRONOUNS

Personal pronouns refer to people, places, or things: *I, you, he, she, it, we, they*.

Subjective personal pronouns refer to people or things who are performing actions or that are the subject being described: *me, you, him, her, it, us, them*. Notice that these pronouns can occur in independent as well as dependent clauses:

<u>I</u> saw what <u>you</u> did, and <u>I</u> know who <u>you</u> are.

Objective personal pronouns refer to people or things that are acted on; they often follow a preposition:

Don't look at <u>me</u>.

Possessive pronouns indicate ownership, or possession: *my, your, his, her, its, our, their*. They function like adjectives in front of a noun:

Every dog has <u>its</u> day.

Absolute possessive pronouns also indicate ownership: *mine, yours, his, hers, its, ours, theirs*. They stand in for both the owner and the thing owned:

The bank account is <u>mine</u>, the property is <u>yours</u>, and the debt is <u>ours</u>.

Relative pronouns introduce subordinate clauses: *who, which, that*. These clauses operate as adjectives, describing the noun or pronoun in the main clause:

Chemical engineers <u>who</u> handle the design and maintenance of chemical processes for large-scale manufacture are usually employed under the title of process engineer.

Fracture mechanics, <u>which</u> was invented during World War I by English aeronautical engineer, A. A. Griffith, seeks to explain the failure of brittle materials.

Griffith calculated <u>that</u> the stress at the tip of a sharp crack approaches infinity.

Demonstrative pronouns point to nouns: *this, these, that, those.*

They may function as adjectives modifying a noun, or they may replace the noun entirely:

Those who ignore history are condemned to repeat it.

Interrogative pronouns begin questions and include words like *who, whose, what, which*, among others:

Who wants to be a millionaire?

Reflexive pronouns name the receiver of an action who or that is identical to the performer of an action and include words like *myself, yourself, himself, herself, itself, ourselves, yourselves, themselves*:

Structural engineers concern themselves with designs that satisfy a given design intent predicated on safety and on serviceability.

Intensive pronouns emphasize a noun or pronoun:

A. A. Griffith himself did not investigate the mechanics of linear-elastic fracture.

Reciprocal pronouns refer to the separate parts of a plural antecedent and include words like *each other, one another*:

Chemical engineering and chemistry can be distinguished from each other by their different objectives.

Indefinite pronouns refer to general (not specific) persons or things.

- **Singular indefinite pronouns**: *another, anyone, anybody, anything, each, either, no one, neither, nobody, one, someone, something*
- **Plural indefinite pronouns**: *both, few, many, several*
- Some indefinite pronouns can be singular or plural depending on the context: *all, most, none, some*

    The stone age, bronze age, and steel age—the history of each is defined by the materials used in that era.

3.1.4 VERBS

The **verb** in a sentence usually expresses an action (*design, implement*), an occurrence (*became*), or a state of being (*is, seemed*). It may be one verb or a phrase made up of a main verb and an auxiliary verb. In English, the verb frequently appears as the second element of a sentence.

Verb Expressing Action
Environmental engineering <u>improves</u> the environment by applying science and engineering principles to provide healthful water, air and land for human habitation and for other organisms.

Verb Expressing Occurrence
It <u>happened</u> one night.

Verb Expressing State of Being
Computer models of designs <u>can be checked</u> for flaws without having to make expensive and time-consuming prototypes.

Verbs may be **regular** (I, you, we, they love; he loves) or **irregular** (I am; you, we, they are; he is).

KINDS OF VERBS

Transitive verbs always take an object:

Engineers <u>use</u> their knowledge of science and mathematics as well as their experience to find suitable solutions to a problem.

ANALYSIS The verb *use* is transitive because it is followed by the object *knowledge*.

Intransitive verbs do not take an object:

Clinical engineers <u>work</u> closely with the IT department and medical physicists.

ANALYSIS The verb *work* is intransitive because it does not take an object.

Linking verbs join a subject to a complement:

Biomedical engineering <u>is</u> the application of engineering principles and techniques to the medical field.

ANALYSIS In this sentence, <u>is</u> is a linking verb that joins the subject to the complement, the noun *application*.

For some, chemical engineering and process engineering <u>seem</u> similar; but they have some important differences.

ANALYSIS In this sentence, *seem* is a linking verb that joins the subject to the complement, the adjective *similar*.

Action verbs, as their name implies, express some form of action:

The bomb <u>exploded</u>.

ANALYSIS In this sentence, *exploded* is an action verb.

Main verbs (as opposed to auxiliary verbs) describe a simple action or state of being:

A mechanical engineer <u>applies</u> physical principles to analyze, design, manufacture, and maintain mechanical systems.

Auxiliary verbs are helping verbs. *Be, do,* and *have* operate as auxiliary verbs in English:

Nanotechnology <u>has been</u> defined as the effort to create a molecular assembler to build molecules and materials.

Modal verbs are verbs like *can, could, may, might, should, would, must,* and *ought.* Modal verbs are used with the present tense form of the verb to express doubt or certainty, necessity or obligation, probability or possibility. They are also often used to make polite requests.

Mechanical engineers <u>must</u> understand and be able to apply concepts from chemistry and electrical engineering.

A mechanical engineer working in thermo-fluids <u>might</u> design a heat sink, an air conditioning system, or an internal combustion engine.

FORMS OF VERBS

Verbs have many forms. They vary according to

- Tense
- Person
- Number
- Mood
- Voice

Verb tense refers to when an action occurs—in the **past**, in the **present**, or in the **future**. Tense expresses the time of an action. To show changes in time, verbs change form and combine with auxiliary verbs.

The **simple tenses** describe a one-time or regular occurrence; the **progressive** form refers to an ongoing event. The **perfect tenses** express completed actions.

SIMPLE TENSES		
TYPE	**EXAMPLE**	**PROGRESSIVE FORM**
Present	I *design*	I *am designing*
Past	I *designed*	I *was designing*
Future	I *will design*	I *will be designing*

PERFECT TENSES		
TYPE	**EXAMPLE**	**PROGRESSIVE FORM**
Present Perfect	I *have designed*	I *have been designing*
Past Perfect	I *had designed*	I *had been designing*
Future Perfect	I *will have designed*	I *will have been designing*

Person of a verb has to do with the subject to which it is attached. Look at the following conjugation of a regular verb in English; note that, in the present tense, regular verbs take an s in the third person singular (with *he, she,* or *it*):

First person singular: I <u>work</u>
Second person singular: you <u>work</u>
Third person singular: he/she/it <u>works</u>
First person plural: we <u>work</u>
Second person plural: you <u>work</u>
Third person plural: they <u>work</u>

Number of a verb has to do with whether the verb is attached to a singular subject or a plural subject. *She works* is singular, but *they work* is plural, and the form of the verb changes to reflect the change in number of the subject.

Mood of a verb refers to whether it is indicative, imperative, or subjunctive.

- **Indicative** mood is used for events that are real or that commonly recur:

  Modern mechanical engineers <u>use</u> various computational tools such as finite element analysis (FEA), computational fluid dynamics (CFD), computer-aided design (CAD), and computer-aided manufacturing (CAM).

- **Imperative** mood is used for commands. The subject in these cases is left out, but is understood to be *you*.

  <u>Remember</u> to join Institute of Electrical and Electronics Engineers (IEEE).

- **Subjunctive** mood is used to express hypothetical situations, things contrary to fact, wishes, requirements, and speculations.

  If I <u>were</u> you, I would learn to question my assumptions.

The **voice** of a verb refers to whether the subject *performs* the action or whether it *receives* the action:

- In the **active** voice, the subject performs the action.

  Customers <u>found</u> that coffee shop to be expensive, filthy, and slow.

  ANALYSIS The subject (*customers*) performs the action (*finding*).

- In the **passive** voice, the subject is acted upon:

  That coffee shop <u>was found</u> to be expensive, filthy, and slow.

  ANALYSIS The subject (*that coffee shop*) receives the action of an implied but unstated performer (*the customers*).

Because the active voice emphasizes the performer of an action, it is more direct and less wordy than the passive voice. On the other hand, because the passive voice stresses the receiver of an action, it is especially useful when the performer of an action is unknown or unimportant.

3.1.5 ADJECTIVES

Adjectives modify or describe nouns or pronouns. They give information about *which one*, *what kind*, or *how many*.

Common Adjective
In 2006, the future for nanotechnology is <u>exciting</u>.

Proper Adjective
Engineering as a profession arose in England during the <u>Victorian</u> era.

> **ESL NOTE**
>
> Words with the suffixes *-ful*, *-ish*, *-less*, and *-like* are usually adjectives.

3.1.6 ADVERBS

An **adverb** can modify a verb, an adjective, another adverb, or a complete sentence. Adverbs give information about *when, where, how*, and *how much*.

Adverb Modifying Verb
A system in which all equalizing processes have gone <u>practically</u> to completion is considered to be in a state of thermodynamic equilibrium.

Thermodynamic processes that develop <u>slowly</u> and thus allow each intermediate step to reach an equilibrium state are called reversible processes.

Adverb Modifying Adjective
Computers are a <u>highly</u> significant cause of a sedentary lifestyle.

Adverb Modifying Adverb
Clinical engineers work <u>very</u> closely with the information technology areas and medical physicists.

Adverb Modifying Sentence
<u>Gradually</u>, undergraduate programs in biomedical engineering are becoming more widespread.

Many adverbs, though not all adverbs, end in -ly, though it is important to realize that many adjectives do as well. Note that adjectives and adverbs may also be **comparative** or **superlative**.

Comparative Adjective
Nikola Tesla argued that Albert Einstein's theory of relativity was much <u>older</u>, dating back to a fellow Serbian some two hundred years earlier.

Superlative Adjective
Nikola Tesla is recognized as being among the <u>most accomplished</u> scientists of the late nineteenth and early twentieth centuries, as well as one of the <u>greatest</u> contributors to the birth of modern technology.

Comparative Adverb

In the late 1880s, Tesla and Edison became adversaries, in part because of Edison's belief that direct current (DC) distributed electric power <u>more efficiently</u> than alternating current, advocated by Tesla.

Superlative Adverb

In the Colorado Springs lab, Tesla "recorded" signals of what he believed were <u>most definitely</u> extraterrestrial radio signals, though his announcements and his data were rejected by the scientific community.

3.1.7 PREPOSITIONS

Prepositions are words that often do express position. They form phrases with nouns and pronouns, and together they often provide information about time and space. Common prepositions include:

about	above	across	after	against	along
among	around	as	at	before	behind
below	beneath	beside	besides	between	beyond
by	concerning	considering	despite	down	during
except	for	from	in	inside	into
like	near	next	of	off	on
onto	opposite	out	outside	over	past
plus	regarding	respecting	round	since	than
through	throughout	till	to	toward	under
underneath	unlike	until	unto	up	upon
with	within	without			

The symbol <u>for</u> torque is τ, the Greek letter *tau*. The concept <u>of</u> torque, also called moment or couple, originated <u>with</u> the work <u>of</u> Archimedes <u>on</u> levers.

Internal-combustion engines produce useful torque only <u>over</u> a limited range <u>of</u> rotational speeds (typically <u>from</u> around 1,000 to 6,000 rpm <u>for</u> a small car).

ESL NOTE The use of prepositions in English is complex. For further discussion of prepositions, see Chapter 8.

3.1.8 CONJUNCTIONS

Conjunctions are words that connect.

COORDINATING CONJUNCTIONS

Coordinating conjunctions join two coordinate or balanced structures. They include the following: *and, or, nor, for* (meaning *because*), *but, yet,* and *so.* When one of these seven words joins two complete sentences, it is working as a coordinating conjunction.

> For U.S. and Canadian–English electronics, a vacuum tube is a device generally used to amplify, <u>or</u> otherwise modify, a signal by controlling the movement of electrons in an evacuated space.

> Slowly, vacuum tubes were replaced by much smaller <u>and</u> less expensive transistors.

> When hot, the filament of a vacuum tube releases electrons into the vacuum, <u>and</u> a cloud of negatively charged electrons results.

SUBORDINATE CONJUNCTIONS

Subordinating conjunctions connect main clauses with subordinate clauses.

after	although	as	as if	as long as	as though
because	before	even if	even though	if	if only
in order that	now that	once	rather than	since	so that
than	that	though	till	unless	until
when	whenever	where	wherever	whereas	while

> <u>Because</u> the processor is integrated onto one or more large-scale integrated circuit packages, the cost of processor power can be greatly reduced.

> <u>While</u> 64-bit microprocessor designs have been in use since the early 1990s, the rise of 64-bit microchips for PCs occurred during the early 2000s.

<u>Although</u> about half of the $44 billion (U.S.) worth of microprocessors that were manufactured and sold in 2003 was spent on CPUs used in desktop or laptop personal computers, that accounts for only about 0.2% of all CPUs sold.

CORRELATIVE CONJUNCTIONS

Correlative conjunctions work in pairs to balance sentence structures.

both . . . and	either . . . or	neither . . . nor
not only . . . but also	whether . . . or . . .	and not . . . so much as . . .

In memory devices, storage cells (conventionally known as capacitors) are also fabricated during front-end surface development, <u>either</u> into the silicon surface <u>or</u> stacked above the transistor.

CONJUNCTIVE ADVERBS

Conjunctive adverbs are used to indicate a connection between main clauses.

also	as a result	for example	furthermore
hence	however	in addition	in other words
instead	moreover	nevertheless	nonetheless
on the contrary	on the other hand	otherwise	similarly
that is	then	therefore	thus

In semiconductor manufacturing, as the number of interconnect levels increases, planarization is required to ensure a flat surface; <u>otherwise</u>, the levels interfere with the ability to pattern.

At one time, DuPont produced ultrapure silicon by reacting silicon tetrachloride with high-purity zinc vapors at 950° C; this technique, <u>however</u>, was plagued with practical problems and was eventually abandoned in favor of the Siemens process.

Notice that the conjunctive adverb need not occur directly after the semicolon. Also, a conjunctive adverb can occur in a sentence by itself, in which case no semicolon is needed at all. Many adverbs can function as conjunctive adverbs. Learn to distinguish between conjunctive adverbs and coordinating conjunctions so that you can avoid comma splice errors.

3.1.9 INTERJECTIONS

An **interjection** is a word that expresses strong feeling. These words are seldom used in formal writing.

<u>Ouch!</u> I hate doing my income taxes.

3.2 PARTS OF SENTENCES

English sentences are usually composed of

* Subjects
* Verbs
* Objects or complements

A sentence is composed of a subject and a predicate. The latter is the name for the verb, its modifiers, and the object or complement in the sentence.

3.2.1 SUBJECTS

The **subject** (S) of a sentence is what the sentence is about. To find the subject of a sentence, ask *who* or *what* of the verb. Subjects can be categorized according to whether they are complete, simple, or compound or whether they occur in a sentence that uses unusual word order.

SIMPLE SUBJECT

The **simple subject** (SS) is the one-word subject with all modifiers, phrases, and clauses stripped away. A simple subject is always either a noun or a pronoun.

               SS

The two leading <u>manufacturers</u>, Hewlett-Packard and Texas Instruments, steadily released more feature-laden calculators during the 1980s and 1990s.

                  SS

More complex scientific <u>calculators</u> support trigonometric, statistical, and other mathematical functions.

In commands (or imperatives), *you* is understood to be the subject.

S
[You] Remember to join the Institute of Electrical and Electronics Engineers (IEEE).

COMPLETE SUBJECT

The **complete subject** (CS) is the actual grammatical subject with all its modifiers, phrases, and clauses.

CS
Robert Noyce is also credited (along with Jack Kilby) with the invention of the integrated circuit or microchip.

CS
A patent for a solid circuit made of germanium, the first integrated circuit, was filed on February 6, 1959.

CS
Minimizing the number of logic elements on the chip, not the number of clock cycles needed to do a computation, is the primary goal of calculator designers.

ANALYSIS What is the primary goal of calculator designers? Not *logic elements*, not *chips*, but minimizing the number of logic elements on the chip.

COMPOUND SUBJECT

A **compound subject** is two or more subjects joined by a coordinating conjunction such as *and* or *or*.

CS   CS
Calculating machines and computers are fundamentally different in that calculators do not qualify as Turing machines.

CS   CS   CS   CS
Abacuses, books of mathematical tables, slide rules, and adding machines have been used in the recent past for serious numeric work.

DELAYED SUBJECTS

Sometimes subjects do not appear before the verb in English sentences. If a sentence begins with an **expletive**, such as *there is* or *there are* (or *there was* or *there were*), then the subject follows the verb. Expletives function as placeholders to begin the sentence.

S
There was some initial <u>fear</u> that basic arithmetic skills would suffer as a result of students' use of calculators for schoolwork.

Subjects sometimes appear at the ends of sentences for dramatic effect.

In use centuries before the adoption of the written Arabic numerals system and widely used by merchants and clerks in China and

S
elsewhere is the <u>abacus</u>.

In questions, the subject often appears between parts of the verb.

S
Do <u>you</u> *know* the answer?

Note: Being able to identify the subject of a sentence is crucial to understanding many common errors in your editing. Look, for example, at sentence fragments, subject–verb agreement, and pronoun case (Chapter 4). If your first language is not English, check out Chapter 8.

3.2.2 PREDICATES

The **predicate** of a sentence includes everything after the complete subject—unless the sentence uses inverted word order. A predicate can express something about the state of the subject, or it can express the action that the subject performs. The predicate includes the verb or verb phrase, complements or objects (direct or indirect), or modifiers. In the following examples, the predicates are italicized:

The chain reaction *is the basis of nuclear power*.

The debate over the use of nuclear power *has often been bitter*.

Silicon *conducts electricity in an unusual way*.

The supervisor *mailed the applicant a description of the job*.

The plant shutdown *left the entire area an economic disaster*.

3.2.3 VERBS, OBJECTS, AND COMPLEMENTS

As you have seen in the preceding section, verbs can be categorized as linking, if followed by a complement; transitive, if followed by an object; or intransitive, if the verb cannot take an object.

LINKING VERBS AND SUBJECT COMPLEMENTS

Linking verbs (LV) must take **subject complements**, a word or words that complete the meaning of the subject (S) with words that rename or describe it.

> S lv   sc
> In the summer of 1958, Kilby was a newly employed engineer at Texas Instruments who did not yet have the right to a summer vacation.
>
> As for their two main advantages over discrete circuits, integrated
>
> S lv sc   sc
> circuits are faster and cheaper.

When a subjective complement renames the subject, it is a noun or pronoun, or a phrase beginning with a noun or pronoun, such as *engineer.* When it describes the noun, it is an adjective, such as *faster* and *cheaper.* Linking verbs are often forms of *to be: be, am, is, are, was, were, being, been.*

Verbs that have no action, such as *appear, become, feel, grow, look, make, prove, remain, stay, seem, smell, sound,* and *taste* are also linking verbs when they are followed by a word or word group that names or describes the subject of the sentence.

TRANSITIVE VERBS AND DIRECT OBJECTS

Transitive verbs (TV) always take a **direct object** (DO). Direct objects name the receiver of an action. A simple direct object must be a noun or pronoun.

> S tv do
> You *will pay* the price.

Transitive verbs are often in the active voice, with the subject (or performer of the action) at the beginning of the sentence and the direct object that receives the action following the verb. Transitive verbs are the only verbs that may appear in the passive voice, with the real subject at the end of the sentence.

Active voice: Starting in the 1930s, researchers such as William Shockley at Bell Laboratories identified semiconductors found in the periodic table as likely materials for solid-state vacuum tubes.

Passive voice: Semiconductors found in the periodic table were identified as likely materials for solid-state vacuum tubes by researchers such as William Shockley at Bell Laboratories, starting in the 1930s.

OR

Passive voice: Semiconductors found in the periodic table were identified as likely materials for solid state vacuum tubes starting in the 1930s.

ANALYSIS The direct object (*semiconductors*) becomes the subject when the sentence is made passive; the original subject is relegated to the end of the sentence, in a prepositional phrase beginning with *by*. Sometimes these *by* phrases are left out in passive constructions.

TRANSITIVE VERBS WITH BOTH INDIRECT AND DIRECT OBJECTS
Sometimes transitive verbs take both direct and **indirect objects** (IO). An indirect object is like a prepositional phrase in which *to* or *for* has been deleted. Indirect objects come before direct objects:

s +v -io do s +v -io do
You *catch* [for] me a fish, and I *will cook* [for] you dinner.

TRANSITIVE VERBS, DIRECT OBJECTS, AND OBJECT COMPLEMENTS
A transitive verb and a direct object are sometimes followed by an **object complement** (OC), a word or group of words that completes the direct object's meaning by describing it or renaming it. If the object complement is used to rename the direct object, it is a noun (or noun phrase) or pronoun.

s +v do oc
Some people *find* nanotechnology a fascinating new frontier of research and development.

ANALYSIS The object complement renames the direct object, *nanotechnology*, as *frontier*.

If the object complement describes the direct object, it is an adjective or adjective phrase.

    s   +v   do   oc
Some people *find* nanotechnology fascinating.

ANALYSIS The object complement describes the direct object, *nanotechnology,* as *fascinating.*

INTRANSITIVE VERBS

An intransitive verb (IV) cannot take either an object or a complement. Intransitive verbs may be followed by adverbs and prepositions or no words at all:

In some processes, intertwined transport and separation unit

    s   iv
operations *combine.*

Although biomedical engineering has been an interdisciplinary

                    s   iv
field for graduate studies, undergraduate programs *have emerged.*

ANALYSIS The verb *combine* in the first sentence is intransitive; it takes no object or complement. Similarly, the verb *have emerged* in the second sentence is intransitive.

When an intransitive verb is followed by an adverb or adverbial phrase, those words modify the verb, an adverb, or an adjective. For example, *slowly but surely* is an adverbial phrase in the sentence *He walked, slowly but surely, to the closet door.*

    s   iv
Bioreactors *are being developed* for use in tissue engineering.

ANALYSIS The intransitive verb *are being developed* takes no object but is followed by an adverbial phrase that modifies it: *for use in tissue engineering.*

Check a dictionary to determine whether a verb is transitive or intransitive. Sometimes a verb can be both, depending on its context.

Transitive: Biomedical engineering (also known as bioengineering) applies engineering principles and techniques to the medical field.

Intransitive: The Rule of Three here does not <u>apply</u>.

ANALYSIS In the first sentence, the verb *applies* takes a direct object that receives the action: *principles*. In the next sentence, [*does*] *apply* takes no object.

3.3 PHRASES AND CLAUSES

Each sentence, no matter how complex, has a **main clause**—the clause containing the main idea, subject, and verb of the sentence. A main clause, also known as an **independent clause**, can always stand alone as a sentence.

Phrases are word groups lacking a subject or verb—or both—and they cannot stand alone (although in recent years fragments are increasingly used for dramatic effect in journalistic writing). Traditionally, phrases are used within sentences, usually as adjectives, adverbs, or nouns.

3.3.1 PREPOSITIONAL PHRASES

Prepositional phrases are groups of words that begin with a preposition, such as *at, across, by, beside, for, from, in, into, off, on, over, to,* or *without*. Prepositional phrases always include a noun or a pronoun, which serves as the object of the preposition.

Prepositional phrases are usually adverbs or adjectives. When a preposition is used as an adjective, it appears right after the noun or pronoun that it modifies:

<u>Before the Industrial Revolution,</u> there were only two kinds <u>of engineers</u>: military engineers and civil engineers.

<u>During the early nineteenth century</u> mechanical engineering <u>in England</u> evolved <u>into a separate field</u> to provide manufacturing machines and the engines to power them.

When a preposition is used as an adverb, it may or may not follow the verb it modifies:

There were only two kinds of engineers <u>before the Industrial Revolution</u>.

ANALYSIS Remember that adverbs answer the questions *how, when, where, why*. Here *before the Industrial Revolution* provides detail on *when*.

3.3.2 VERBAL PHRASES

Verbal phrases are formed from parts of the verb, but they function as nouns, adjectives, or adverbs. There are three kinds of verbal phrases: infinitives, participial phrases, and gerunds.

INFINITIVE PHRASES

The **infinitive** consists of *to* plus the base form of the verb: *to be, to do, to design*. The infinitive often functions as a noun in a sentence, but it can also be an adjective or an adverb. An **infinitive phrase** is the infinitive together with its complements, objects, or other modifiers:

> <u>To design a heat sink, an air conditioning system, or an internal combustion engine</u> requires a thorough understanding of thermodynamics.

> Systems designed by mechanical engineers can be as small as a nano-sized gear or as large as the structure of a supertanker used <u>to carry oil around the world</u>.

PARTICIPIAL PHRASES

Participial phrases are constructed from present participles and past participles and modify nouns or pronouns:

- **Present participles** are formed with *-ing* and act like adverbs and adjectives.

  The molecules of a gas at high temperature move within a closed space, <u>constantly colliding with walls and other molecules</u>.

  An automobile <u>traveling at 60 km/h</u> has more kinetic energy than it has traveling at 30 km/h.

  ANALYSIS The participial phrase in the first sentence tells *how* the molecules move; the participial phrase in the second sentence provides descriptive detail about *automobile*.

- **Past participles** are forms of regular verbs that end in *-ed* or the equivalent forms of irregular verbs. They, too, operate like adjectives in a sentence.

  While elastic deformation refers to the ability of the material to regain its original shape after the external load is removed, permanent material deformation, <u>called plastic deformation</u>, occurs with large loads.

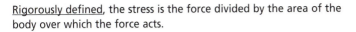

<u>Rigorously defined</u>, the stress is the force divided by the area of the body over which the force acts.

3.3.3 GERUNDS

Similar in appearance to a participial phrase, the gerund plays the role of noun. A **gerund** is a single word with -*ing* used as a noun. A **gerund phrase** is a single word with -*ing* accompanied by its objects, complements, and modifiers; it is a group of words acting as noun.

In the iron-core type transformer, the <u>winding</u> is wrapped around an iron bar.

The <u>splitting</u> of an atom produces a great amount of energy.

Jarvik changed his artificial heart design in 1974 by <u>fitting his model with a highly flexible three-layer diaphragm made of smooth polyurethane</u>.

<u>Reversing the rotation of the electrohydraulic heart pump</u> reverses the direction of the hydraulic flow.

> **ESL NOTE** Gerunds and infinitives, when used as objects, sometimes pose problems for ESL students. (See Chapter 8.)

3.3.4 APPOSITIVE PHRASES

Appositives, or appositive phrases, are used to rename nouns or pronouns. They function like nouns.

The history of thermodynamics begins with Robert Boyle, <u>an Irish physicist and chemist</u>, and Robert Hooke, <u>an English scientist</u>, who together in 1656 invented the air pump.

Thermodynamics, <u>a branch of physics that studies the effects of changes in temperature, pressure, and volume on physical systems at the macroscopic scale</u>, analyzes the collective motion of particles using statistics.

3.3.5 ABSOLUTE PHRASES

An **absolute phrase** modifies a clause or a sentence the same way that an adverb clause does. In fact, the absolute phrase is a reduction of an adverb clause: the subordinating conjunction and the verb have been deleted.

> The laws of thermodynamics once defined, researchers could begin to describe how systems respond to changes in their surroundings.

> ANALYSIS The absolute in this example could be rewritten as *When the law of thermodynamics was defined*.

3.3.6 SUBORDINATE CLAUSES

Subordinate clauses (SC), like main clauses, contain a subject and a verb, but they cannot stand alone as complete sentences; instead, they function in sentences as adjectives, adverbs, or nouns. A subordinate clause is constructed with a relative pronoun or subordinating conjunction.

ADJECTIVE CLAUSES

An **adjective clause** modifies a noun or a pronoun. Adjective clauses are constructed with relative pronouns (for example, *who, whom, whose, which*, or *that*) or with a word such as *when* or *where*:

> In 1965, Lotfi A. Zadeh published his seminal work on fuzzy set in which he detailed the mathematics of fuzzy set theory.

> He is also credited, along with John R. Ragazzini, in 1952, as having pioneered the development of the Z-transform method, which is now standard in digital signal processing.

> Fuzzy logic, a component of fuzzy set theory, focuses on reasoning that is approximate rather than precisely deduced from classical logic.

ADVERBIAL CLAUSES

Adverbial clauses act as adverbs by modifying adjectives, adverbs, or verbs, usually answering the questions: When? Where? Why? How? Under what conditions? or To what degree? These clauses are constructed with subordinating conjunctions (for example, *after, although, as, as if, because, even though, if, provided that, rather than, since, so that, though, whether*):

> Although it is widely accepted and has a broad track record of successful applications, fuzzy logic is controversial in some circles.

> Fuzzy logic is specifically designed to deal with imprecision of facts, while probability deals with chances of something happening (although the result is still considered to be precise).

ANALYSIS Notice that the parenthetical text is also an adverb—an adverbial clause within another adverbial clause.

NOUN CLAUSES

Noun clauses function as nouns; they can be used as subjects, objects, or complements. Noun clauses are constructed with relative pronouns such as *how, that, which, who, whoever, whom, whomever, what, whatever, when, where, whether, whose, why*.

> Experimental discoveries showed <u>that semiconductor devices could perform the functions of vacuum tubes</u>.

> The law of the conservation of energy states <u>that the total amount of energy (including potential energy) in a closed system remains constant</u>.

3.4 SENTENCE TYPES

There are two approaches to the classification of sentences. They can be categorized according to their structure or their function.

3.4.1 SENTENCE STRUCTURE

Sentences can be simple, compound, complex, or compound-complex, depending on the number and type of clauses they contain. In order to be a sentence, there must be at least one independent (or main) clause. A subordinate clause, like the main clause, has a subject and a verb, but it cannot stand alone as a sentence.

SIMPLE SENTENCE

A **simple sentence** is one independent clause with no subordinate clauses. Therefore, a simple sentence has just one subject and verb.

> s   v   do
> Marvin Minsky co-founded the Artificial Intelligence Lab at MIT.

COMPOUND SENTENCE

A **compound sentence** is made up of two or more independent clauses (ic), with no subordinate clauses. The main clauses may be joined by a comma and a coordinating conjunction (*and, or, nor, for, but, yet*, or *so*) or with a semicolon.

> ic
> <u>Artificial intelligence systems are now in routine use in many professional</u>

ic

fields, and they are being built into many consumer products, computer software applications, and video strategy games.

ic

According to advocates of strong AI, machines to become self-aware,

ic

but these same machines may or may not exhibit human-like thought processes.

COMPLEX SENTENCE

A **complex sentence** is made up of one independent clause (ic) with at least one subordinate clause (sc).

ic

"Weak" artificial intelligence programs cannot be called "intelligent"

sc

because they cannot really think.

sc

While the brain gets its power from performing many parallel operations,

ic

a computer excels by performing operations very quickly.

ic

The human brain can perform a huge number of operations per second,

sc

since it has roughly 100 billion neurons operating simultaneously, connected by roughly 100 trillion synapses.

COMPOUND-COMPLEX SENTENCE

A **compound-complex sentence** is made up of at least two independent clauses and at least one subordinate clause.

ic   *sc*   *ic*   *sc*

Give me whatever money you can spare, and I will see to it that you get a tax receipt.

ANALYSIS This sentence has two main clauses and a subordinate clause in each of them.

3.4.2 SENTENCE FUNCTION

Note that sentences can be declarative statements; imperatives used either as requests or commands; interrogative sentences used to form questions; or exclamatory sentences used to show exclamation.

Declarative
To err is human.

Imperative
Be true to your school.

Interrogative
What vision do I see before me?

Exclamatory
You must listen to me!

4

GRAMMATICAL SENTENCES

4

Grammatical Sentences

4   Grammatical Sentences

The following sections catalog a number of common errors made by writers in English. Use these sections to revise your work and to respond to questions and suggestions about your use of language. If you discover that you have a tendency to make any one of these errors with some frequency, review the relevant section carefully to internalize the information you find. Doing so will improve your sense of grammar for the next piece of writing you complete.

4.1 SUBJECT–VERB AGREEMENT

Every sentence has a subject (stated or implied) and a verb. **Subject–verb agreement** refers to the relationship between the subject and the verb. In the present tense, verbs must agree with subjects in two ways:

- In **number**. Number means the subject can be singular (for example, *I*) or plural (for example, *we*).
- In **person**. Person can be first person (*I, we*), second person (*you*) or third person (*she, he, it,* or *they*).

The word *engineer* comes from the Latin root *ingeniosus*, meaning "skilled,"

If the verb is a regular verb and the subject is in the third-person singular, use the -s (or -es) form of the verb. Although it ends in -s, *comes* is a singular verb.

PRESENT TENSE FORMS OF *WORK*		
	SINGULAR	**PLURAL**
First Person	I *work*	we *work*
Second Person	you *work*	you *work*
Third Person	she/he/it *works*	they *work*

Notice how the following irregular verbs achieve subject–verb agreement:

PRESENT-TENSE FORMS OF *DO*	
I *do*	we *do*
you *do*	you *do*
she/he/it *does*	they *do*

PRESENT-TENSE FORMS OF *HAVE*	
I *have*	we *have*
you *have*	you *have*
she/he/it *has*	they *have*

The verb *to be* has different forms for the present and past tense:

PRESENT-TENSE FORMS OF *BE*	
I *am*	we *are*
you *are*	you *are*
she/he/it *is*	they *are*

PAST-TENSE FORMS OF *BE*	
I *was*	we *were*
you *were*	you *were*
she/he/it *was*	they *were*

If you have been speaking and writing English for a long time, problems with subject–verb agreement should be obvious to your ear or eye. For example, the sentence *The word "engineer" do not come from the word "engine"* immediately sounds or looks incorrect. It is obvious that the subject and verb do not agree. The sentence should be *The word "engineer" does not come from the word "engine."*

However, some subject–verb agreement problems are more difficult to spot. A number of English sentence constructions make the subject difficult to identify—often the subject is located far from the verb—and, as a result, it is easy to make verb-agreement mistakes. The following section reviews some of the most common subject–verb agreement problems.

4.1.1 WORDS BETWEEN SUBJECT AND VERB

Modifying words between the sentence subject and verb can be mistaken for the subject. When evaluating any sentence for subject—verb agreement, ignore all modifying elements, such as prepositional phrases, so that only the sentence subject and verb remain. Then, determine whether the subject and verb agree. Try drawing an arrow to connect the subject with the verb, as has been done in the following example:

Incorrect: An individual engineering <u>task</u>, such as designs for bridges, electric power plants, and chemical plants, *require* approval by a professional or chartered engineer.

Correct: An individual engineering <u>task</u>, such as designs for bridges, electric power plants, and chemical plants, *requires* approval by a professional or chartered engineer.

ANALYSIS The subject of the preceding sentence is *task*, not *tasks*, *bridges*, or *plants*. The sentence verb is *require*, which is incorrect. Because the subject is in the third-person singular, the correct verb form is *requires*, not *require*.

Incorrect: <u>Engineers</u>, using a prototype, scale model, simulation, destructive test, nondestructive test, or stress test, *attempt* to ensure that products will perform as expected.

Correct: <u>Engineers</u>, using a prototype, scale model, simulation, destructive test, nondestructive test, or stress test, *attempt* to ensure that products will perform as expected.

ANALYSIS The subject of the sentence is *engineers*, not *prototype*, *model*, or *test*. The sentence verb is *attempts*, which is incorrect. The subject *engineers* is in the third-person plural. To be correct, the verb should be in the form *attempt*, not *attempts*.

Phrases beginning *along with, as well as, in addition*, and *together with* do not change the number of the subject because they are not really part of the subject:

Incorrect: <u>Paul Winchell</u>, along with the more well-known Robert Jarvik, *were* the early developers of the artificial heart in the early 1960s.

Correct: <u>Paul Winchell</u>, along with the more well-known Robert Jarvik, *was* the early developer of the artificial heart in the early 1960s.

ANALYSIS *Paul Winchell* is the main subject of the sentence. If the writer had wanted to emphasize both the Winchell *and* Jarvik, the sentence could have been structured as follows: *Paul Winchell and the more well-known Robert Jarvik were the early developers of the artificial heart in the early 1960s.*

4.1.2 SUBJECTS WITH *AND*

A **compound subject** contains two or more independent subjects joined by *and*. The compound subject requires a plural verb:

Correct: In the *Star Trek* series, <u>Captain Jean-Luc Picard</u> and <u>Joseph Sisko</u> both *have* artificial hearts.

Correct: <u>Paul Winchell</u>, who patented the first design in 1963, <u>and Robert Jarvik</u>, who used that design in his *Jarvik-7, are* the early developers of the artificial heart.

However, when the parts of the subject refer to a single person or idea, they require a singular verb:

Correct: <u>Rice and tomato</u> *is* a San Francisco treat.

Correct: <u>Spaghetti and clam sauce</u> *has* been a favorite in our house for years.

The pronouns *every* and *each* are singular and require singular verbs, even if the subjects they precede are joined by *and*.

Correct: Every woman and man *has* a right to vote.

4.1.3 SUBJECTS WITH *OR* OR *NOR*

When a compound subject is joined by *or* or *nor*, make the verb agree with the part of the subject nearer to the verb:

Correct: Neither biomedical engineering <u>nor fields</u> such as clinical engineering *are* nearly as well known as electrical engineering or mechanical engineering.

Correct: Either aeronautical engineering <u>or</u> astronautical <u>engineering</u> *is* a field of aerospace engineering you might investigate.

4.1.4 INDEFINITE PRONOUNS

An **indefinite pronoun** does not refer to a specific person or thing. The following are indefinite pronouns:

all	anybody	anyone	anything
each	either	everybody	everyone
everything	neither	no one	nobody
none	nothing	one	some
somebody	someone	something	

Even though many indefinite pronouns seem to refer to more than one person or thing, most require a singular verb. Note especially that *each, every,* and words ending in *-body* and *-one* are singular.

Correct: <u>Everybody</u> attending the IEEE conference *wants* to visit the windmill museum.

Correct: <u>Somebody</u> at the windmill museum *wants* to know if windmills for generating electricity are legal in residential areas.

Correct: <u>Everyone</u> at the windmill museum *agrees* that wind-generated electricity is an underused resource.

NEITHER *AND* NONE

When used alone, the indefinite pronouns *neither* and *none* require singular verbs.

> *Correct:* Neither *is* correct.

> *Correct:* Of the IEEE members who visited the windmill museum, none *regrets* the time spent there.

When prepositional phrases with plurals follow the indefinite pronouns *neither* and *none*, in some cases a plural or singular verb may be used. However, it is best to treat *neither* and *none* consistently as singular.

> *Correct:* Neither of those wind-turbine designs developed at NREL *has* been adopted for commercial use.

> *Correct:* None of these bills introduced in Congress *provides* economic incentives to use wind power to generate electricity.

SINGULAR OR PLURAL INDEFINITE PRONOUNS THAT CAN BE SINGULAR

A few indefinite pronouns, such as *all, any, more, most, none,* and *some,* can be singular or plural, depending on the noun or pronoun to which they refer.

> *Singular:* All of the world's petroleum energy *is* being used up rapidly.

> *Plural:* All of the renewable energy resources *are* being rapidly developed.

ANALYSIS In the first example, *energy* is a singular noun, so the singular verb *is* is required. In the second example, *resources* is a plural noun, so the plural verb *are* is required.

4.1.5 COLLECTIVE NOUNS

A **collective noun** names a class or a group of people or things. Some examples of collective nouns include *board, committee, family, group, jury,* and *team.* Use a singular verb with the collective noun when you want to communicate that the group is acting as a unit:

> *Correct:* The board *agrees* that it needs a new product line.

> *Incorrect:* The first-year engineering class *have* a record enrollment.

> *Correct:* The first-year engineering class *has* a record enrollment.

ANALYSIS The *class* is considered a single unit, and individual action is not important to sentence meaning. Therefore, a singular verb is required.

Use a plural verb when you want to communicate that members of the group are acting independently:

Correct: The <u>board</u> *have* a variety of conflicting ideas about which new product areas to pursue.

ANALYSIS Each member of the board has his or her separate ideas about products to develop.

Sometimes it is possible to better capture the idea of individual action by recasting the sentence with a plural noun:

Correct: The <u>board members</u> *have* a variety of conflicting ideas about which new product areas to pursue.

NUMBER AS A SINGULAR OR PLURAL NOUN
If the collective noun *number* is preceded by *the*, treat it as a singular noun:

Correct: <u>The number</u> of renewable energy resources being developed for commercial use *is* slowly increasing.

If *number* is preceded by *a*, treat it as a plural:

Correct: <u>A number</u> of renewable energy resources *are* undesirable to some because of esthetic and environmental concerns.

UNITS OF MEASUREMENT
Use a singular verb when the unit of measurement is used collectively—that is, when the thing described by the noun cannot be counted:

Correct: <u>One-half</u> of the recoverable oil in the world *is* located in the Middle East, according to a U.S. Geological Survey.

Use a plural verb when the unit of measurement refers to individual persons or things that can be counted:

<u>Two-thirds</u> of all the world's oil reserves *are* located in the Middle East, according to several oil companies and the U.S. Department of Energy.

4.1.6 SUBJECT AFTER VERB

Most often the verb follows the subject in sentences. However, in certain cases, the verb may come before the subject, making it difficult to evaluate subject–verb agreement.

EXPLETIVE CONSTRUCTIONS
Expletive constructions include phrases such as *there is, there are*, or *it is, it was*. When these phrases appear at the beginning of a sentence, the verb often precedes the subject:

Correct: There *are* enough petroleum <u>reserves</u> to continue current production rates for 50 to 100 years, according to a 2005 U.S. Geological Survey estimate.

INVERTED SENTENCE ORDER
To achieve sentence variety, you may from time to time wish to invert traditional subject–verb order. Ensure that when you do this, you check that the subject and the verb agree.

Incorrect: Most troubling of current energy theories *are* the Hubbert peak <u>theory</u>, also known as "peak oil," which held in 1956 that oil production would peak in the continental United States between 1965 and 1970, and worldwide in 2000.

Correct: Most troubling of current energy theories *is* the Hubbert peak <u>theory</u>, also known as "peak oil," which held in 1956 that oil production would peak in the continental United States between 1965 and 1970, and worldwide in 2000.

ANALYSIS The subject is not *theories*. Instead, it is *theory*. It requires the singular verb *is*.

Incorrect: Most troubling in the current energy debate *is* the Hubbert peak <u>theory</u> and the more recent <u>estimate</u> of the Association for the Study of Peak Oil and Gas that world oil production will peak around the year 2010 and natural gas, somewhere between 2010 to 2020.

Correct: Most troubling in the current energy debate *are* the Hubbert peak <u>theory</u> and the more recent <u>estimate</u> of the Association for the Study of Peak Oil and Gas that world oil production will peak around the year 2010 and natural gas, somewhere between 2010 to 2020.

4.1.7 SUBJECT COMPLEMENT

A **subject complement** is a noun or adjective that follows a linking verb and renames or describes the sentence subject, as in *HVAC is an acronym that stands for "heating, ventilation and air-conditioning."* Because of its relationship to the subject, the complement can often be mistaken for the subject and result in subject–verb agreement errors:

Incorrect: A cleanroom's primary <u>use</u> *are* environments that require a low level of pollutants such as dust, airborne microbes, aerosol particles, and chemical vapors.

Correct: A cleanroom's primary <u>use</u> *is* environments that require a low level of pollutants such as dust, airborne microbes, aerosol particles, and chemical vapors.

ANALYSIS The subject of the sentence is *use*, which is singular. If the subject is singular, the sentence requires a singular verb. The plural *environments* is the complement.

4.1.8 SUBJECT–VERB AGREEMENT WITH RELATIVE PRONOUNS

WHO, WHICH, THAT
A **relative pronoun** such as *who, which,* or *that* usually introduces an adjective clause that modifies the subject. The relative pronoun must agree with its antecedent. The antecedent is the noun or pronoun to which the relative pronoun refers. Thus, the verb must agree with the antecedent:

Correct: Marion King Hubbert, who is best known for the Hubbert curve, was a geophysicist who made important contributions to the fields of geology and geophysics.

ANALYSIS The singular noun *Hubbert* is the antecedent of *who*. The verb *is* must then be singular.

ONE OF THE *AND* ONLY ONE OF THE *CONSTRUCTIONS*
Subject–verb agreement mistakes are often made with relative pronouns when the sentence contains *one of the* or *only one of the*. Generally, with *one of the* constructions, use a plural verb:

Incorrect: Dust is <u>one of the</u> important environment pollutants that *causes* problems in cleanrooms.

Correct: Dust is <u>one of the </u>important environment pollutants that cause problems in cleanrooms.

ANALYSIS The antecedent of *that* is *Dust*, not *pollutants*. Since the antecedent is plural, to agree, the verb *cause* must also be plural.

Generally, with *only one of the* constructions, use a singular verb:

Incorrect: Dust is <u>only one of the</u> important environment pollutants that *cause* problems in cleanrooms.

Correct: Dust is <u>only one of the</u> important environment pollutants that causes problems in cleanrooms.

ANALYSIS The word *only* makes the antecedent of *that* is *one*, not *Dust*. Since the antecedent is singular, to agree, the verb must also be singular: *is*.

4.1.9 PLURAL FORM, SINGULAR MEANING

Some words ending in -ics or -s are singular in meaning, even though they appear plural in appearance. These words include the following:

athletics	economics	ethics
physics	politics	statistics
mathematics	measles	mumps
news		

Nouns such as the preceding generally require a singular verb:

Incorrect: <u>Physics</u> *are* the science of the natural world, dealing with the fundamental constituents of the universe, the forces they exert on one another, and the results of these forces.

Correct: <u>Physics</u> *is* the science of the natural world, dealing with the fundamental constituents of the universe, the forces they exert on one another, and the results of these forces.

When nouns such as *mathematics, physics,* and *statistics* refer to a particular item of knowledge, as opposed to the collective body of knowledge, they are treated as plural:

Correct: International Energy Agency <u>statistics</u> *show* that oil production is on the decline in 33 of the 48 largest oil-producing countries.

4.1.10 TITLES AND WORKS AS WORDS

A work referred to by its title is treated as a singular entity, even if the title includes a plural word:

Incorrect: <u>*The Invaders*</u>, a late-1960s science-fiction television series, *feature* an architect who learns of an alien invasion underway and attempts to defeat the aliens.

Correct: <u>*The Invaders*</u>, a late-1960s science-fiction television series, *features* an architect who learns of an alien invasion underway and attempts to defeat the aliens.

Incorrect: <u>*Sliders*</u> *focus* on a group of travelers who "slide" between parallel worlds using a wormhole which they call the "Einstein-Rosen-Podolsky bridge."

Correct: <u>*Sliders*</u> *focuses* on a group of travelers who "slide" between parallel worlds using a wormhole which they call the "Einstein-Rosen-Podolsky bridge."

4.2 OTHER PROBLEMS WITH VERBS

The verb communicates vital information in any sentence, often by indicating what is occurring, what has occurred, or what will occur. For writers, there are several major potential trouble spots relating to verbs. If you have read the previous section, you'll know that one common problem is making sure that a sentence's subject and verb agree. Unfortunately, there are other areas where writers frequently experience problems with verb usage. Some of these potential trouble spots include:

- Irregular verbs forms (such as *drink, drank, drunk*)
- *Lie* and *lay*
- *-s* (or *-es*) endings with verbs
- *-ed* endings
- Omitted verbs

- Verb tenses
- Subjunctive mood
- Active versus passive voice

Fortunately, information, models, and guidelines in this section will help you to avoid common verb problems in these areas. You might flip through the section to obtain a general sense of how its contents can help you meet your particular writing goals.

> **ESL NOTE** If your first language is not English, consult Chapter 8, which offers advice on working with verbs in English.

4.2.1 IRREGULAR VERBS

With the exception of the verb *be*, English verbs have five forms. These forms are shown in the chart below, in brief sample sentences using the regular verb *talk* and the irregular verb *ring*:

Base (Simple) Form: Today I (talk, write).

Past Tense: Yesterday I (talked, wrote).

Perfect: I (have talked, have written) many times in the past.

Progressive: I (am talking, am writing) this moment.

-s Form: She/he/it (talks, writes).

With regular verbs, such as *talk*, the past tense and past participle verb forms are created by adding *-ed* or *-d* to the **base**, or simple, form of the verb. As you can see in the chart above, this pattern is not followed for the irregular verb *write*. In fact, many writers mix up the past tense and past participle of irregular verb forms, creating non-standard English sentences:

Nonstandard: Engineers *have wrote* many reports.

Standard: Engineers *have written* many reports.

Nonstandard English is language other than edited English. You may encounter and use it in very informal speaking contexts, and also in much modern literature, but it is not acceptable in formal, academic writing and speaking contexts. Standard English should be used in all formal writing and speaking situations.

STANDARD ENGLISH VERB FORMS
Past tense verbs are used to communicate action that happened completely in the past. These verbs do not require helping verbs such as *is* or *are*.

In 1971, Marion King Hubbert *predicted* that global oil production would peak between 1995 and 2000.

The **past participle** always requires a helping verb. For the perfect tenses, these helping verbs could be *has*, *have*, or *had*. For the passive voice, the helping verbs are *be, am, is, are, was, were, being,* or *been*.

Past tense: In 2004, the Association for the Study of Peak Oil and Gas *predicted* that world oil production <u>would peak</u> around the year 2010.

Past participle: The Association for the Study of Peak Oil and Gas *has predicted* that world oil production <u>will peak</u> around the year 2010.

When you want to choose the standard English verb form for formal writing, first check to see if the verb is irregular. See if it is listed on the chart of common irregular verbs on pages 124 to 128. If the verb in your sentence does not require a helping verb, select the appropriate past-tense verb. If the verb does require a helping verb, use the past-participle form listed in the chart. Of course, regular verbs will follow the pattern outlined above for the regular verb *talk*.

Incorrect: The world *was <u>shook</u>* in the late 18th century by Thomas Malthus's prediction that human population growth would outstrip agriculture's ability to support it.

Correct: The world *was <u>shaken</u>* in the late 18th century by Thomas Malthus's prediction that human population growth would outstrip agriculture's ability to support it.

Incorrect: In the old days, people <u>sung</u> the national anthem at the start of every baseball game.

Correct: In the old days, people <u>sang</u> the national anthem at the start of every baseball game.

Incorrect: According to NASA, Apollo 13's number 2 oxygen tank was <u>broke</u> when Mission Control requested that the crew perform a "cryo stir."

Correct: According to NASA, Apollo 13's number 2 oxygen tank was <u>broken</u> when Mission Control requested that the crew perform a "cryo stir."

COMMON IRREGULAR VERBS

The three-column table below lists many of the more common irregular verbs in the base (simple), past-tense, and past-participle forms. When composing a sentence or editing your work, find the relevant verb in the first column, then determine if you have used the correct past-tense or past-participle form.

BASE (SIMPLE) FORM	PAST TENSE	PAST PARTICIPLE
arise	arose	arisen
awake	awoke, awaked	awaked, awoken
be	was, were	been
bear (to carry)	bore	borne
bear (to give birth)	bore	born
beat	beat	beaten, beat
become	became	become
begin	began	begun
bend	bent	bent
bet	bet	bet
bid	bid	bid
bind	bound	bound
bite	bit	bitten, bit
blow	blew	blown
break	broke	broken

bring	brought	brought
build	built	built
burst	burst	burst
buy	bought	bought
catch	caught	caught
choose	chose	chosen
cling	clung	clung
come	came	come
cost	cost	cost
creep	crept	crept
cut	cut	cut
deal	dealt	dealt
dig	dug	dug
dive	dived, dove	dived
do	did	done
drag	dragged	dragged
draw	drew	drawn
dream	dreamed	dreamed
drink	drank	drunk
drive	drove	driven
eat	ate	eaten
fall	fell	fallen
feed	fed	fed
feel	felt	felt
fight	fought	fought
find	found	found
flee	fled	fled

fling	flung	flung
fly	flew	flown
forbid	forbade, forbad	forbidden
forget	forgot	forgotten
freeze	froze	frozen
get	got	got, gotten
give	gave	given
go	went	gone
grow	grew	grown
hang (execute)	hanged	hanged
hang (suspend)	hung	hung
have	had	had
hear	heard	heard
hide	hid	hidden
hit	hit	hit
hold	held	held
hurt	hurt	hurt
keep	kept	kept
know	knew	known
lay (put)	laid	laid
lead	led	led
leave	left	left
lend	lent	lent
let (allow)	let	let
lie (recline)	lay	lain
light	lighted, lit	lighted, lit
lose	lost	lost

make	made	made
mean	meant	meant
pay	paid	paid
prove	proved	proved, proven
put	put	put
quit	quit	quit
read	read	read
ride	rode	ridden
ring	rang	rung
rise (get up)	rose	risen
run	ran	run
say	said	said
see	saw	seen
seek	sought	sought
send	sent	sent
set (place)	set	set
shake	shook	shaken
shine	shone, shined	shone, shined
shoot	shot	shot
show	showed	shown, showed
shrink	shrank	shrunk, shrunken
sing	sang	sung
sink	sank	sunk
sit (be seated)	sat	sat
slay	slew	slain
sleep	slept	slept
slide	slid	slid

sling	slung	slung
speak	spoke	spoken
spend	spent	spent
spin	spun	spun
spring	sprang	sprung
stand	stood	stood
steal	stole	stolen
sting	stung	stung
stink	stank, stunk	stunk
strike	struck	struck, stricken
strive	strove	striven
swear	swore	sworn
sweep	swept	swept
swim	swam	swum
swing	swung	swung
take	took	taken
teach	taught	taught
tear	tore	torn
tell	told	told
think	thought	thought
throw	threw	thrown
understand	understood	understood
wake	woke, waked	waked, woken
wear	wore	worn
win	won	won
wring	wrung	wrung
write	wrote	written

4.2.2 *LIE* AND *LAY*

The forms of the verbs *lie* and *lay* are frequently confused. *Lie* means "to recline." It is an intransitive verb, which means it cannot be followed by a direct object:

In Figure 9, World Trade Center perimeter columns, several stories high and still connected, *lie* amongst all the debris on the ground.

The simple past tense of the verb *lie* is *lay*:

Much of the World Trade Center site *lay* atop an old landfill, which had extended from the west side of Lower Manhattan 700 feet out into the Hudson.

The past participle of *lie* is *lain*:

Some argue that information about the impending attacks on the World Trade Center had *lain* unheeded on the desks of key U.S. government officials for months.

The present participle of *lie* is *lying*:

In Figure 9, World Trade Center perimeter columns, several stories high and still connected, are *lying* amongst all the debris on the ground.

Lay means "to put down." It is a transitive verb, which means it must be followed by a direct object:

Architects chose to lay *much* of the World Trade Center foundations atop an old landfill, which had extended from the west side of Lower Manhattan 700 feet out into the Hudson.

The simple past tense of *lay* is *laid*:

Architects laid *much* of the World Trade Center foundations atop an old landfill, which had extended from the west side of Lower Manhattan 700 feet out into the Hudson.

The past participle of *lay* is *laid*:

Much of the World Trade Center foundations *were laid* atop an old landfill, which had extended from the west side of Lower Manhattan 700 feet out into the Hudson.

The present participle of *lay* is *laying*:

Construction workers were *laying* the foundations of the World Trade Center in 1965.

LIE AND LAY

BASE (SIMPLE) FORM	PAST TENSE		PRESENT PARTICIPLE
lie	*lay*	*lain*	*lying*
lay	*laid*	*laid*	*laying*

Here are some additional examples of the use of lie and lay:

Incorrect: After the collapse of the Twin Towers, debris <u>laid</u> in all directions for many blocks.

Correct: After the collapse of the Twin Towers, debris <u>lay</u> in all directions for many blocks.

ANALYSIS Here the meaning is *lie* as in "to recline" or "to rest." *Laid* is the past-tense form of *lay*, which means "to put down." The correct past-tense form of *lie* is *lay*.

Incorrect: Osama bin Laden and members of al-Qaeda had been <u>lying</u> plans for the attack on the Twin Towers as early as 2001.

Correct: Osama bin Laden and members of al-Qaeda had been <u>laying</u> plans for the attack on the Twin Towers as early as 2001.

ANALYSIS *Lying* is the present participle of the verb *lie*, which means "to recline" or "to rest." Here the meaning is *lay*, which means "to put down" or "to place." The correct present participle of *lay* is *laying*.

Incorrect: The information about the impending attacks on the World Trade Center had <u>laid</u> unheeded on his desk for months.

Correct: The information about the impending attacks on the World Trade Center had <u>lain</u> unheeded on his desk for months.

ANALYSIS *Laid* is the past participle of the verb *lay*, which means "to put down." An additional meaning of *lie* is "to rest on a surface." This verb better reflects the meaning of the sentence. The past participle of *lie* is *lain*.

Other verb pairs that are easy to confuse are *sit* and *set* and *rise* and *raise*.

4.2.3 *-S* ENDINGS

Use the *-s* (or *-es*) form of a verb in the present when the subject is third-person singular. Often, people who speak English dialects or who are just learning English omit verb endings such as *-s* (or *-es*) that are required in standard English.

- *First and second-person singular and plural:* I tell, you tell, we tell, you tell
- *Third-person plural:* they tell, visitors tell
- *Third-person singular:* she tells, he tells, it tells, Mother tells, everybody tells

As you can see, the third-person singular includes nouns such as *Mother*; the pronouns *she, he,* and *it*; and the indefinite pronoun *everybody.*

Incorrect: Pre-production <u>involve</u> all the planning until the day the camera operator <u>take</u> out the camera.

Correct: Pre-production <u>involves</u> all the planning until the day the camera operator <u>takes</u> out the camera.

ANALYSIS The sentence subject *pre-production* is third-person singular, so the verb must end in *-s*. In the subordinate clause, the subject is *camera operator,* which is also third-person singular and thus requires the *-s* ending, as in *takes.*

If the subject is not third-person singular, do not add an *-s* (*-es*) verb ending:

Incorrect: I <u>drafts</u> a script for the video.

Correct: I <u>draft</u> a script for the video.

Incorrect: The actors <u>eats</u> a lot of cheese and crackers between filming.

Correct: The actors <u>eat</u> a lot of cheese and crackers between filming.

ANALYSIS In the first sentence the subject is *I,* which is first-person singular; thus the *-s* ending is not required. In the second sentence, the subject is *actors,* which is third-person plural and again does not require the *-s* verb ending.

If the subject is third-person singular, use *has, does,* or *doesn't*:

Incorrect: That *expressway* usually <u>have</u> heavy traffic jams in the morning.

Correct: That *expressway* usually <u>has</u> heavy traffic jams in the morning.

Incorrect: <u>Do</u> *anybody* know what time it is?

Correct: <u>Does</u> *anybody* know what time it is?

Incorrect: The stage manager don't think it is necessary to alter the backdrop.

Correct: The stage *manager* <u>doesn't</u> think it is necessary to alter the backdrop.

4.2.4 *-ED* ENDINGS

For regular verbs, the past tense and past participle are formed by adding *-ed* or *-d* to the base verb, as in *talked* and *have talked,* or *used* and *have used.* However, some speakers omit *-ed* endings from past-tense and past-participle regular verbs and also omit the endings from these words in their writing. Often it can be difficult to remember the *-ed* ending, since in some verbs it is not pronounced:

asked, learned, licensed, passed, practiced, supposed to

Make sure to add the *-ed* or *-d* ending to the base (simple) verb for all regular past-tense or past-participle verbs.

PAST TENSE

The past tense indicates an action that has happened at a particular time in the past. To form the past tense of a regular verb, add *-ed* or *-d* to the base (simple) verb form:

Incorrect: Last year, the *firm* <u>help</u> firms save over $3.5 million by reducing chemical recovery boiler shutdowns.

Correct: Last year, the *firm* <u>helped</u> firms save over $3.5 million by reducing chemical recovery boiler shutdowns.

Incorrect: In their annual reports, the *firms* <u>report</u> no capital costs necessary for the 5- to 20-percent increased chemical recovery boiler capacity.

Correct: In their annual reports, the *firms* <u>reported</u> no capital costs necessary for the 5- to 20-percent increased chemical recovery boiler capacity.

PAST PARTICIPLES

Past participles can be used in three ways:

- To form one of the perfect tenses, in which case, the past participle follows *have, has*, or *had*. In the following example, *had developed* uses the past-perfect tense (*had* is followed by the past participle, *developed*):

  The scientist had <u>developed</u> a new technique for destroying cancer cells.

- To form the passive voice, in which case, the past participle follows *be, am, is, are, was, were, being*, or *been*. In the following example, *were created* uses the passive voice: The real subject, *the social club*, is the actor. In the passive voice, a form of *be* (here *were*) is followed by the past participle (here *created*):

  The decorations were <u>created</u> by the social club.

- To act as an adjective modifying nouns. In the following example, the past participle *weakened* modifies the noun *soldier*:

  The medics carried the <u>weakened</u> soldier to the field hospital.

4.2.5 OMITTED VERBS

In formal writing contexts, do not omit required linking verbs or helping (auxiliary) verbs.

LINKING VERBS

A linking verb is a main verb that links a subject with a subject complement (a word or words that renames or describes the subject). Linking verbs indicate a condition or state of being. They are often forms of the verb *be*, such as *am, was*, or *been*.

Incorrect: After tracing the call, the police discovered that the caller a fellow employee.

Correct: After tracing the call, the police discovered that the caller <u>was</u> a fellow employee.

Some linking verbs appear in contractions, such as in *I'm* for *I am* or *it's* for *it is*. Especially in formal writing, do not leave out linking verbs.

HELPING (AUXILIARY) VERBS

A helping verb, also known as an auxiliary verb, is a form of *be, do, can, have, may, will, shall, could, would, might,* or *must* that combines with a main verb to express tone, mood, or voice:

Incorrect: Housing prices <u>escalated</u> in recent months.

Correct: Housing prices <u>have</u> escalated in recent months.

Like linking verbs, helping verbs can be contracted as in *she's going* or *they've been warned*. Especially in formal writing, do not leave out helping verbs.

ESL NOTE
If English is not your first language, consult the section on omitted verbs, in the following pages.

4.2.6 TENSE

Tense indicates the time or duration of an action. The following section outlines some common writing problems associated with verb tenses and suggests strategies for avoiding these problems and ways of correcting them. One major problem connected with verb tense—shifts from one verb tense to another—is covered in detail in Chapter 6. Before exploring tense problems in detail, it is important to understand English verb tenses. There are three basic verb tenses:

- Present
- Past
- Future

Each tense has three forms, also known as aspects:

- Simple
- Perfect
- Progressive

SIMPLE TENSES

The three simple tenses divide time into present, past, and future:

- The **simple present tense** indicates something that happens regularly or in the present:

  I <u>repair</u> the socket.

- The **simple past tense** indicates a completed action that happened at a particular time in the past:

  She <u>repaired</u> the socket yesterday.

- The **simple future tense** indicates an action that will happen sometime in the future:

  We <u>will repair</u> the socket tomorrow.

The following table provides the simple forms for the regular verb *cook*, the irregular verb *take*, and the verb *be*:

SIMPLE TENSE			
SIMPLE PRESENT			
Singular		**Plural**	
I	cook, take, am	we	cook, take, are
you	cook, take, are	you	cook, take, are
she/he/it	cooks, takes, is	they	cook, take, are
SIMPLE PAST			
I	cooked, took, was	we	cooked, took, were
you	cooked, took, were	you	cooked, took, were
she/he/it	cooked, took, was	they	cooked, took, were
SIMPLE FUTURE			
I, you, she/he/it, we, they	will cook, take, be		

PERFECT TENSES

The perfect tense can be present, past, or future:

- The **present perfect tense** indicates action that has happened in the past:

  Marine engineers <u>have developed</u> turbulence models to simulate water flow around ship hulls.

- The **past perfect tense** indicates an action that has been completed at some time in the past before another past action:

  Until recently, marine engineers <u>had used</u> expensive, often inaccurate experimental measurements to design vessels subjected to sloshing.

- The **future perfect tense** indicates an action that will take place before some time in the future:

  In the next decade, research in computational fluid dynamics <u>will have enabled</u> the shipbuilding industry to design hull forms that run faster and generate less noise.

PERFECT TENSE	
PRESENT PERFECT	
I, you, we, they	*have cooked, taken, been*
she/he/it	*has cooked, taken, been*
PAST PERFECT	
I, you, she/he/it, we, they	*had cooked, taken, been*
FUTURE PERFECT	
I, you, she/he/it, we, they	*will have cooked, taken, been*

PROGRESSIVE FORMS

The simple and perfect verb tenses have progressive forms. The **progressive** forms indicate that an action, occurrence, or state of being is ongoing, or progressive. Progressive verb forms are created by combining the present participle and forms of the verb *be* as a helping, or auxiliary, verb.

PROGRESSIVE FORMS

PRESENT PROGRESSIVE

I	*am cooking, taking, being*
she/he/it	*is cooking, taking, being*
you, we, they	*are cooking, taking, being*

PAST PROGRESSIVE

I, she/he/it	*was cooking, taking, being*
you, we, they	*were cooking, taking, being*

FUTURE PROGRESSIVE

I, you, she/he/it, we, they	*will be cooking, taking, being*

PRESENT PERFECT PROGRESSIVE

I, you, we, they	*have been cooking, taking, being*
she/he/it	*has been cooking, taking, being*

PAST PERFECT PROGRESSIVE

I, you, she/he/it, we, they	*had been cooking, taking, being*

FUTURE PERFECT PROGRESSIVE

I, you, she/he/it, we, they	*will have been cooking, taking, being*

SPECIAL USES OF THE PRESENT TENSE

Along with indicating that something happens or can happen in the present, the present tense has several special uses in formal academic writing.

- Use the **literary present tense** to discuss events in fiction, movies, television shows, and other such works:

  Incorrect: In the first episode of *Star Trek: Deep Space Nine*, the crew <u>discovered</u> a wormhole, which <u>provided</u> nearly immediate travel to and from the distant Gamma Quadrant.

  Correct: In the first episode of *Star Trek: Deep Space Nine*, the crew <u>discovers</u> a wormhole, which <u>provides</u> nearly immediate travel to and from the distant Gamma Quadrant.

- Use present tense for general truths and scientific principles:

  Incorrect: According to mechanical advantage, a machine <u>reduced</u> the force necessary to move a load.

  Correct: According to mechanical advantage, a machine <u>reduces</u> the force necessary to move a load.

- However, if a scientific principle has been disproved, use the past tense:

  Correct: According to the early Greeks, the brain <u>pumped</u> blood through the body.

- When summarizing, paraphrasing, or directly quoting an author's views as expressed in a non-literary work, use the present tense:

  Incorrect: In his book *The Existential Pleasures of Engineering*, Samuel C. Florman <u>contended</u> that "the wealthy, the clever, and daring" hold the real power in the world—not technologists.

  Correct: In his book *The Existential Pleasures of Engineering*, Samuel C. Florman <u>contends</u> that "the wealthy, the clever, and daring" hold the real power in the world—not technologists.

  Incorrect: In his *Opticks*, Newton <u>wrote</u> that "The changing of bodies into light, and light into bodies, is very conformable to the course of Nature, which seems delighted with transmutations."

  Correct: In his *Opticks*, Newton <u>writes</u> that "The changing of bodies into light, and light into bodies, is very conformable to the course of Nature, which seems delighted with transmutations."

Obviously, Sir Isaac is no longer alive. The present tense is used to introduce an author's view, even if an author is dead.

Exception: If you are following the American Psychological Association style for in-text citation, use the past tense when the date and author's name are provided:

G. Temple (1995) <u>described</u> the patient's way of transforming words and numbers into images.

PAST PERFECT TENSE

The **past perfect tense** indicates an action that has been completed at some time in the past, before another action. The past perfect is formed by using the helping verb *had* before the past participle of a verb:

In the mid-1800s, many *had suspected* that electric currents possessed magnetic effects.

Several mistakes involving the past perfect tense are common:

- Using the simple past tense when the sentence's meaning and grammar require the past perfect tense:

Incorrect: In 1800, Alessandro Volta, an Italian physicist, announced that he <u>found</u> a new source of electricity, which is now known as the battery.

Correct: In 1800, Alessandro Volta, an Italian physicist, announced that he <u>had</u> found a new source of electricity, which is now known as the battery.

ANALYSIS Volta's discovery is a completed action that took place in the past before his announcement of that discovery; therefore, the past perfect tense is required.

Incorrect: By the time Faraday <u>invented</u> his rotator, he was too busy with other matters to give much attention to electromagnetism.

Correct: By the time Faraday <u>had</u> invented his rotator, he was too busy with other matters to give much attention to electromagnetism.

ANALYSIS Because the action of inventing the rotator had been completed, the past perfect tense is required.

- Overusing the past perfect tense. Do not use the past perfect tense if two past actions in the sentence happened at the same time:

Incorrect: In his 1831 report to the Royal Society on electromagnetic induction, Faraday <u>had</u> presented his theories explaining how magnetism could produce electricity.

Correct: In his 1831 report to the Royal Society on electromagnetic induction, Faraday <u>presented</u> his theories explaining how magnetism could produce electricity.

TENSES WITH INFINITIVES AND PARTICIPLES

An **infinitive** is a verbal consisting of the simple verb form and, usually, *to*. However, things get complicated when different tenses are involved:

- *Present tense infinitive:* Use the present tense infinitive (*to exercise*, *to fly*) to indicate action that takes place at the same time as, or later than, the action of the sentence's main verb:

  Incorrect: Before Heinrich Hertz, physicists who sought to prove Maxwell's theory had tried to <u>have designed</u> a transmitter of electromagnetic waves, a detector, and some method for confirming the wave-like nature of the phenomena.

  Correct: Before Heinrich Hertz, physicists who sought to prove Maxwell's theory had tried <u>to design</u> a transmitter of electromagnetic waves, a detector, and some method for confirming the wave-like nature of the phenomena.

  ANALYSIS The action of the infinitive (*to design*) would take place at the same time or later than the action of the sentence verb (*had tried*).

- *Present perfect infinitive:* The perfect form of the infinitive is created by placing the helping verb *to have* in front of the past participle (*to have exercised*, *to have flown*). Use the present perfect infinitive to indicate action that takes place before the action of the main sentence verb:

  Incorrect: I would like to <u>finish</u> the design on time, but I had better things to do.

  Correct: I would like to <u>have finished</u> the design on time, but I had better things to do.

  ANALYSIS The liking occurs in the present; finishing the design would have occurred in the past.

PARTICIPLES

The tense of the participle is determined by the tense of the main sentence verb.

- *Present participle:* Use the present participle (*risking*, *making*) when the action occurs at the same time as that of the main sentence verb:

  Correct: Applying his mathematical skills and his ingenuity to physics, James Clerk Maxwell was able to make major contributions to science.

ANALYSIS The present participle is required because Maxwell made his contributions concurrently with his use of mathematics.

- *Past participle or present perfect participle:* Use the past participle (*risked, made*) or the present perfect participle (*having risked, having made*) to indicate action that occurred before that of the main sentence verb:

  Correct: Made of cedar, the canoe floated well.

  ANALYSIS The making of the canoe took place before the canoe floated.

  Correct: Having invested all his money in a dot.com company, Yuri lost it all when profits in the high-technology sector dramatically declined.

  ANALYSIS Yuri's investment took place before his unfortunate financial loss.

4.2.7 MOOD

The **mood** of a verb indicates the manner of action. There are three moods of a verb: the indicative, the imperative, and the subjunctive. By using one of three verb moods, a writer can show how he or she views a thought or action. For instance, through his or her choice of mood, a writer can indicate whether he or she is expressing an opinion or a wish.

- The **indicative mood** states a fact or opinion or asks a question:

  The library *needs* our assistance.

- The **imperative mood** gives an order or a direction:

  Be at the library by eight.

- The **subjunctive mood** is used to express contrary-to-fact-or-wish conditions; in *that* clauses expressing demands, recommendations, requests; and with expressions such as *be that as it may*. Frequently, writers use the subjunctive mood incorrectly.

FORMING THE SUBJUNCTIVE

To create the subjunctive mood with *that* clauses, use the base (simple) form of a verb (*come, decide, produce*). Do not change the verb to indicate whether the subject is singular or plural.

> It is crucial that I *be* [not *am*] at the interview on time.

> The boss requested that he *produce* [not *produces*] more work.

If you are employing the subjunctive to express a contrary-to-fact clause starting with *if* or to express a wish, use *were*, the past tense form of *be*. Do not use *was*.

> I wish that I *were* [not *was*] younger.

APPROPRIATE USES OF THE SUBJUNCTIVE

The subjunctive mood is used in the following writing situations:

- Contrary-to-fact clauses beginning (usually) with *if* or *unless*. In the subjunctive, use *were* to express a contrary-to-fact clause starting with *if* or *unless*:

  Incorrect: If she <u>was</u> the president, she would honor the Kyoto Protocol.

  Correct: If she <u>were</u> the president, she would honor the Kyoto Protocol.

  Incorrect: The United States would approve the Kyoto Protocol if George W. Bush <u>was</u> not the president.

  Correct: The United States would approve the Kyoto Protocol if George W. Bush <u>were</u> not the president.

  ANALYSIS The *if* clauses express conditions that do not exist— they are contrary to fact; therefore, they require the subjunctive.

The subjunctive is not used when the *if* clause expresses a condition that does or may exist:

  Not contrary to fact: If she is elected president, she <u>will</u> approve the Kyoto Protocol.

- Contrary-to-fact clauses expressing a wish. In the subjunctive, use *were* to express a wish:

Formal writing: I wish that U.S. commercial network news reporting <u>were</u> more balanced.

Note: In informal speech, people often use the indicative mood in place of the subjunctive.

Informal speech: I wish that U.S. commercial network news reporting <u>was</u> more balanced.

In a formal writing context, when you need to express a wish, always use the subjunctive.

- *That* clauses after verbs such as *ask, insist, recommend, request, require, suggest,* and *urge.* The following verbs can indicate a requirement or a request: *ask, insist, recommend, request, require, suggest,* and *urge.* These verbs often come before a clause starting with *that.* These sentences require the subjunctive. To form the subjunctive, use the base (simple) form of the verb whether the subject is singular or plural:

  Incorrect: The memorandum *<u>requires</u> that* all employees <u>are</u> there.

  Correct: The memorandum *<u>requires</u> that* all employees be there.

  Incorrect: It *<u>is required</u> that* the examination candidate <u>presents</u> himself or herself at precisely 9:30 A.M.

  Correct: It *<u>is required</u> that* the examination candidate <u>present</u> himself or herself at precisely 9:30 A.M.

  ANALYSIS The sentence is expressed in the subjunctive because the requirement is not yet a reality.

- In idiomatic expressions such as <u>suffice it to say</u> or <u>be that as it may</u>. The subjunctive mood is used in idioms and expressions such as the following: *as it were, be that as it may, come rain or shine, far be it from me, God be thanked, suffice it to say.*

  Suffice it to say, they concur.

4.2.8 VOICE

Voice refers to whether the sentence subject is the actor performing the action communicated by the sentence verb or the receiver of that action. In the active voice the subject (S) performs the action, while in the passive voice the subject is acted upon. Only transitive verbs (TV)—those verbs requiring a direct object (DO)—can be active or passive.

$$S \quad +v \quad\quad do$$

Active voice: The <u>guard</u> *took* the prisoner to his cell in the Guantanamo Bay detainment camp.

$$S \quad\quad +v$$

Passive voice: The <u>prisoner</u> *was taken* to his Guantanamo camp cell by the guard.

The passive voice consists of the helping, or auxiliary, verb *be* and the past participle, as in *was chosen.* For more information on the active and passive voice, see Chapters 6 and 7.

CHANGING A SENTENCE FROM THE PASSIVE TO THE ACTIVE VOICE
In most cases, use the active voice in your academic writing. The active voice tends to be clearer and more concise, direct, powerful, and dramatic. To change a sentence from the passive to the active voice, identify the sentence actor (who or what performs the main sentence action) and make that actor the sentence subject:

Passive voice: Every spring, gardens <u>are planted by</u> millions of North American homeowners.

Active voice: Every spring, millions of North American homeowners <u>plant</u> gardens.

WHEN THE PASSIVE VOICE IS APPROPRIATE
Although the active voice is often preferable, the passive voice does have important functions. It is appropriate when the actor is unknown or unimportant, or when responsibility for an action should not or cannot be assigned.

The passive voice is often used in technical and scientific writing in which a process, as opposed to a person, is important:

The solution is pumped through a special filter where harmful impurities *are extracted.*

When writing about historical events, you can use the passive voice to avoid assigning blame for an action, to emphasize that a certain group was acted upon, or to avoid having to state an obvious or uninteresting subject:

NASA claims that out of its cumulative $2.4 trillion budget, less than 0.8 percent has been spent on the entire space program.

ANALYSIS It's obvious that the U.S. government and NASA have done the spending.

ESL NOTE Do not avoid the passive voice. It is, in many cases, appropriate. For information on how to transform the active voice to the passive, consult Chapters 6 and 7.

4.3 PROBLEMS WITH PRONOUNS

A **pronoun** is a word that replaces a noun or another pronoun. Three major types of pronoun problems occur frequently in writing:

- *Antecedent agreement problems:* The pronoun does not agree with the noun or pronoun to which it refers.
- *Reference problems:* It is not clear to which noun or pronoun the pronoun refers.
- *Case problems:* The **case** of a pronoun is its form in a particular sentence context—whether the pronoun functions as a subject, object, or a possessive. Writers sometimes confuse pronoun case. Two common pronoun case difficulties involve the following:
  - *Personal pronouns:* These problems occur when writers use *I*, for example, instead of *me* or vice versa, or *he* instead of *him* or vice versa.
  - *Whether to use who or whom in sentences*

The following four sections provide guidance in identifying, avoiding, and—if necessary—correcting these types of pronoun errors.

PRONOUN–ANTECEDENT AGREEMENT

The **antecedent** is the word the pronoun replaces. (*Ante* in Latin means *before*.) A pronoun must agree with its antecedent. If the antecedent is singular, the pronoun that refers to it must also be singular:

The microbiologist adjusted his microscope.

Similarly, if the antecedent is plural, the pronoun must be plural:

The choir <u>members</u> opened <u>their</u> song books.

ESL NOTE Pronouns such as *he, she, his, her* and *its* agree in gender with their antecedents, not with the words they modify: Lorna *traveled with* her [not *his*] *chauffeur to Saskatoon*.

MAKING PRONOUNS AGREE WITH ANTECEDENTS THAT ARE INDEFINITE PRONOUNS

Indefinite pronouns do not refer to any specific person, thing, or idea:

another	anybody	anyone	anything
each	either	everybody	everyone
everything	neither	nobody	none
no one	nothing	one	somebody
someone	something		

In formal English, treat indefinite pronouns as singular even though they may seem to have plural meanings:

<u>Anyone</u> who knows the answer should enter it using <u>his or her</u> [not *their*] keyboard.

INDEFINITE-PRONOUN AGREEMENT PROBLEMS

When editing your writing, you may find that you have used a plural pronoun to refer to a singular indefinite pronoun. In such instances, apply the following strategies for correcting the pronoun agreement problem.

- Change the plural pronoun to a singular, such as *he or she*:

  Incorrect: When the airplane hit severe turbulence, everyone feared for <u>their</u> safety.

  Correct: When the airplane hit severe turbulence, everyone feared for <u>his or her</u> safety.

- Make the pronoun's antecedent plural:

  Incorrect: When the airplane hit severe turbulence, everyone feared for <u>their</u> safety.

  Correct: When the airplane hit severe turbulence, the <u>passengers</u> feared for their safety.

- Recast the sentence to eliminate the pronoun agreement problem:

  Incorrect: When the airplane hit severe turbulence, everyone feared for <u>their</u> safety.

  Correct: When the airplane hit severe turbulence, safety became the common concern among all those on board.

Because the use of *his or her* can be awkward and wordy, especially if used repeatedly, you might consider the second and third correction strategies as preferable alternatives.

GENERIC NOUNS

A **generic noun** names a typical member of a group, such as a typical *classroom teacher* or a typical *dentist*. Generic nouns might appear to be plural; however, they are singular:

Correct: Each Olympic <u>athlete</u> must sacrifice if <u>he or she</u> *plans* [not *they plan*] to win a gold medal.

If a plural pronoun incorrectly refers to a generic noun, there are three major ways to remedy the error:

Incorrect: Although the <u>ordinary worker</u> complains about overwork, they feel powerless to cut back.

Correct: Although the <u>ordinary worker</u> complains about overwork, *he or she* feels powerless to cut back.

Correct: Although <u>ordinary workers</u> complain about overwork, *they* feel powerless to cut back.

Correct: The <u>ordinary worker</u> complains about overwork but feels powerless to cut back.

COLLECTIVE NOUNS

A **collective noun** names a group of people or things. Examples of collective nouns include the following words:

audience	army	choir
class	committee	couple
crowd	faculty	family
group	jury	majority
number	pack	team

- If the collective noun refers to a unit, use the singular pronoun:

  The <u>audience</u> stood and applauded to show *its* approval.

- If parts of the collective noun act individually, use a plural pronoun:

  The <u>audience</u> folded *their* collapsible chairs and placed them in a storage room.

- Often it is a good idea to emphasize that the antecedent is plural by adding a word, such as *members*, describing individuals within the group:

  The audience <u>members</u> folded *their* collapsible chairs and placed them in a storage room.

Whether you treat the collective noun as singular or plural, ensure that you consistently treat references within the sentence as singular or plural, respectively.

COMPOUND ANTECEDENTS

A **compound antecedent** is formed by *and, or, nor, either . . . or,* or *neither . . . nor.*

- If the common antecedents are joined by *and,* use a plural pronoun whether the antecedents are plural or singular:

  At the Sorbonne, Marie and Pierre Curie focused <u>their</u> [not *his and her*] studies on radioactive materials and, in 1898, announced <u>their</u> [not *his and her*] discovery: that the pitchblende contained traces of some unknown radioactive component which was far more radioactive than uranium.

In late 1947, <u>John Bardeen and Walter Brattain</u> at Bell Labs announced that *they* had been able to use semiconductors to create triodes.

- If the common antecedents are joined by *or, nor, either . . . or, neither . . . nor*, make the pronoun agree with the nearest antecedent:

  Either the president or the <u>Democrats</u> will have *their* way.

  Note: With a compound antecedent such as the preceding, place the plural noun last to prevent the sentence from sounding awkward.

- If one common antecedent is masculine and the other is feminine, rewrite the sentence to avoid any gender problem:

  Original: He incorrectly thinks that either Pierre or Marie Curie was awarded for her research on radium.

  Rewrite: He incorrectly thinks that the judges awarded the Nobel Prize either to Pierre Curie's or to Marie Curie's research on radium.

4.3.1 PRONOUN REFERENCE

A **pronoun** is a word that replaces a noun or another pronoun. Using pronouns allows you to avoid repeating nouns in speech and writing:

When <u>Marie Curie</u> was hired to teach physics and chemistry at the <u>Sorbonne,</u> <u>she</u> became the first woman to teach <u>there</u>.

ANALYSIS The noun or pronoun that the pronoun replaces is its antecedent. Here the pronoun *she* clearly relates to the antecedent *Marie Curie*, and the demonstrative pronoun *there* clearly relates to the antecedent *Sorbonne*.

However, when the relationship between the antecedent and the pronoun is ambiguous, implied, vague, or indefinite, your intended meaning becomes unclear or it can be completely lost to the reader.

AMBIGUITY AND PRONOUN REFERENCE

When it is possible for a pronoun to refer to either one of two antecedents, the sentence is ambiguous:

Ambiguous: <u>Franz</u> told <u>his father</u> that <u>his</u> car needed a new transmission.

ANALYSIS In this sentence the second possessive pronoun *his* could refer to *Franz* or *his father*.

To eliminate the ambiguity, either repeat the clarifying antecedent or rewrite the sentence:

Option 1: Franz told his father that his father's car needed a new transmission.

Option 2: Franz said to his father, "Dad, your car needs a new transmission."

PROBLEMS WITH IMPLIED ANTECEDENTS

The reader should be able to clearly understand the noun antecedent of any pronoun you use. This antecedent must be stated and not implied or merely suggested:

Incorrect: Before the raging fire spread too close to nearby farms, <u>they</u> were ordered to leave their homes.

Correct: Before the raging fire spread too close to nearby farms, the <u>residents</u> were ordered to leave their homes.

ANALYSIS Although in the original sentence it is implied that the occupants of farms were the ones ordered to leave, it is not explicitly stated. The pronoun *they* has no clear antecedent.

Make sure that antecedents refer to nouns present in, or near, the sentence:

Incorrect: In Samuel C. Florman's *The Civilized Engineer*, <u>he</u> narrates how he chose engineering as his profession.

Correct: In <u>his</u> *The Civilized Engineer*, <u>Samuel C. Florman</u> narrates how he chose engineering as his profession.

ANALYSIS In the original sentence, it is not clear who is doing the narrating. The wording allows the possibility that it is not *Samuel C. Florman*, but perhaps a contributor to his book.

VAGUE USE OF THIS, THAT, WHICH, AND IT

Pronouns such as *this*, *that*, *which*, and *it* should refer clearly to a specific noun antecedent and not large groups of words expressing ideas or situations:

Incorrect: The international figure skating organization agreed to a major overhaul of the judging process; however, <u>it</u> took time.

Correct: The international figure skating organization agreed to a major overhaul of the judging process; however, <u>the change</u> took time.

Incorrect: A spot forecast may state that a temperature range for a specific canyon in the forest will be between 25 and 30 degrees, the humidity between 12 and 14 percent, and the winds 15 kilometers an hour. <u>This interests</u> fire fighters.

Correct: A spot forecast may state that a temperature range for a specific canyon in the forest will be between 25 and 30 degrees, the humidity between 12 and 14 percent, and the winds 15 kilometers an hour. <u>All of these data interest</u> fire fighters.

PROBLEMS WITH INDEFINITE USE OF IT, THEY, OR YOU

The pronouns *it*, *they*, and *you* can cause problems with clarity in writing. They must have clear definite antecedents.

- Do not use the pronoun *it* indefinitely; for example, "In this book [article, chapter, and so on] it says . . .":

Incorrect: <u>In</u> Chapter 1 of the book <u>it</u> states that in the 1800s and early 1900s, higher education focused on forming "gentlemen" through the liberal arts.

Correct: Chapter 1 of the book states that in the 1800s and early 1900s, higher education focused on forming "gentlemen" through the liberal arts.

- Never use *they* without a definite antecedent:

Incorrect: In a typical Hollywood movie, <u>they</u> manipulate the audience's emotions.

Correct: In a typical Hollywood movie, <u>the director, screenwriters, and actors</u> manipulate the audience's emotions.

- In formal writing, the use of *you* is acceptable when you are addressing the reader directly:

  Correct: If <u>you</u> do not want the beeper on, select OFF; if you want it loud, select HIGH.

- However, do not use *you* as an indefinite pronoun in formal writing:

  Incorrect: In ancient Greece <u>you</u> dropped a mussel shell into a certain jar to indicate that a defendant was guilty.

  Correct: In ancient Greece <u>one</u> dropped a mussel shell into a certain jar to indicate that a defendant was guilty.

4.3.2 PRONOUN CASE (*I* VS. *ME*, ETC.)

Case refers to the form a noun or pronoun takes according to the function of that noun or pronoun in a sentence. In English there are three cases:

- The **subjective case** indicates that the pronoun functions as a subject or a subject complement.
- The **objective case** indicates that the pronoun functions as the object of a preposition or a verb.
- The **possessive case** indicates that the pronoun shows ownership.

PRONOUN CASES		
SUBJECTIVE	**OBJECTIVE**	**POSSESSIVE**
I	me	my
we	us	our
you	you	your
she/he/it	her/him/it	her/his/its
they	them	their

The remainder of this section helps you to clearly distinguish between the subjective and the objective case and explains how to avoid common pronoun case errors. The final part of the section explains common uses of pronouns in the possessive case.

SUBJECTIVE CASE PRONOUNS

The subjective case (*I, we, you, she/he/it, they*) must be used when the pronoun functions as a subject or as a subject complement. In the following example, the pronoun functions as a subject and therefore the subjective case is used:

<u>Tony and I</u> *split* the cost of the video.

A **subject complement** is a noun or adjective that follows a linking verb and renames or describes the sentence subject. Because the use of pronouns in the subjective case sounds quite different from the way you might use pronouns in informal speech, subjective case pronouns as subject complements frequently cause writing difficulties:

Incorrect: The students who did the most work are Ivan and <u>her</u>.

Correct: The students who did the most work are Ivan and <u>she</u>.

In all formal writing, ensure that you use the subjective pronoun case when the pronoun is part of the subjective complement:

Correct: The woman Anatole married is <u>she</u>.

ANALYSIS *She* is the complement of the subject *woman.* If the construction sounds too unnatural, recast the sentence.

She is the woman Anatole married.

OBJECTIVE CASE PRONOUNS

Use an objective case pronoun (*me, us, you, her/him/it, them*) if the pronoun functions as

- A direct object:

  The instructor asked her to read the poem.

- An indirect object:

  The proctor provided Sam, Duncan, and me with pencils.

- The object of a preposition:

Just between you and me, the Russian's routine was superior.

PRONOUNS IN COMPOUND SUBJECTS AND OBJECTS
Choosing the correct pronoun can be tricky when the pronoun occurs in a compound subject or a compound object:

Compound subject: She and I went to the multiplex to see a movie.

Compound object: The park proposal surprised her and me.

The fact that the subject or object is compound does not affect the case of the pronoun. However, often a compound structure causes a writer to confuse pronoun case. To determine whether you have selected the correct pronoun case, try mentally blocking out the compound structure, except for the pronoun in question. Then, decide if the pronoun case you have selected is correct:

Incorrect: ~~My two roommates and~~ **me** did the driving.

Correct: ~~My two roommates and~~ **I** did the driving.

Here are some additional examples:

Incorrect: After class, the instructor gave study materials to Rachel and I.

Correct: After class, the instructor gave study materials to Rachel and me.

ANALYSIS Because the pronoun is the object of the sentence, the objective case is required.

Incorrect: In spite of many difficulties, Beth and me set up the project successfully.

Correct: In spite of many difficulties, Beth and I set up the project successfully.

ANALYSIS Because the pronoun is the subject of the sentence, the subjective case is required.

Avoid using a reflexive pronoun such as *myself* or *himself* when you are uncertain about the pronoun case:

Incorrect: The professional organization sent the entry forms to Marley and <u>myself</u>.

Correct: The professional organization sent the entry forms to Marley and <u>me</u>.

ANALYSIS *Marley* and *me* are the indirect objects of the verb *sent*.

PRONOUNS AS APPOSITIVES

An **appositive** is a noun or noun phrase that renames a noun, noun phrase, or pronoun. When a pronoun functions as an appositive, it has the same function, and hence case, as the noun or pronoun it renames:

Incorrect: Three members of the project team—Clara, Michael, and <u>me</u>—won a trophy.

Correct: Three members of the project team—Clara, Michael, and <u>I</u>—won a trophy.

ANALYSIS *Clara, Michael, and I* is an appositive for the sentence subject *three members*. As a result, the subjective case of the pronoun is required.

Incorrect: Let's you and <u>I</u> take the weekend off and go to Big Bend National park.

Correct: Let's you and <u>me</u> take the weekend off and go to Big Bend National park.

ANALYSIS *You and me* is an appositive to *us* [*let's* is a contraction of *let us*]. *Us* is the objective of the verb *let*; therefore, the objective case of the pronoun is required.

PRONOUN CASE WITH *WE* OR *US* AND NOUNS

When you need to decide whether *we* or *us* should come before a noun or noun phrase, mentally block out the noun so that only the pronoun remains. Then, decide which pronoun case is correct:

Incorrect: **Us** ~~piano players~~ play scales every day.

Correct: **We** ~~piano players~~ play scales every day.

ANALYSIS *We* is correct since a subjective case of the pronoun is required.

Follow the same procedure when considering pronouns that function as sentence objects:

Incorrect: Our supervisor has little patience for <u>we</u> drafting interns.

Correct: Our supervisor has little patience for <u>us</u> drafting interns.

ANALYSIS You wouldn't say *Our supervisor has little patience for we*. *Us* functions as an object, and the objective case is most appropriate.

COMPARISONS USING **THAN** OR **AS** WITH PRONOUNS

Pronoun problems also arise when you make comparisons using *than* or *as*. However, the case of the pronoun is determined by its function in the implied part of the sentence, which has been omitted. To determine the correct pronoun case in a sentence that uses *than* or *as* to make a comparison, supply the implied or missing part of the sentence. Then, decide if the pronoun case is correct:

Incorrect: The groom is a full meter taller *than* <u>her</u>.

Correct: The groom is a full meter taller *than* <u>she</u> (is).

Incorrect: Charles Kuen Kao, a pioneer in the use of fiber optics in telecommunications, holds many more patents *than* <u>me</u>.

Correct: Charles Kuen Kao, a pioneer in the use of fiber optics in telecommunications, holds many more patents *than* <u>I</u> (hold patents).

Incorrect: Our supervisor has the same expectations for you *as* she has for <u>I</u>.

Correct: Our supervisor has the same expectations for you *as* she has for <u>me</u>.

POSSESSIVE CASE AND GERUNDS

A **gerund** is a form of a verb that ends in *-ing* and is used as a noun; for example, *Fencing is my favorite sport*. Use a pronoun in the possessive case (*my, our, your, her/his/its, their*) to modify a gerund or gerund phrases:

Incorrect: The physical trainer disapproved of <u>him</u> *eating* bacon before workouts.

Correct: The physical trainer disapproved of <u>his</u> *eating* bacon before workouts.

Nouns as well as pronouns can modify gerunds. The possessive is formed by adding an apostrophe and -s to the end of the noun:

Correct: <u>Wayne's</u> *smoking* is the cause of his bad breath.

4.3.3 *WHO* AND *WHOM*

Who and *whom* are pronouns. *Who* is the subjective case; it is used only for subjects and subject complements. *Whom* is the objective case; it is used only for objects.

- As interrogative pronouns, *who* and *whom* are used to open questions.
- As relative pronouns, *who* and *whom* are used to introduce subordinate clauses.

INTERROGATIVE PRONOUNS: WHO AND WHOM
To decide whether to use *who* or *whom*, first determine the pronoun's function within the question. Does the interrogative pronoun function as a subject or subject complement, or as an object? Determine whether the subjective or objective case is required by recasting the question as a statement. Then, temporarily substitute a subjective case pronoun, such as *he*, or an objective case pronoun, such as *him*.

Incorrect: <u>Whom</u> commanded the coalition forces during the war in Afghanistan?

Correct: <u>Who</u> commanded the coalition forces during the war in Afghanistan?

ANALYSIS Possible answers are *he commanded the force* and *him commanded*; the latter will obviously not work. *Who* is the subject of the verb *command*. Therefore, the subjective case is required and *who* is the correct interrogative pronoun.

Incorrect: <u>Who</u> did the human resources manager interview?

Correct: <u>Whom</u> did the human resources manager interview?

ANALYSIS The choices are *The human resources manager interviewed her* and *The human resources manager interviewed she*; the latter will not work. *Whom* is the direct object of the verb *did interview*. Therefore, the objective case is required, and *whom* is the correct interrogative pronoun.

RELATIVE PRONOUNS: WHO AND WHOM

Use *who* and *whoever* as relative pronouns for subjects, and use *whom* and *whomever* for objects. When deciding which pronoun to use, you must determine whether the relative pronoun functions as a subject or an object within the subordinate clause. A good technique to employ when making this decision is to mentally block off the subordinate clauses you are considering:

Incorrect: The Nobel Prize in Physics is presented to <u>whomever</u> has made the most significant contribution to the field of physics over the course of her or his career.

Correct: The Nobel Prize in Physics is presented to <u>whoever</u> has made the most significant contribution to the field of physics over the course of her or his career.

ANALYSIS To determine the correct relative pronoun, mentally block off *The Nobel Prize in Physics is presented to*. You can see that the relative pronoun of the subordinate clause is the subject of the verb *has*, and thus, the relative pronoun *whoever* is the correct choice.

Incorrect: We don't know <u>who</u> the board has selected to be the next CEO.

Correct: We don't know <u>whom</u> the board has selected to be the next CEO.

ANALYSIS Mentally block out *We don't know*, and focus exclusively on the subordinate clause. The relative pronoun is the object of the verb *selected*. Thus, the objective case is required, and the correct pronoun is *whom*.

Ignore interrupting expressions such as *I know, they think*, or *she believes*, which often come after *who* or *whom* in a subordinate clause:

Incorrect: Alfred Nobel intended the prize to be awarded to only those individuals <u>whom</u> the committee thought had "conferred the greatest benefit on mankind."

Correct: Alfred Nobel intended the prize to be awarded to only those individuals <u>who</u> the committee thought had "conferred the greatest benefit on mankind."

ANALYSIS Mentally block out *Alfred Nobel intended the prize to be awarded to only those individuals* and *the committee thought*. The relative pronoun of the remaining subordinate clause is the subject of the verb *had conferred*. Therefore, the subjective case is required, and the correct relative pronoun is *who*.

4.4 ADVERBS AND ADJECTIVES

Adjectives and adverbs are modifiers. Adjectives modify nouns and pronouns. Adverbs can modify:

- Verbs

  He left the examination *early*.

- Adjectives

  Her cheeks were *slightly* red.

- Other adverbs

  They left the coffee shop *very* late.

Many adverbs end in -ly (*walk **quickly***); however, some do not (*walk **often***). As well, a number of adjectives end in -ly.

That is a *lovely* vase.

Problems can occur when adjectives are incorrectly used as adverbs or vice versa. The best way to decide whether a modifier should be an adjective or an adverb is to determine its function in the sentence. If you are in doubt about whether a word is an adjective or an adverb, you might also consult a dictionary.

4.4.1 ADVERBS

In modifying a verb, another adverb, or an adjective, an adverb answers questions such as the following: Why? When? Where? How? The following are some common misuses of adjectives in situations where adverbs are required.

- Incorrect use of an adjective to modify a verb:

  Incorrect: The system ran <u>smooth</u> and <u>efficient</u>.

  Correct: The system ran <u>smoothly</u> and <u>efficiently</u>.

- Incorrect use of the adjective *good* when the adverb *well* is required:

  Incorrect: The CEO indicated to the human resources personnel that she wants new hires to demonstrate that they can write <u>good</u>.

  Correct: The CEO indicated to the human resources personnel that she wants new hires to demonstrate that they can write <u>well</u>.

For more detail on the correct uses of *good* and *well*, see the usage glossary in Chapter 7.

- Incorrect use of an adjective to modify an adjective or adverb:

  Incorrect: The museum at Niagara Falls has a <u>real</u> unusual collection of artifacts.

  Correct: The museum at Niagara Falls has a <u>really</u> unusual collection of artifacts.

4.4.2 ADJECTIVES

Usually adjectives come before the nouns they modify:

A <u>space</u> elevator is a <u>hypothetical</u> structure designed to transport material from the <u>planet</u> surface into space.

However, adjectives can also function as subject complements that follow a linking verb. The subject complement renames or describes the sentence subject:

Many engineers believe that space elevators are <u>impossible</u>.

Linking verbs communicate states of being as opposed to actions:

Incorrect: From a distance, a space elevator would look <u>strangely</u>.

Correct: From a distance, a space elevator would look <u>strange</u>.

Incorrect: She feels <u>happily</u>.

Correct: She feels <u>happy</u>.

Incorrect: Using current materials, the "beanstalk" type space elevator would perform <u>bad</u>.

Correct: Using current materials, the "beanstalk" type space elevator would perform <u>badly</u>.

ANALYSIS The verbs *taste, feel,* and *look* communicate states of being, so adjectives are required.

Some verbs, such as *look, feel, smell,* and *taste,* may or may not be linking verbs. When the word after the verb modifies the subject, the verb is a linking verb, and this modifying word should be an adjective. However, when the word modifies the verb, it should be an adverb.

• If the word modifies a noun, use the adjective:

The space elevator looked <u>strange</u>.

ANALYSIS *Strange* modifies the subject, *elevator*. Here, *looked* is a linking verb and the modifier is an adjective. The adjective *strange* indicates the state of being strange.

- If the word modifies something other than a noun, use the adverb:

  The engineer looked *curiously* at her designs for a geosynchronous orbital tether (also known as a "beanstalk").

  ANALYSIS *Curiously* modifies the verb, *looked*. The adverb *curiously* answers how the act of looking was done.

4.4.3 COMPARATIVES AND SUPERLATIVES

Most adjectives and adverbs have three forms:

- Positive
- Comparative
- Superlative

FORMING COMPARATIVES AND SUPERLATIVES

ADJECTIVES

Positive	Comparative	Superlative
With one- and most two-syllable adjectives		
red	redder	reddest
crazy	crazier	craziest
With longer adjectives		
intoxicating	more intoxicating	most intoxicating
selfish	less selfish	least selfish
A few adjectives are irregular.		
good	better	best
bad	worse	worst
Some have no comparative or superlative form.		
unique	–	–
pregnant	–	–

ADVERBS		
Positive	**Comparative**	**Superlative**
With adverbs ending in -ly		
selfishly	more selfishly	most selfishly
gracefully	less gracefully	least gracefully
With other adverbs		
fast	faster	fastest
hard	harder	hardest
A few adverbs are irregular.		
well	better	best
badly	worse	worst
Some have no comparative or superlative form.		
really	–	–
solely	–	–

COMPARATIVE AND SUPERLATIVE FORMS
When comparing two entities, use the comparative form:

Incorrect: Which is the least of the two evils?

Correct: Which is the <u>lesser</u> of the two evils?

When comparing three or more entities, use the superlative:

Incorrect: Of the three playwrights, I feel that William Shakespeare is the <u>greater</u>.

Correct: Of the three playwrights, I feel that William Shakespeare is the <u>greatest</u>.

DOUBLE COMPARATIVES OR SUPERLATIVES
Avoid combining comparatives and superlatives:

Incorrect: Of the two scientists, I believe that Isaac Newton is the <u>more</u> greater.

Correct: Of the two scientists, I believe that Isaac Newton is the greater.

Incorrect: The painting is probably one of the most <u>beautifulest</u> in the museum.

Correct: The painting is probably one of the most <u>beautiful</u> in the museum.

COMPARATIVES OR SUPERLATIVES WITH ABSOLUTE CONCEPTS

Absolute concepts by their very nature do not come in degrees and cannot be compared. Some examples of absolute concepts are *favorite, unique, perfect, pregnant, impossible, infinite,* or *priceless.* If three diamonds are perfect, of two of them one cannot be more perfect than the other. Similarly, of the three, one is not most perfect.

Incorrect: The cat looked <u>more pregnant</u> than she did last week.

Correct: The cat's pregnancy was more obvious this week.

Incorrect: That painting by da Vinci is <u>very</u> unique.

Correct: That painting by da Vinci is unique.

Incorrect: The bizarre comedy was the most <u>unique</u> I have ever seen.

Correct: The bizarre comedy was the most <u>unusual</u> I have ever seen.

4.4.4 DOUBLE NEGATIVES

A **double negative** is a nonstandard English construction in which negative modifiers such as *no, not, neither, none, nothing,* and *never* are paired to cancel each other. Double negatives should be avoided in all formal writing:

Incorrect: The government <u>never</u> does nothing to solve the problems affecting the poor.

Correct: The government <u>never</u> does <u>anything</u> to solve the problems affecting the poor.

Correct: The government does <u>nothing</u> to solve the problems affecting the poor.

In standard English, the modifiers *barely, hardly,* and *scarcely* are considered negative modifiers. These words should not be paired with words such as *no, not,* or *never*.

Incorrect: They could <u>not barely</u> hear the tiny girl speak.

Correct: They could <u>barely</u> hear the tiny girl speak.

Two negatives are acceptable in a sentence only if they create a positive meaning:

She was <u>not disappointed</u> with her ten-game hitting streak.

4.5 SENTENCE FRAGMENTS

A sentence is one complete independent clause that contains a subject and verb. A **sentence fragment**, on the other hand, is part of a sentence that is set off as if it were a whole sentence by a beginning capital letter and a final period or other end punctuation. However, the fragment lacks essential requirements of a grammatically complete and correct sentence. For example, the fragment may

- Lack a verb:

  Just Phil and I.

- Lack a subject:

  Pacing the hallway.

- Be a subordinate clause commencing with a subordinating word:

  When I fly a kite.

Sentence fragments give readers a fragment of a thought as opposed to a complete thought, and they interfere with writing clarity. In any type of academic writing, sentence fragments are considered a serious writing error, and they must be eliminated.

4.5.1 TESTING FOR SENTENCE FRAGMENTS

Fragments can be spotted easily when they appear in isolation, but fragments are more difficult to identify when they are near complete sentences. If you suspect a group of words is a sentence fragment, consider these questions:

- Does the word group have a verb?
  - YES. Consider the next question.
  - NO. *The word group is a fragment and must be revised to include a verb.*
- Does the word group have a subject?
  - YES. Consider the next question.
  - NO. *The word group is a fragment and must be revised to include a subject.*
- Does the word group start with a subordinating word, making it a subordinate clause?
  - YES. *The word group is a sentence fragment and must be revised to create a complete sentence that is an independent clause.*
  - NO. If you answered yes to the two previous questions and no to this one, the word group is a complete sentence and does not require revision for sentence completeness.

Be sure to consider all three questions when reviewing your sentence, because the fragment could be missing more than one essential sentence element. If the questions indicate that you have a sentence fragment, use the following strategies to transform it into a complete sentence.

4.5.2 ELIMINATING SENTENCE FRAGMENTS

To fix the sentence fragment and make it a complete sentence, do one of the following:

- Attach the sentence fragments to an independent clause, or a clause that contains the essential element lacking in the fragment (e.g., a subject or a verb):

  Just Phil and I were pacing the hallway.

- Compose an independent clause from the fragment:

  At the emergency ward, the parents were pacing the hallway.

- Drop the subordinating word:

  ~~When~~ I fly a kite.

> **ESL NOTE**
>
> Subjects may not be omitted in English, except in the case of imperative sentences. Verbs may not be omitted either. If English is not your first language, consult Chapter 8 for further information.

4.5.3 SOURCES OF SENTENCE FRAGMENTS

Subordinate clauses, phrases, predicates, and certain transitions are common sources of fragments.

SUBORDINATE CLAUSES

A **subordinate clause** contains a subject and a predicate, or verb, but the clause begins with a subordinating word or phrase, such as *after, although, if,* or *until,* or a relative pronoun such as *that, which, what, who.* Therefore, the clause is not independent. You can make a subordinate clause into an independent clause in one of two ways:

- Merge the subordinate clause with a nearby sentence:

   Incorrect: Carrier was slow to turn its "Weathermaker," the first practical home air conditioner, into a full-scale manufacturing effort. <u>Because</u> Carrier's main business in 1928 was still large applications such as factories and theaters.

   Correct: Carrier was slow to turn its "Weathermaker," the first practical home air conditioner, into a full-scale manufacturing effort <u>because</u> Carrier's main business in 1928 was still large applications such as factories and theaters.

- Delete the subordinating element of the clause:

   Incorrect: Carrier was slow to turn its "Weathermaker," the first practical home air conditioner, into a full-scale manufacturing effort. <u>Because</u> Carrier's main business in 1928 was still large applications such as factories and theaters.

   Correct: Carrier was slow to turn its "Weathermaker," the first practical home air conditioner, into a full-scale manufacturing effort. Carrier's main business in 1928 was still large applications such as factories and theaters.

PHRASES

A **phrase** is a group of words that does not have either a subject or a verb, and, therefore, cannot stand alone as an independent clause or sentence. Look at these examples:

> to go kayaking
> for the umpteenth time
> with great trepidation

Major types of phrases include noun phrases, adjective phrases, adverb phrases, and prepositional phrases.

You can address phrase fragment problems in two ways:

- Incorporate the phrase into a nearby sentence:

Incorrect: Our community library has an amazing array of resources. <u>For</u> every local citizen to use.

Correct: Our community library has an amazing array of resources, <u>which is there for</u> every local citizen to use.

ANALYSIS The phrase *for every local citizen to use* has been turned into a prepositional phrase functioning as an adjective to modify *resources*.

Incorrect: As a subscriber to the paranormal, he took part in the smudging. <u>A ceremony</u> using smoke to purify the psychic energy field, or aura, around a person.

Correct: As a subscriber to the paranormal, he took part in the smudging, <u>a ceremony</u> using smoke to purify the psychic energy field, or aura, around a person.

ANALYSIS The writer has used the phrase beginning *A ceremony using smoke* as an appositive to rename *smudging*.

- Turn the phrase into a complete sentence by adding a subject, predicate (verb), or both:

Incorrect: Smoke jumpers land with heavy gear, including two parachutes, puncture-proof Kelvar suits, freeze-dried food, fire shelters, and personal effects. <u>Followed by</u> cardboard boxes containing chain saws, shovels, and axes, which are heaved out of the airplane.

Correct: Smoke jumpers land with heavy gear, including two parachutes, puncture-proof Kelvar suits, freeze-dried food, fire shelters, and personal effects. <u>The jumpers are followed</u> by cardboard boxes containing chain saws, shovels, and axes, which are heaved out of the airplane.

ANALYSIS A clear subject and verb have been added to the phrase beginning *Followed by*.

COMPOUND PREDICATES

A **predicate** is the part of the sentence that contains the verb. It indicates what the subject is doing or experiencing, or what is being done to the subject. A **compound predicate** contains two or more predicates with the same subject:

Incorrect: In 1915, Willis Haviland Carrier and several partners formed the Carrier Engineering Corporation. <u>And dedicated</u> themselves to improving the technology of air conditioning.

Correct: In 1915, Willis Haviland Carrier and several partners formed the Carrier Engineering Corporation <u>and dedicated</u> themselves to improving the technology of air conditioning.

ANALYSIS The fragment starting *And dedicated themselves* has been made part of the compound predicate. Note that no comma is required between compound elements of this predicate.

EXAMPLES INTRODUCED BY TRANSITIONS

When you introduce examples, illustrations, and explanations to support your discussion, watch out for the following transition words, which can create fragments:

also	and	as an illustration	besides
but	equally important	especially	for example
for instance	furthermore	in addition	in particular
including	like	mainly	namely
or	specifically	such as	that is
to illustrate			

Sometimes a fragment introduced by any one of the foregoing words or phrases can be attached to the sentence before it to create a complete sentence:

Incorrect: Any review of the development of air conditioning technology must include a discussion of the technology's early pioneers. <u>Such as</u> Willis Haviland Carrier, Alfred Wolff, Fred Wittenmeier, L. Logan Lewis, Alfred Mellowes, Thomas Midgley, and Frederick McKinley Jones.

Correct: Any review of the development of air conditioning technology must include a discussion of the technology's early pioneers, <u>such as</u> Willis Haviland Carrier, Alfred Wolff, Fred Wittenmeier, L. Logan Lewis, Alfred Mellowes, Thomas Midgley, and Frederick McKinley Jones.

However, in some instances you may find it necessary to change the fragment containing examples into a new sentence:

Incorrect: Jan Morris's travel pieces cover many interesting cities. For instance, exploring Beirut, visiting Chicago, and discovering "The Navel City" of Cuzco.

Correct: Jan Morris's travel pieces cover many interesting cities. For instance, <u>she explores</u> Beirut, <u>visits</u> Chicago, and <u>discovers</u> "The Navel City" of Cuzco.

ANALYSIS The writer corrected the fragment beginning *For instance* by including the subject *she*, creating a complete sentence.

FRAGMENTS IN LISTS
Occasionally, list elements are fragmented. This type of writing problem can usually be corrected through the use of a colon or dash:

Incorrect: During my rare vacations, I work on my three R's. Reading, rest, and running.

Correct: During my rare vacations, I work on my three R's: reading, rest, and running.

4.5.4 ACCEPTABLE FRAGMENTS

Professional writers may use sentence fragments intentionally for emphasis or effect.

- The following italicized fragment creates emphasis:

  Once I imagined what would happen if I reached down and lifted one of the carp "scientists" out of the pond. Before I threw him back into the water, he might wiggle furiously as I examined him. I wondered how this would appear to the rest of the carp. To them, it would be a truly unsettling event. They would first notice that one of their "scientists" had disappeared from their universe. *Simply vanished, without leaving a trace.* Wherever they would look, there would be no evidence of the missing carp in their universe.

  —*Michio Kaku, Hyperspace: A Scientific Odyssey Through Parallel Universes, Time Warps, and the 10th Dimension*

- The following one is used for transition:

  And now for the bad news.

- The following fragment acts as an exclamation:

  Not bloody likely!

- The following one answers a question:

  And should we go along with this position? Under no circumstances.

- The following fragment is typical of those used in advertising:

  Proven effective.

Many instructors do not accept sentence fragments, even intentional ones, in formal writing. Fragments may be acceptable in less-formal writing contexts, such as an informal personal essay or an article for a campus newspaper. Even in contexts where they are permitted, do not overuse sentence fragments.

4.6 COMMA SPLICES AND FUSED SENTENCES

Incorrectly joining two or more independent clauses within a sentence is a common writing error. An independent clause, or main clause, contains at least a subject and a verb, and the clause can stand on its own as a separate grammatical unit. When two independent clauses appear in a single sentence, they must be joined in one of two ways:

- Using a comma and one of the seven coordinating conjunctions: *and, but, for, nor, or, so, yet*
- With a semicolon or other acceptable punctuation such as a dash or a colon

Fused sentences (also known as run-on sentences) or comma splices occur when two independent clauses are incorrectly joined within the same sentence.

4.6.1 IDENTIFYING FUSED SENTENCES OR COMMA SPLICES

FUSED SENTENCES

In a **fused sentence**, no punctuation or coordinating conjunction appears between the two independent clauses:

Incorrect: Nuclear fusion in stars and supernovae creates new natural elements <u>it</u> is this reaction that is harnessed in fusion power.

Correct: Nuclear fusion in stars and supernovae creates new natural elements. It is this reaction that is harnessed in fusion power.

ANALYSIS *Nuclear fusion in stars and supernovae create new natural elements*, and *it is this reaction that is harnessed in fusion power*— these are both independent clauses.

COMMA SPLICES

In comma splices, the independent clauses are joined (or spliced) with commas and no coordinating conjunction.

Incorrect: Nuclear fusion in stars and supernovae creates new natural elements, <u>it</u> is this reaction that is harnessed in fusion power.

Correct: Nuclear fusion in stars and supernovae creates new natural elements; <u>it</u> is this reaction that is harnessed in fusion power.

Often writers use conjunctive adverbs in place of coordinating conjunctions and, in doing so, create comma splice errors. A **coordinating conjunction** is one of these seven words: *and, but, or, nor, for,* so, and yet. A **conjunctive adverb**, on the other hand, is a word such as *furthermore, however,* or *moreover.*

Incorrect: Nuclear fusion in stars and supernovae creates new natural elements, <u>moreover</u>, it is this reaction that is harnessed in fusion power.

Correct: Nuclear fusion in stars and supernovae creates new natural elements; <u>moreover,</u> it is this reaction that is harnessed in fusion power.

Use the following checklist to determine if a sentence is fused or is a comma splice.

- Does the sentence contain two independent clauses?
  - NO. Neither of the errors applies.
  - YES. *Proceed to the next question.*
- Are the independent clauses joined by a comma and a coordinating conjunction?
  - YES. The clauses are correctly joined.
  - NO. *Proceed to the next question.*
- Are the independent clauses joined by a semicolon or other acceptable punctuation, such as a colon or a dash?
  - YES. The clauses are correctly joined.
  - NO. *Use one of the revision strategies provided in the next section to correct the fused sentence or comma splice.*

4.6.2 STRATEGIES FOR CORRECTING FUSED SENTENCES OR COMMA SPLICES

You have four major options for correcting fused sentences or comma splices.

REVISION WITH COORDINATING CONJUNCTIONS
Add a comma and a coordinating conjunction: *and, but, for, nor, or, so, yet:*

Incorrect: Nuclear fusion in stars and supernovae creates new natural elements, it is this reaction that is harnessed in fusion power.

Correct: Nuclear fusion in stars and supernovae creates new natural elements, <u>and</u> it is this reaction that is harnessed in fusion power.

REVISION WITH SEMICOLONS OR COLONS

Add a semicolon or other appropriate punctuation, such as a colon or a dash:

> *Incorrect:* Nuclear fusion in stars and supernovae creates new natural elements, it is this reaction that is harnessed in fusion power.

> *Correct:* Nuclear fusion in stars and supernovae creates new natural elements; it is this reaction that is harnessed in fusion power.

> *Correct:* Nuclear fusion in stars and supernovae creates new natural elements; <u>moreover,</u> it is this reaction that is harnessed in fusion power.

Use a semicolon without a conjunction if the relationship between the two independent clauses is very clear:

> *Incorrect:* The results of our engineering project were disappointing, our attempt to design a fully autonomous humanoid robot soccer player failed miserably.

> *Correct:* The results of our engineering project were disappointing; our attempt to design a fully autonomous humanoid robot soccer player failed miserably.

Use a semicolon and a comma with independent clauses that are joined with a conjunctive adverb or transitional phrase, such as the following:

also	as a result	besides	consequently
conversely	for example	for instance	furthermore
in addition	in fact	meanwhile	moreover
nonetheless	next	on the other hand	otherwise
similarly	subsequently	then	therefore
thus			

> *Incorrect:* Margaret Atwood is Canada's foremost living novelist, she is among the country's leading poets.

Correct: Margaret Atwood is Canada's foremost living novelist; <u>furthermore</u>, she is among the country's leading poets.

REVISION WITH SUBORDINATE CLAUSES

Revise the sentence to subordinate one of the clauses.

Incorrect: Nuclear fusion in stars and supernovae creates new natural elements, it is this reaction that is harnessed in fusion power.

Correct: <u>While</u> nuclear fusion in stars and supernovae creates new natural elements, it is this reaction that is harnessed in fusion power.

The foregoing option is often the most effective, because it shows what you want to emphasize and the logic connecting the clauses:

Incorrect: The Deep Blue team beat chess master Garry Kasparov in 1996, another "grand challenge" project was achieved.

Correct: <u>When</u> the Deep Blue team beat chess master Garry Kasparov in 1996, another "grand challenge" project was achieved.

Incorrect: The DARPA Grand Challenge began offering a $2 million prize to any autonomous robot that could complete a 132-mile course in the Mojave Desert, no vehicle has completed the challenge.

Correct: <u>Since</u> the DARPA Grand Challenge began offering a $2 million prize to any autonomous robot that could complete a 132-mile course in the Mojave Desert, no vehicle had completed the challenge as of 2005.

Use a colon if the first independent clause introduces the second:

Incorrect: The requests are thorough and varied, a chicken or rabbit will be skinned, boned, quartered, shredded, turned into patties, prepared for stew, the liver for this, the kidney for that.

Correct: The requests are thorough and varied: a chicken or rabbit will be skinned, boned, quartered, shredded, turned into patties, prepared for stew, the liver for this, the kidney for that.

REVISION BY SEPARATING SENTENCES

Turn each independent clause into a separate complete sentence:

Incorrect: Nuclear fusion in stars and supernovae creates new natural elements, it is this reaction that is harnessed in fusion power.

Correct: Nuclear fusion in stars and supernovae creates new natural elements. <u>It</u> is this reaction that is harnessed in fusion power.

Incorrect: Two hours before the start, the teams were given GPS coordinates for the race route to load into their robots, the robots were not allowed to contact humans in any way once the race had started.

Correct: Two hours before the start, the teams were given GPS coordinates for the race route to load into their robots. <u>The</u> robots were not allowed to contact humans in any way once the race had started.

Incorrect: In the 2005 DARPA Grand Challenge, there were entries from Stanford and Carnegie Mellon, they finished within minutes of each other.

Correct: In the 2005 DARPA Grand Challenge, there were entries from Stanford and Carnegie Mellon. They finished within minutes of each other.

5

PUNCTUATION

5

Punctuation

5 Punctuation

This section on punctuation follows the one on grammar because so many issues of punctuation depend on a knowledge of how the language is put together grammatically. Refer to this section for advice on problem areas you encounter as you compose and revise.

5.1 COMMAS

The following section provides rules and guidance for instances in which you must use commas and instances in which commas prevent confusion.

5.1.1 INDEPENDENT CLAUSES WITH COORDINATING CONJUNCTIONS

When two or more **independent clauses** (clauses that can stand on their own as sentences) are linked by **coordinating conjunctions** (*and, or, for, but, so, nor,* and *yet*), place a comma before the coordinating conjunction:

> The Law of Conservation of Energy states that energy can neither be created nor destroyed, and no exceptions to this law have been discovered to date.

> ANALYSIS Two independent clauses are joined by *and*.

A single multiplication in the Mark I took 6 seconds, a division took 15.3 seconds, and a logarithm or a trigonometric function took over 1 minute.

ANALYSIS Three independent clauses are joined, the last two by *and*.

5.1.2 COMPOUND ELEMENTS

Do not use a comma to separate **compound predicates**, or other compound elements of a sentence that are not complete independent clauses.

COMPOUND PREDICATES

Robots are used to perform tasks too dangerous or difficult for humans, or can be used to automate repetitive tasks that can be performed more cheaply.

ANALYSIS The words following the comma do not form an independent clause.

The word *robot* first appeared in a Czech science fiction play in 1921, and was probably invented by the painter Josef Čapek.

ANALYSIS *And* links two elements, but the element after *and* is not an independent clause. Instead, *and* joins two parts of a compound predicate: *appeared* and *invented*.

COMPOUND PHRASES

A robot may act under the direct control of a human or under the control of a pre-programmed computer.

ANALYSIS *Or* links two prepositional phrases, both of which are introduced by *under*.

COMPOUND WORDS

Large and complex robots are used in manufacturing such as car production.

ANALYSIS *And* links two adjectives, *Large* and *complex*.

Developing a robot with a natural human or animal gait is extraordinarily difficult.

ANALYSIS *Or* links two objects of the preposition *with*.

Human and animal bodies use many different muscles in their movement, making the development of robots with natural gaits a difficult and expensive task.

ANALYSIS *And* links an initial pair of adjectives, *Human* and *animal*, and then another, *difficult* and *expensive*.

5.1.3 INTRODUCTORY ELEMENTS

Introductory groups of words in sentences are often made up of the following:

- Adverbial clauses
- Prepositional phrases
- Verbal phrases
- Absolute phrases

In each case, the introductory group of words is followed by a comma, which signals to the reader that the main part of the sentence is about to begin.

COMMAS WITH INTRODUCTORY ADVERBIAL CLAUSES
An **introductory adverbial clause** is a construction with a subject and a verb that introduces a main clause; for example *Whenever I feel in need of cheering up*, I go to the park. Introductory adverbial clauses indicate *where, when, why, how*, or *under what conditions* the main sentence action takes place. After such clauses, use a comma to indicate that the main part of the sentence is about to start:

When Jack Kilby presented the findings of his research to the management of Texas Instruments‸ he proved that circuit components could be mass-produced in a single piece of semiconductor material.

When Seymour Cray's company released its first computer (Cray-1) in 1976‸ no other computer at that time could match its speed and performance.

COMMAS WITH INTRODUCTORY PREPOSITIONAL PHRASES
After any introductory prepositional phrase, use a comma to indicate that the main part of the sentence is about to start:

In 1986‸ Apple Computer bought a Cray X-MP and announced that they would use it to design the next Apple Macintosh.

<u>After several years of development</u>‸ Cray's company released its first product (Cray-1) in 1976.

COMMAS WITH INTRODUCTORY VERBAL PHRASES

Verbals include participles, gerunds, and infinitives. Place a comma after short and long introductory verbal phrases:

Participle
<u>Later known as the diode</u>‸ the "kenotron" allowed electric current to flow in only one direction, enabling the rectification of alternating current.

<u>Holding 1,093 patents in the United States alone</u>‸ Edison can be considered one of the most prolific inventors in history.

Gerund
<u>By using wax-coated cardboard cylinders</u>‸ Alexander Graham Bell and others were able to solve the problems of low sound quality that characterized Thomas Edison's initial phonograph.

Infinitive
<u>To enable light bulbs to last much longer</u>‸ Edison used a high-resistance lamp in a very high vacuum.

COMMAS WITH INTRODUCTORY ABSOLUTE PHRASES

An absolute phrase modifies the entire sentence. Follow an introductory absolute phrase with a comma:

<u>All things considered</u>‸ the Internet is as an important milestone in the history of communication as are the printing press, radio, and television.

5.1.4 ITEMS IN A SERIES

A series in a sentence could be three or more words, phrases, or clauses that have the same grammatical form and are of equal importance. Place a comma after each item in the series. An item might be one word, a phrase, or a clause.

Series of Words
<u>Resistors, capacitors, inductors, transformers, and even diodes</u> are all considered passive electronic devices.

ANALYSIS Although the comma before *and* is considered optional, use this comma in engineering, scientific, and technical documents. It helps to prevent misreading.

Series of Phrases

The urban transportation planning model requires the estimation of <u>trip generation</u>, <u>trip distribution</u>, <u>mode choice</u>, and <u>route assignment</u>.

ANALYSIS A series of noun phrases, each an object of the preposition *of*.

Transportation engineering design includes <u>sizing the transportation facilities</u>, <u>determining the subgrade and wear-surface materials and their thicknesses used in pavement</u>, and <u>designing vertical and horizontal alignment of the roadway</u>.

ANALYSIS A series of gerund phrases, each acting as a direct object of the verb *includes*.

Series of Predicates

Jack St. Clair Kilby <u>was born in1923</u>, <u>received his bachelor of science degree from the University of Illinois at Urbana-Champaign in 1947</u>, <u>obtained a master's degree from the University of Wisconsin in 1950</u>, <u>began work as an engineer at Texas Instruments in 1958</u>, <u>created the first integrated circuit in 1959</u>, and <u>went on to receive a Nobel Prize in 2000</u>.

ANALYSIS There are six predicates in this sentence!

Series of Dependent Clauses

The vacuum tube is a device <u>that is used to amplify a signal by controlling the movement of electrons in an evacuated space</u>, <u>that has been replaced by the much smaller and less expensive transistor</u>, but <u>that is still used in specialized applications such as audio systems and high power RF transmitters</u>.

ANALYSIS Three dependent clauses (specifically, adjective clauses), each modifying *device*.

Series of Independent Clauses

Vacuum tubes are bulky and consume a great amount of power; they produce great amounts of waste heat; and they easily fail or burn out.

5.1.5 COORDINATE AND CUMULATIVE ADJECTIVES

Coordinate adjectives are two or more adjectives that separately and equally modify the noun or pronoun. The order of these adjectives can be changed without affecting the meaning of the sentence.

Coordinate adjectives can be joined by *and*. Separate coordinate adjectives with commas:

> Grace Hopper was well-known for her <u>lively, colorful, irreverent</u> speaking style at computer-related events.

> ANALYSIS You can tell these adjectives are coordinate since they can easily be linked with *and* (*lively* and *colorful* and *irreverent*).

> His résumé included various, short-term, landscaping jobs.

> An exhibit of <u>authentic, early Inca</u> art was on display at the Royal Ontario Museum.

> The music festival featured <u>many Canadian folk</u> acts.

Cumulative adjectives modify the adjective after them and a noun or pronoun. Cumulative adjectives increase meaning from word to word as they progress toward the noun or pronoun. They are not interchangeable and cannot be joined by *and*. Do not use a comma between cumulative adjectives:

> The IBM-developed Mark I at Harvard University was the <u>first large-scale automatic digital</u> computer in the United States.

> ANALYSIS *digital* modifies *computer*, *automatic* modifies *digital computer*, and *large-scale* modifies *automatic digital computer*, and finally *first* modifies *large-scale automatic digital computer*. The order of the adjectives cannot be changed, nor can the coordinating conjunction *and* be placed between the adjectives.

5.1.6 RESTRICTIVE AND NONRESTRICTIVE ELEMENTS

Adjective clauses, adjective phrases, and appositives can modify nouns and pronouns. These modifying elements may be either restrictive or nonrestrictive.

WHAT IS A RESTRICTIVE ELEMENT?

A **restrictive element** limits, defines, or identifies the noun or pronoun that it modifies. The information in a restrictive element is essential to a sentence's meaning. *Do not set off a restrictive element with commas.* Traditionally, *that* is used instead of *which* for restrictive clauses, but *which* is often used in these cases as well:

> The Uncompahgre Ute Indians of Colorado used an ingenious invention <u>that allowed the creation of light by piezoelectricity thousands of years before the modern world learned of the concept</u>.

ANALYSIS It was not just any invention but an invention enabling the creation of light using piezoelectricity. Omitting the restrictive element would greatly alter the meaning of the sentence, turning it into a uselessly general statement.

> Ultrasonic transducers <u>that could transmit sound waves through air</u> were used in early television remote controls.

ANALYSIS Once again, it's not just any ultrasonic transducer, but specifically one that can transmit sound waves through air. Notice how restrictive elements *restrict* the universe of what they modify to a smaller subset.

WHAT IS A NONRESTRICTIVE ELEMENT?

A **nonrestrictive element** adds nonessential, or parenthetical, information about an idea or term that is already limited, defined, or identified; hence, a nonrestrictive element is set off with a comma or commas. While *which* as opposed to *that* is occasionally used in restrictive clauses, *that* is quite awkward in nonrestrictive clauses:

> Piezoelectric materials, <u>which are more complicated than purely elastic materials because of electromechanical coupling</u>, are widely used in micro- and nano electro-mechanical systems.

ANALYSIS The essential meaning of the sentence—piezoelectric materials are widely used in micro- and nano electro-mechanical systems—is not lost if the information about their behavior is removed; thus, this information is nonrestrictive.

> Ultrasonic time-domain reflectometers, <u>which send an ultrasonic pulse through a material and measure reflections from discontinuities</u>, are used to evaluate structural safety.

CONTEXT

Often, to decide whether a modifying element is restrictive or non-restrictive, you can simply read the sentence without the element. If the sentence doesn't make sense without the element, the element is restrictive and does not require commas; if it does make sense, the element is nonrestrictive and does require commas. In some cases, however, you will need to know the context in which a sentence appears to decide whether a restrictive or nonrestrictive element is required:

> Drivers, who drink excessively, should be banned from driving.

> Drivers who drink excessively should be banned from driving.

ANALYSIS The first example, the nonrestrictive version, makes the bizarre claim that all drivers drink excessively. The commas make the clause nonrestrictive and produce this meaning: "Oh, by the way, all drivers drink excessively." The second example, the restrictive one, limits the universe of drivers to that subset of drivers who drink excessively.

ADJECTIVE CLAUSES

Adjective clauses are dependent clauses, appearing at the beginning, middle, or end of sentences that modify nouns or pronouns. Adjective clauses can begin with:

- Relative pronouns: *who, whose, whom, that, which*
- Relative adverbs: *when, where*

Adjective clauses can be nonrestrictive or restrictive. Nonrestrictive adjective clauses are set off with commas. Use *which*, rather than *that*, with nonrestrictive adjective clauses.

Nonrestrictive Clause
Tofino‸ which is well over 200 kilometers from Victoria‸ attracts many tourists during November for storm watching.

ANALYSIS The information about Tofino's approximate distance from Victoria is not essential to the main idea of the sentence. Commas are required to set off the nonrestrictive clause.

A restrictive adjective clause should not be set off by commas since removing the clause would change the meaning of the sentence.

Restrictive Clause
Drivers who drink excessively should be banned from driving.

ANALYSIS The *who* clause is restrictive. It limits the *drivers* to those *who drink excessively*. Commas are not required.

Use *that*, rather than *which*, with restrictive adjective clauses:

It is common belief that Edison did not use mathematics, but his notebooks reveal that he used sophisticated mathematical analyses of Ohm's and Joule's Laws.

PHRASES THAT FUNCTION AS ADJECTIVES

Verbal and prepositional phrases can function as adjectives and may be nonrestrictive or restrictive. Whether they appear at the beginning, middle, or end of sentences, nonrestrictive adjective phrases must be set off with commas to indicate that the information is nonessential.

Nonrestrictive Adjective Phrase

Charles Kao‸ while working at Standard Telecommunications Laboratories in England, realized that the problem with glass as a light transmitter was a result of impurities rather than silica itself.

ANALYSIS The adjective phrase beginning *while* provides nonessential information about *Charles Kao*. The phrase is not needed to identify *Charles Kao*. Consequently, the phrase is nonrestrictive and must be set off by commas.

A restrictive adjective phrase should never be set off by commas since removing the phrase would change the sentence's meaning.

Restrictive Adjective Phrase

Shoppers <u>using debit cards to make small purchases</u> are becoming more common today.

ANALYSIS *Using debit cards to make small purchases* restricts the meaning of *shoppers*. Since it is a restrictive phrase, it is not set off by commas.

APPOSITIVES

An **appositive** renames or extends the meaning of a nearby noun or noun phrase. Most appositives are nonrestrictive and therefore must be set off with commas.

Nonrestrictive Appositive

Seymour Roger Cray‸ <u>a U.S. electrical engineer and supercomputer architect</u>, founded the company Cray Research in 1989.

The ILLIAC IV, <u>a specialized one-of-a-kind machine that rarely operated near its maximum performance except on very specific tasks</u>, was the only machine able to perform at the same level as the Cray-1 in 1976.

ANALYSIS The appositive *a specialized one-of-a-kind machine that rarely operated near its maximum performance except on very specific tasks* is not essential information. Since it is a nonrestrictive appositive, commas should be used to set it off.

Some appositives, however, are restrictive and do not require commas.

Restrictive Appositive
Charles Kao and a <u>colleague George Hockham</u> published a paper in 1966 predicting that optical fibers could be made pure enough to carry signals for miles.

ANALYSIS The appositive following *George Hockman* restricts the meaning. It is not any *colleague* that is meant; rather, it is a specific colleague, George Hockman.

In 1951, Columbia University <u>physicist Charles Townes</u> was searching for a way to alter the length of radio waves in order to probe the structure and behavior of molecules.

ANALYSIS The appositive *Charles Townes* is restrictive. It is not just any Columbia University professor that was doing this research. Thus, no commas should be used.

5.1.7 CONCLUDING ADVERBIAL CLAUSES

Adverbial clauses introducing a sentence should always be punctuated with a comma. However, when adverbial clauses conclude a sentence and their meaning is essential to the sentence, they are not set off by commas. Adverbial clauses that begin with the following subordinated conjunctions are usually essential: *after, as soon as, before, because, if, since, unless, until,*

Water boils at sea level <u>when it reaches a temperature of 100 degrees Celsius</u>.

ANALYSIS The concluding adverb clause commencing *when* is essential to the meaning of the sentence, so it is not set off with a comma.

Place a comma before adverbial clauses that contain nonessential information. Often, adverbial clauses beginning with the following subordinating conjunctions are nonessential: *although, even though, though*

> He missed the turn for the expressway‸ <u>even though signs for the on-ramp were well posted</u>.

5.1.8 TRANSITIONS, PARENTHETICAL EXPRESSIONS, ABSOLUTE PHRASES, CONTRASTS

TRANSITIONAL EXPRESSIONS

Transitional expressions are words or groups of words that function as links between or within sentences. A transitional expression can appear at the beginning, end, or within a sentence. Examples of transitional expressions are phrases such as *therefore, however, for example, in addition*, and *on the contrary*.

If a transitional expression appears between independent clauses in a compound sentence, place a semicolon before it and, most often, a comma after the transitional expression:

> A single optical fiber can accommodate multiple light streams‸in fact‸ more than a thousand distinct streams can ride along a single glass fiber at the same time.

> Lasers play an important role in medicine; <u>for example</u>, medical lasers simultaneously make surgical incisions and cauterize blood vessels, minimizing bleeding.

Except when they are used to join two independent clauses, punctuate transitional expressions with commas wherever they occur in a sentence:

> <u>However</u>, all laser light has the same highly organized nature.

> A DVD player, <u>for example</u>, uses laser light to read digital information from a spinning disc.

PARENTHETICAL EXPRESSIONS

Parenthetical expressions contain additional information the writer inserts into the sentence for such purposes as to explain, qualify, or give his or her point of view. If parenthetical expressions do not appear in parentheses, they are set off with commas:

> <u>Typically</u>, light traveling along ordinary glass fiber loses about 99 percent of its energy after just 30 feet.

In most writing situations, <u>such as this one</u>, commas are used to set off parenthetical expressions.

ABSOLUTE PHRASES

An **absolute phrase** contains a noun subject and a participle that modify an entire sentence. Set off absolute phrases with commas:

<u>The war being over</u>, the refugees returned home.

<u>Their profits steeply declining</u>, many computer companies laid off employees.

EXPRESSIONS OF CONTRAST

Expressions of contrast include words such as *not, nor, but*, or *unlike*. Set off expressions of contrast with commas:

One further benefit of fiber is that even with long distances, fiber cables, <u>unlike some types of electrical transmission lines</u>, experience no crosstalk.

In practical fibers, a resin buffer layer surrounded by a jacket layer contributes strength to the fiber, <u>but not to transmission capabilities</u>.

5.1.9 NOUNS OF DIRECT ADDRESS, YES AND NO, INTERROGATIVE TAGS, INTERJECTIONS

Use commas to set off the following:

- Nouns of direct address:

  Your back flip, <u>Olga</u>, is of Olympic caliber.

- The words *yes* and *no*:

  <u>No</u>, you can not rappel down the face of the university administration tower.

- Interrogative tags:

  You did turn off the burglar alarm, <u>didn't you</u>?

- Mild interjections:

  <u>Then</u>, incidents like that are inevitable.

5.1.10 HE SAID, ETC.

Use commas with speech tags such as *he said, she wrote*, or phases like *according to* to set off direct quotations:

"So nimble are these little electrons," wrote John Ambrose Fleming, "that however rapidly we change the rectification, the plate current is correspondingly altered, even at the rate of a million times per second."

According to Jack Kilby in 1958, "Extreme miniaturization of many electrical circuits could be achieved by making resistors, capacitors and transistors and diodes on a single slice of silicon."

"The future of integrated electronics is the future of electronics itself," Gordon Moore wrote in 1965.

5.1.11 DATES, ADDRESSES, TITLES, NUMBERS

DATES

When the date is within the sentence, use commas following the day and the year in month-day-year dates:

> On August 14, 1945, Japan surrendered.

When the date is inverted or when just the month and year are given, commas are not required:

> Rear Admiral Grace Murray Hopper was born on December 9, 1906 and died January 1, 1992.

> January 2002 was unseasonably warm.

ADDRESSES

Use a comma between the city and state or city and country. When a sentence continues on after the city and state or city and country, also use a comma after the state or country:

> Nikola Tesla did most of his research in New York, New York, during the 1880s and 1890s.

In a complete address, separate all items except the postal code:

> Nelson, a division of Thomson Canada Limited, is located at 1120 Birchmount Road, Scarborough, Ontario M1K 5G4.

> Harcourt College Publishers was located at 301 Commerce Street, Fort Worth, Texas 76102.

TITLES

When an abbreviated title follows a name, place a comma after the name and a second comma after the title:

> Philip Bacho, Ph.D., taught the course on writing scripts.

NUMBERS

U.S. writers insert a comma (but no space) after every three digits, starting from the right:

256
1,024
1,048,576

Canadian readers Canada follows the international system of metric measurement, which does not use commas in numbers. Instead, spaces are used to separate sets of three digits. Four-digit numbers may be grouped together.

4673
233 971
6 299 381

5.1.12 TO PREVENT CONFUSION

Frequently, a comma is essential to ensure that readers clearly understand your intended meaning. Omitting or misplacing a comma can easily lead to misreadings:

<u>When connecting</u>, the modem makes a series of noises indicating the sequence of events required to establish communication with a server.

ANALYSIS Without the comma, this sentence could be misread as *connecting the modem*.

In many writing situations, commas are required to prevent reader confusion. Occasionally commas are required to help readers group units of meaning as the writer intended:

<u>Those who can</u>, run every chance they get.

A comma is used to indicate that an understood word or words have been omitted:

Michael Faraday is famous for his work with magnetism and electricity; Hans Christian Ørsted, for his discovery of electromagnetism.

5.1.13 COMMON PROBLEMS WITH THE COMMA

Be able to state a rule, such as the rules in the chapter, whenever you use a comma in your writing.

BETWEEN VERB AND SUBJECT OR OBJECT

Do not use a comma to separate a subject from its verb or a verb from its object:

Incorrect: Alpha Centauri, is the brightest star system in the southern constellation of Centaurus.

Correct: Alpha Centauri is the brightest star system in the southern constellation of Centaurus.

ANALYSIS *Centauri*, the subject of the sentence, should not be separated from the main verb *is*.

Incorrect: Zefram Cochrane's apparent place of origin, was changed from Alpha Centauri to Montana in the feature film.

Correct: Zefram Cochrane's apparent place of origin was changed from Alpha Centauri to Montana in the feature film.

ANALYSIS *place*, the subject of the sentence, should not be separated from the main verb *was changed*.

Incorrect: That Zefram Cochrane and his Star Trek colleagues wished to usher in a new visionary era for humanity with the development of the warp drive, is much in doubt.

Correct: That Zefram Cochrane and his Star Trek colleagues wished to usher in a new visionary era for humanity with the development of the warp drive is much in doubt.

ANALYSIS That Zefram Cochrane and his Star Trek colleagues wished to usher in a new visionary era for humanity with the development of the warp drive is a big noun clause functioning as the subject. You could substitute "This" for it. However large it is, don't separate this subject from its verb is.

COMMON INCORRECT USES OF THE COMMA

Never use a comma in the following situations.

• To separate an adjective from a following noun:

An optical fiber is a thin, transparent⸏ fiber used for the transmission of light.

- To separate an adverb from a following adjective:

  By using a laser source with an extremely ⸗narrow spectrum, real-world applications can achieve data rates of up to 40 gigabits per second.

- After *like* or *such as*:

  Glass optical fibers are almost always made from silica, but for longer-wavelength infrared applications, other materials are used, such as ⸗ fluorozirconate, fluoroaluminate, and chalcogenide glasses.

- After a coordinating conjunction (*and, but, for, nor, or, so, yet*):

  Optical fiber exhibits exceptionally low loss over long distances between amplifiers or repeaters, and ⸗ its inherently high data-carrying capacity enables thousands of electrical links to be replaced by a single high-bandwidth fiber.

- After *although*:

  Although ⸗ either transparent plastic or glass fibers can be used to manufacture optical fiber, glass is always used in long-distance telecommunications applications in order to lower the rate of optical attenuation.

- Before *than*:

  Optical fiber is generally used in systems that require higher bandwidths ⸗ than electrical cabling can provide.

- To set off an indirect quotation:

  In an Asiaweek.com interview, Charles K. Kao, considered the father of fiber optics, said that ⸗ optical fiber might indeed be the root of what allowed the Internet to take off.

- With an exclamation point or a question mark:

  "Help! My Internet connection is too fast! ⸗" the tortoise cried.

- Before a parenthesis:

  Typically, plastic optical fiber has much higher rate of attenuation than does glass fiber ⸗ (the amplitude of the signal in plastic optical fiber decreases more rapidly).

5.2 SEMICOLONS

A semicolon is used to separate major elements of a sentence that are of equal grammatical rank.

5.2.1 INDEPENDENT CLAUSES WITH NO COORDINATING CONJUNCTION

An **independent clause** expresses a complete thought and can stand on its own as a sentence. When related independent clauses appear in a sentence (as in a compound sentence), they are usually linked by a comma and a coordinating conjunction (*and, but, for, nor, or, so,* and *yet*). The conjunction indicates the relationship between the clauses.

When the relationship between independent clauses is clear without the conjunction, you can instead link the two clauses with a semicolon:

> A teacher affects eternity; no one can tell where his
> influence stops.

> —*Henry Adams*

Use a semicolon if a coordinating conjunction between two independent clauses has been omitted. If you use a comma, you will create a grammatical error known as a comma splice:

> An engineering education must include a strong foundation in mathematics and science; engineering students should then apply this knowledge to a specific field.

> Materials science is closely related to many other areas of engineering and science; as such, it is one of the most multidisciplinary of all engineering fields.

Strategies for revising comma splice errors can be found in Chapter 4. You may wish to consider other alternatives to using a semicolon.

5.2.2 INDEPENDENT CLAUSES WITH TRANSITIONAL EXPRESSIONS

Transitional expressions can be conjunctive adverbs or transitional phrases.

COMMON CONJUNCTIVE ADVERBS			
accordingly	also	anyway	besides
certainly	consequently	conversely	finally

further	furthermore	hence	however
incidentally	indeed	instead	likewise
meanwhile	moreover	namely	nevertheless
next	nonetheless	now	otherwise
similarly	specifically	still	subsequently
then	thereafter	therefore	thus
undoubtedly			

TRANSITIONAL PHRASES

after all	as a matter of fact	as an illustration	as a result
at any rate	at the same time	equally important	even so
for example	for instance	in addition	in conclusion
in fact	in other words	in short	in spite of
in summary	in the first place	in the same way	of course
on the contrary	on the other hand	to be sure	to illustrate

When a transitional expression comes between independent clauses, place a semicolon before the expression and a comma after it:

She is an authority on the West Nile virus; <u>furthermore</u>, we need someone with her expertise.

If the transitional expression is in the middle of or at the end of the second independent clause, the semicolon is placed between the independent clauses:

In about 80% of cases, the West Nile infection causes no symptoms; in the rest, <u>on the other hand</u>, the virus causes mild flu-like symptoms known as West Nile fever.

Do not confuse the punctuation for transitional expressions with that used with coordinating conjunctions (*and, but, for, nor, or, so,* and

yet). When a coordinating conjunction links two independent clauses, it is preceded by a comma.

5.2.3 SERIES WITH INTERNAL PUNCTUATION

Usually, commas separate items in a series. However, when series items contain commas, a semicolon is placed between items to make the sentence easier to read:

> ENIAC contained 17,468 vacuum tubes, 7,200 crystal diodes, 1,500 relays, 70,000 resistors, 10,000 capacitors, and around 5 million hand-soldered joints; weighed 27 tons; measured 8 feet by 3 feet by 100 feet; took up 1800 square feet; and consumed 150 kilowatts of electricity.

> ANALYSIS Without semicolons, the sentence is difficult to read. The first element in the series contains a comma of its own. If only commas are used, it is difficult for the reader to distinguish between the five predicates. The semicolons help the reader to group information accurately.

INCORRECT USES OF THE SEMICOLON

Never use a semicolon in the following writing situations.

- Between independent clauses joined by *and, but, for, nor, or, so,* or *yet:*

> San Antonio-based Datapoint Corporation later decided not to use the
> *developing, and*
> chip it had been ~~developing; and~~ Intel marketed it as the 8008 in 1972 as the world's first 8-bit microprocessor.

- Between a subordinate clause and the remainder of the sentence:

> Because it used four of the accumulators controlled by a special multiplier
> *unit, the*
> ~~unit; the~~ ENIAC could perform 385 multiplication operations per second.

- Between an appositive and the word to which it refers:

> To the top of the metal layer of the substrate of an integrated circuit is
> *photoresist: a*
> applied a layer of ~~photoresist; a~~ chemical that hardens when exposed to light (often ultraviolet).

- To introduce a list:

  Integrated circuits are fabricated in a layer process which includes these key

  process ~~steps; imaging~~ *steps: imaging*, deposition, and etching

5.3 COLONS

The colon is used most often to indicate a formal, emphatic introductory word, phrase, or clause that follows it. The colon also has
technical and scientific uses.

5.3.1 BEFORE A LIST, AN APPOSITIVE, OR A QUOTATION

Use a colon before:

- A list:

  Stress can be applied to materials in three common ways: <u>axial,
  torsional, and flexural</u>.

- An appositive:

  In the 1930s, researchers realized that complex calculations could be
  achieved by means of the binary numbering system: <u>a counting system
  that uses only ones and zeros</u>.

- A clause introduced by an independent clause:

  As the ENIAC was being developed, the prediction that frequent tube
  failures would render the machine useless turned out to be partially
  correct: <u>every day, several tubes burned out, rendering it nonfunctional
  about half the time</u>.

  In 1874 a German scientist named Ferdinand Braun discovered rectification:
  <u>the fact that current tends to flow through a semiconductor crystal in
  only one direction</u>.

- A quotation:

  Jack Kilby, an electrical engineer at Texas Instruments, summed up his
  ideas about the integrated circuit he had created in 1958 in these
  words: "<u>Extreme miniaturization of many electrical circuits could be
  achieved by making resistors, capacitors and transistors and diodes on a
  single slice of silicon</u>."

5.3.2 BETWEEN INDEPENDENT CLAUSES

Use a colon between independent clauses if the first clause introduces the second:

In the 1960s, semiconductor devices began replacing vacuum tubes in most applications: they use electronic conduction in the solid state as opposed to the gaseous state or thermionic emission in a high vacuum.

Microprocessor development has generally followed Moore's Law: the law holds that the complexity of an integrated circuit, in relation to minimum component cost, doubles every 24 months.

5.3.3 CONVENTIONAL USES

The colon is conventionally used

- After the salutation of a formal letter:

  Dear Dr. Hughes:

- To indicate hours and minutes:

  6:31 a.m. (or p.m.)

- Between numbers in ratios:

  The ratio of men over fifty was 5:1.

- Between the title and subtitle of a book:

  Invention by Design: How Engineers Get from Thought to Thing

- To separate the city from the publisher and date in a bibliographic entry:

  Toronto: Nelson, 2003.

- Between Bible chapters and verses:

  Psalms 23:1–3

5.3.4 INCORRECT USES OF THE COLON

Except in documentation, a complete independent clause must precede a colon. Do not use a colon in the following writing situations.

- Between a verb and its complement or object:

  contained 17,468
  ENIAC contained: 17,468 vacuum tubes, 7,200 crystal diodes, 1,500 relays, 70,000 resistors, 10,000 capacitors and around 5 million hand-soldered joints; weighed 27 tons; was roughly 8 feet by 3 feet by 100 feet; took up 1,800 square feet; and consumed 150 kW of power.

- Between a preposition and its object:

  into gasoline
  When used to supply energy, petroleum is converted into: gasoline, fuel oils, lubricants, kerosene, and jet fuels.

- After *for example, such as*, and *including* or *included*:

  included boreal
  The content of the botanist's lecture included: boreal forests, a Carolinian forest, and an Amazonian rainforest.

5.4 APOSTROPHES

Apostrophes are used for possessives, contractions, and some plurals.

5.4.1 POSSESSIVE NOUNS

An apostrophe (') appears as part of a noun to indicate that the word is possessive. Often ownership is obvious, as in *Mishka's hockey stick* or the *instructor's briefcase*. Sometimes, ownership is not as explicit, as in the *journey's end* or the *river's tributaries*. To test if a noun is possessive, see if you can state it as an *of* phrase, as in *the end of the journey* or the *tributaries of the river*. According to this test, both nouns, *journey's* and *river's*, are possessive.

- Add -'s if the noun does not end in -s:

  The medical robot's equipment includes microphones, cameras, and databases that allow doctors and patients to communicate almost as if they were in the same room.

- Add -'s if the noun is singular and ends in -s:

  When loads are applied to a truss's joints, forces are transmitted only in the direction of each of its members.

- Do not add -s to a singular noun if doing so would make the word difficult to pronounce:

  <u>Moses'</u> mountain journey is an important part of his legacy.

- Add only an apostrophe if the noun is plural and ends in -s:

  <u>Workers' rights</u> were neglected by the military regime.

- Add -s or -s' to the last noun only of compound subjects to show joint possession:

  John Markoff and John M. Broder's March 2004 article in the *New York Times* reported that none of 15 robot vehicles completed the 142-mile race in the Mojave Desert sponsored by DARPA.

- Add -s or -s' to all nouns when each noun individually possesses something:

  Robot artwork such as James Seawright's House Plants and Ken Rinaldo's Autotelematic Spider Bots feature interaction with viewers and each other.

- Add -s or -s' to the last element of compound nouns to show possession:

  His brother-in-law's collaboration enabled Charles Townes to make progress in the development of the laser.

5.4.2 POSSESSIVE INDEFINITE PRONOUNS

An indefinite pronoun refers to a general or nonspecific person or thing. Examples of indefinite pronouns are *somebody, anything*, and *anyone*. Add -'s to the end of the indefinite pronoun to make it possessive:

It is <u>anybody's</u> guess how some nanorobots will be used in medicine.

<u>Someone's</u> laptop was stolen from the reference library.

5.4.3 CONTRACTIONS

The apostrophe takes the place of missing letters in contractions:

<u>Who's</u> going to do it doesn't matter.

ANALYSIS *Who's* written in full is *Who is*, and *doesn't* written in full is *does not*.

The apostrophe can also indicate that the first two digits of years have been left out:

In the '50s, much important theoretical and experimental work leading to the laser was done.

Did you enjoy That *'70s Show*?

However, -s without an apostrophe is added to years in a decade:

Important theoretical and experimental work leading to the laser was done in the 1950s.

5.4.4 PLURALS OF NUMBERS, LETTERS, ETC.

In general usage, an -'s is used to pluralize numbers, letters, and phrases.

- Numbers:

  Computers use the binary numbering system consisting of 1's and 0's.

- Letters:

  The living rooms in the condominiums all formed *L*'s.

- Word as words:

  Science fiction is usually based on *what if*'s.

Notice that -s is not italicized when used with italicized words or letters.

- Abbreviations:

  DARPA's rules for robot racing in the Mojave Desert specify that contestants are not allowed to control or send signals to the robots once the race starts.

However, according to most engineering and technical style guides, no apostrophe is needed with plurals of numbers, years, and abbreviations:

He has trouble writing 6s.

In the mid-1970s, much of the groundwork was laid for fiber-optic communications.

Typically, DVDs can hold up to 133 minutes of high-resolution video, with soundtrack available in up to 8 languages and subtitles in up to 32 languages.

5.4.5 INCORRECT USES OF THE APOSTROPHE

Do not use an apostrophe with:

- Nouns that are not possessive:

  clients
  The clients; had expected us to pick up the tab for dinner.

- The possessive pronouns *his*, *hers*, *its*, *ours*, *theirs*, and *whose*:

  its
  That is it's first time out of the box.

  ANALYSIS Here, *its* is the possessive form. *It's* is the contraction for *it is*.

5.5 QUOTATION MARKS

Quotation marks are used not just for quotations but for other instances such as titles and words treated as words. The following section also provides rules on punctuating and formatting quotations.

5.5.1 DIRECT QUOTATIONS

Direct quotations are the exact words copied from a print source or transcribed from what a person says. Direct quotations must be enclosed within quotation marks:

"I do not know with what weapons World War III will be fought, but World War IV will be fought with sticks and stones," Einstein wrote in a letter to President Harry Truman.

Of his vision of the new information world, Vannevar Bush wrote, "Wholly new forms of encyclopedias will appear, ready-made with a mesh of associative trails running through them, ready to be dropped into the memex and there amplified."

On the other hand, **indirect quotations** paraphrase or summarize what has appeared in a print source or what a person has said. Indirect quotations are not placed within quotation marks:

In his July 1945 *Atlantic Monthly* article "As We May Think," Vannevar Bush maintained that great things surely must come from a world full of cheap electronic devices.

QUOTING LONGER PASSAGES BY A SINGLE SPEAKER

If you are directly quoting passages by a single speaker, start each new paragraph with quotation marks, but do not use closing quotation marks until the end of the quoted material.

MARKING A CHANGE IN SPEAKER WITHIN DIALOG

Start a new paragraph to signal a change in the speaker. The following is a dialog between two fictional characters:

> "I said me, not you."
>
> "Oh. You got a car outside?"
>
> "I can walk."
>
> "That's five miles back to where the van is."
>
> "People have walked five miles."
>
> —Alice Munro, "Friend of My Youth"

5.5.2 LONG QUOTATIONS

PROSE

A "long" quotation of prose is any passage that is more than four typed or handwritten lines. Start the quotation ten spaces from the left margin. You do not need to enclose the longer quotation within quotation marks because the indented format establishes for the reader that the quotation is taken exactly from a source. Usually, longer quotations are introduced by a sentence ending with a colon:

> In an essay on lasers and fiber optics, Charles H. Townes reflects on what led him to the initial idea of the maser:
>
>> The laser invention happened because I wanted very much to be able to make an oscillator at frequencies as high as the infrared in order to extend the field of microwave spectroscopy in which I was working. I had tried several ideas, but none worked very well. At the time I was also chairman of a committee for the Navy that was examining ways to obtain very short-wave oscillators. In 1951, on the morning before the last meeting of this committee in Washington, I woke up early worrying over our lack of success. I got dressed and stepped outside to Franklin Park, where I sat on a bench admiring the azaleas and mulling over our problem [7].
>
> (Source: http://www.greatachievements.org/?id=3717)

Placing the source number within brackets is the citation style prescribed by the IEEE and CBE. (See Chapter 12.) If the direct quotation had included additional paragraphs, each new paragraph would need to be indented an additional three spaces.

POETRY

A "long" quotation of poetry is more than three lines of the poem. Start the quotation 10 spaces from the left margin. You do not need to enclose the longer quotation within quotation marks because the indented format establishes for the reader that the quotation is taken exactly from the poem. Use quotation marks within the quotation only if they are part of the poem.

> Here is a snippet of computer-generated poetry from the technofile's Computer Poetry Corner:
>
> *The blue giants*
>
> *Run loosely to the death,*
>
> *When lively, plastic, sparkling*
>
> *Loud priests noisily.*

Guidelines vary on formatting poetry and long quotations. Check your style guide produced by the associated organization—for example, the American Psychological Association.

5.5.3 QUOTATIONS WITHIN QUOTATIONS

Single quotation marks enclose quotations within quotations:

> According to Charles H. Townes, "When Schawlow and I first distributed our paper on how to make a laser, a number of friends teased me with the comment, 'That's an invention looking for an application. What can it do?'"
>
> Source: http://www.greatachievements.org/?id=3717

Two different quotation marks appear at the end of the quotation. The single quotation mark completes the interior quotation, while the double quotation mark completes the main quotation.

5.5.4 QUOTATIONS AND TITLES

Use quotation marks around titles of works that are included within other works, such as poems, short stories, newspaper and magazine articles, radio programs, television episodes, and chapters and other subdivisions of books:

> For many, Vannevar Bush's article "As We May Think," published in the July 1945 edition of the *Atlantic Monthly*, is an essential work in the development hypertext.

The titles of plays, books, and films and the names of magazines appear in italics if you are typing your manuscript. They are underlined if your manuscript is handwritten.

5.5.5 QUOTATIONS AND WORDS AS WORDS

Use quotation marks to set off words used as words. However, it is acceptable to use quotation marks for this purpose:

> Norbert Wiener coined the term "cybernetics" in his 1948 book *Cybernetics or Control and Communication in the Animal and the Machine*.

Use italics for terms when they are being defined:

> *Cybernetics* is the science of communication, control, and feedback in living organisms, in machines, and in combinations of the two.

Use quotation marks when words are used in an unusual, non-standard way:

> A computer program that has numerous functional problems is said to be "buggy" and prone to "crashing."

5.5.6 QUOTATIONS WITH OTHER PUNCTUATION

The following section provides rules for using punctuation with quotation marks.

COMMAS AND PERIODS

Place commas and periods inside quotation marks:

> "[N]o more than 10 to 20 million [will be] killed," Turgidson remarks to the president about the outcome of a pre-emptive nuclear war. "Tops!"

Also follow the foregoing punctuation rule in the following cases:

- With single quotation marks
- For titles of works
- For words used as words

SEMICOLONS AND COLONS

Place semicolons and colons outside quotation marks:

> In the 1964 Stanley Kubrick movie, *Dr. Strangelove or: How I Learned to Stop Worrying and Love the Bomb*, Jack D. Ripper is obsessed with "precious bodily fluids"; he apparently has had a problem with impotence.

> As the bank's head economist, she asserts that the economy will soon "take off"; several of her colleagues at other banks strongly disagree.

QUESTION MARKS AND EXCLAMATION POINTS

If the question mark or exclamation point is part of the quoted material, place the question mark or exclamation point *inside* the quotation marks. No other end punctuation is needed.

Part of the Quoted Material

At the end of the1964 Stanley Kubrick movie, *Dr. Strangelove or: How I Learned to Stop Worrying and Love the Bomb*, Dr. Strangelove jumps out of his wheelchair and exclaims, "Mein Führer, I can walk!"

In the same movie, the George C. Scott character reassures the president about the outcome of a preemptive nuclear war: "Now I'm not saying we wouldn't get our hair mussed, but I am saying no more than 10 to 20 million killed. Tops!"

If the question mark or exclamation point applies to the entire sentence, place a question mark or exclamation point *outside* the quotation marks.

Applies to the Entire Sentence

What do you think of Napoleon's view that "history is a set of lies agreed upon"?

Pay particular attention to your use of question marks and exclamation points with quotation marks. If the quotation is itself a question, place the question mark before the closing quotation mark; place a sentence period after the closing bracket in the parenthetical citation:

> At one point in *Dr. Strangelove or: How I Learned to Stop Worrying and Love the Bomb*, Jack D. Ripper says, "Mandrake, do you recall what Clemenceau once said about war?" [5].

5.5.7 INTRODUCING QUOTED MATERIAL

You have three major punctuation options when using a group of words to introduce a quotation:

- Colon
- Comma
- No punctuation

WHEN TO USE THE COLON

Use the colon if the quotation has been formally introduced. A formal introduction is a complete independent clause:

> In a *Wall Street Journal* ad appearing July 31, 1979, Edward Teller stated that he was the only one whose health was affected by the nuclear accident at Three-Mile Island: "It was not the reactor. It was Jane Fonda. Reactors are not dangerous."

WHEN TO USE THE COMMA

Use a comma if a quotation is introduced with or followed by an expression such as *she said* or *he uttered*:

> In the *Wall Street Journal* ad, Teller wrote, "On May 7, a few weeks after the accident at Three-Mile Island, I was in Washington. I was there to refute some of that propaganda that Ralph Nader, Jane Fonda and their kind are spewing to the news media in their attempt to frighten people away from nuclear power."

> "You might say that I was the only one whose health was affected by that reactor near Harrisburg. No, that would be wrong. It was not the reactor. It was Jane Fonda. Reactors are not dangerous," Teller went on to say.

WHEN A QUOTATION IS BLENDED INTO A SENTENCE

If a quotation is worked into your own sentence, punctuate as if the quotation marks were not there. Use punctuation according to how the quotation fits into the grammatical structure of your sentence.

> In 1991, in recognition of his "lifelong efforts to change the meaning of peace as we know it," Teller was awarded one of the first Ig Nobel Prizes for Peace, a parody of the Nobel Prize.

> Each year in early autumn about the time the Nobel Prizes are announced, the Ig Nobel Prizes are announced for ten achievements that "first make people laugh, and then make them think."

WHEN A QUOTATION BEGINS A SENTENCE

Use a comma to set off a quotation at the beginning of a sentence:

"I have harnessed the cosmic rays and caused them to operate a motive device," Nikola Tesla once remarked in a newspaper interview.

However, a comma is not needed if the opening quotation ends with a question mark or an exclamation point:

"Space elevators! Impossible! The cost for the carbon nanotubes alone would be $500 million!" shouted the old professor.

WHEN A QUOTED SENTENCE IS INTERRUPTED BY EXPLANATORY WORDS

Use commas to set off the explanatory words:

"Excuse me, sir," replied a younger professor, "but already there is a consortium of space elevator companies called the LiftPort Group that is planning to use carbon nanotubes in the construction of a 100,000 km space elevator. They have already successfully experimented with high-altitude robotic lifters."

WHEN TWO SUCCESSIVE QUOTED SENTENCES ARE INTERRUPTED BY EXPLANATORY WORDS

Use a comma before the explanatory words within the quotation marks of the first quotation. End the explanatory words with a period:

"The scientists from Franklin to Morse were clear thinkers and did not produce erroneous theories," Nikola Tesla once remarked. "The scientists of today think deeply instead of clearly. One must be sane to think clearly, but one can think deeply and be quite insane."

5.5.8 INCORRECT USES OF QUOTATION MARKS

Do not use quotation marks around indirect quotations:

Incorrect: Albert Einstein was convinced that God does not "gamble with the universe."

Correct: Albert Einstein was convinced that God does not gamble with the universe.

ANALYSIS Einstein is not reported as ever having said this. He is believed to have said that "God does not play dice with the universe."

Do not use quotation marks to set off the title of your engineering document. When you place the title on the cover, title page, or at the top of page 1, don't use quotation marks. However, when you refer to your engineering document by its title (for example, in e-mail), you should use quotation marks (or italics if your style guidelines so require).

5.6 OTHER MARKS

5.6.1 PERIODS

Periods are commonly used to indicate the end of a sentence and within abbreviations.

ENDING SENTENCES

Use the period after statements, indirect questions, and mild commands:

> At the end of the nineteenth century, scientists, inventors, and entrepreneurs such as Nikola Tesla, Charles Steinmetz, and George Westinghouse championed AC as the power-supply medium of choice.

> Has anybody considered the negative social and environmental costs of obtaining an extremely cheap source of energy such as that potentially provided by nuclear fusion?

> One of the panelists at the symposium asked whether anyone had considered the negative social and environmental costs of obtaining an extremely cheap source of energy such as that potentially provided by nuclear fusion.

After a direct question, use a question mark. However, if the question is indirect, use a period to end the sentence.

MILD COMMAND

After a strong command, use the exclamation point.

> Please, call an ambulance now!

However, after a **mild command**—an imperative or declarative sentence that is not an exclamation—use a period:

> Please pick up the groceries.

ABBREVIATIONS

Use periods in abbreviations such as the following:

a.m. (or A.M.)	p.	B.A.	Dr.	Inc.
p.m. (or P.M.)	etc.	M.A.	Ms.	Ltd.
B.C. (or B.C.E.)	e.g.	M.B.A.	Mrs.	Dec.
A.D. (or C.E.)	i.e.	Ph.D.	Mr.	St.

Canadian readers Do not use periods with Canada Post abbreviations, such as SK, ON, and NB.

Widely recognized abbreviations for organizations, companies, and countries do not require periods:

CBS CSIS NFB UK USA IBM UN NBA CFL

If you are in doubt about whether or not an abbreviation requires a period, consult a good dictionary or encyclopedia. To check the abbreviation of a name of a company, you might consult that company's web site.

DECIMAL VALUES

Each GPS satellite is equipped with an atomic clock that it can keep time to within 0.000000003 of a second.

ABBREVIATIONS AT THE END OF SENTENCES

Do not add a second period if the sentence ends with an abbreviation's period:

He always wanted to complete his M.A.

5.6.2 QUESTION MARKS

FOLLOWING A DIRECT QUESTION

Use a question mark after any direct questions:

Exactly how does Global Positioning System (GPS) manage to show you your exact position on Earth any time, anywhere, in any weather?

Use a period after an indirect question:

You may be asking yourself how Global Positioning System (GPS) manages to show you your exact position on Earth any time, anywhere, in any weather.

FOLLOWING QUESTIONS IN A SERIES THAT ARE NOT COMPLETE SENTENCES

Use a question mark to end each question in a series, even if series questions are not complete sentences:

How many of the twentieth century's greatest engineering achievements can you find around your house or apartment? A telephone? Cell phone? Television? Radio? CD player? Microwave oven?

5.6.3 EXCLAMATION POINTS

Use the exclamation point with an emphatic declaration or a strong command:

The transfer of motion from one end of a conductor occurs at the speed of light, 186,000 miles per second!

Do not overuse the exclamation point:

Overuse: The Mark I measured 51 feet by 8 feet by 2 feet! It weighed about 5 tons, and its basic calculating units required a 50-foot shaft driven by a 5-horsepower electric motor! It could store 72 numbers, each 23 decimal digits long! It could do 3 additions or subtractions in 1 second! A multiplication took 6 seconds, a division took over 15 seconds, and a logarithm or a trigonometric function took over 60 seconds!

Correction: The Mark I measured 51 feet by 8 feet by 2 feet. It weighed about 5 tons, and its basic calculating units required a 50-foot shaft driven by a 5-horsepower electric motor. It could store 72 numbers, each 23 decimal digits long! It could do 3 additions or subtractions in 1 second. A multiplication took 6 seconds, a division took over 15 seconds, and a logarithm or a trigonometric function took over 60 seconds!

ANALYSIS Every sentence ends with an exclamation point, and as a result, the punctuation mark loses its effectiveness in communicating emphasis. Communicate strong impressions through the powerful use of words, not through overuse of the exclamation point.

5.6.4 DASHES

The dash marks a strong break in the continuity of a sentence. It can be used to add information, emphasize part of a sentence, or to set part of the sentence off for clarity.

To make a dash using your computer, enter two unspaced hyphens (–). Do not leave a space before the first hyphen or after the second hyphen. Many computer programs automatically format dashes when you enter two consecutive hyphens.

Use dashes to

- Enclose a sentence element that interrupts the flow of thought, or to set off parenthetical material that deserves emphasis:

  In 1951 Columbia University physicist Charles Townes was a leading expert in spectroscopy—the science of matter's interactions with electromagnetic energy—searching for a way to alter the length of radio waves in order to probe the structure and behavior of molecules.

- Set off appositives that contain commas:

  In 1962, three research organizations—General Electric, IBM, and MIT's Lincoln Laboratory—simultaneously developed a gallium arsenide laser whose basic workings would be used much later in CD and DVD players.

- Show a dramatic shift in tone or thought:

  Soon Townes and his students were building the first laser—not generating laser beams, at this point, but simply demonstrating its principles.

- Restate or amplify:

  The information-carrying capacity of a transmission medium is directly related to frequency—the number of wave cycles per second, or hertz.

  The pulses of some lasers are so brief—a few quadrillionths of a second—that they can freeze the movements of molecules in a chemical reaction.

- Prepare a list:

  In the storage room are all the paint supplies—paints, paint thinner, canvas drop cloths, brushes, rollers, and rolling tins.

If dashes overused, they can lose their effectiveness and create a disjointed effect in your writing. Limit the number of dashes in a sentence to two paired dashes or one unpaired dash; limit sentences containing dashes to one or two per paragraph.

5.6.5 PARENTHESES

Parentheses are used to set off helpful, nonessential, additional information. While dashes usually call attention to the information they enclose, parentheses often de-emphasize the information they enclose.

Use parentheses to:

- Enclose supplemental information, such as definitions, examples, digressions or asides, and contrasts:

Devices that switch messages from one fiber to another (for example, from one router to another on the Internet) still must convert the data back and forth from light to electricity.

Urban transportation planning requires an understanding of trip origination (how many trips for what purpose), trip destination (where people are going), mode choice (what transportation mode is used), and route assignment (which routes are used).

- Enclose letters or numbers that label items in a series:

In one basic clock cycle of 200 microseconds, the ENIAC could (1) write a number to a register, (2) read a number from a register, or (3) add or subtract two numbers.

Do not overuse parentheses. Including too much parenthetical information can make your writing seem choppy and awkward. Often, you can integrate information from parentheses into your sentences so they flow more smoothly:

William Shockley (February 13, 1910–August 12, 1989) was an American physicist and co-inventor of the transistor (with John Bardeen and Walter Houser Brattain), for which he was awarded the Nobel Prize in physics (in 1956).

William Shockley (February 13, 1910–August 12, 1989) was an American physicist and coinventor of the transistor (with John Bardeen and Walter Houser Brattain) for which he was awarded the Nobel Prize in physics (in 1956).

5.6.6 BRACKETS

Brackets are used to enclose any words you have inserted into quoted material. You may need to add or change a word so a quotation will fit more smoothly into the structure of your essay sentences, or to clarify information or ideas for readers. Also, square brackets can be used to indicate an error in the original quoted material.

TO ADD OR SUBSTITUTE PUNCTUATION, CAPITALIZATION,
OR CLARIFYING INFORMATION TO A QUOTATION

"[N]o more than 10 to 20 million [will be] killed," Turgidson remarks to the president about the outcome of a preemptive nuclear war. "Tops!"

ANALYSIS In the first part of this quotation from the movie, what was the second part of a fragmentary statement in the movie has been turned into a complete sentence with the [N] and with the completion of the verb phrase [will be].

TO INDICATE ERRORS IN ORIGINAL MATERIAL
The Latin word *sic* means "so" or "thus." The word *sic* is placed in square brackets immediately after a word in a quotation that appears erroneous or odd. *Sic* indicates that the word is quoted exactly as it stands in the original:

"Growing up on the small island [*sic*] of Nanaimo, British Columbia, Diana Krall has made a name for herself as a jazz singer."

ANALYSIS [Sic] indicates to the reader that the writer who is quoting the sentence realizes the author of the original article is wrong in calling Nanaimo an "island," when in fact it is a city.

5.6.7 ELLIPSIS MARKS

An ellipsis mark consists of three spaced periods (. . .). The ellipsis is used to indicate that you have omitted material from the original writer's quoted words.

WHEN DELETING MATERIAL FROM A QUOTATION
Nikola Tesla wrote that "[e]re many generations pass, our machinery will be driven by a power obtainable at any point in the universe. . . . Throughout space there is energy. Is this energy static or kinetic? If static our hopes are in vain; if kinetic—and this we know it is, for certain—then it is a mere question of time when men will succeed in attaching their machinery to the very wheelwork of nature."

Notice the bracketed *e*: this indicates a change to the original (it was a capital E).

An ellipsis is not required at the beginning of a quotation. Do not place an ellipsis at the end of the quotation, unless you have omitted content from the final quoted sentence.

WHEN DELETING A FULL SENTENCE FROM THE MIDDLE OF A QUOTED PASSAGE

Use a period before the three ellipsis points if you need to delete a full sentence or more from the middle of a quoted passage:

> In the *New York Times*, October 19, 1931, Nikola Tesla was quoted as saying, "If Edison had a needle to find in a haystack, he would proceed at once with the diligence of the bee to examine straw after straw until he found the object of his search. . . . I was a sorry witness of such doings, knowing that a little theory and calculation would have saved him ninety per cent of his labor."

WHEN QUOTING POETRY

Use a full line of spaced dots to indicate that you have omitted a line or more from the quotation of a poem:

> Death, be not proud, though some have called thee
> Mighty and dreadful, for thou art not so;
>
> ....
>
> From rest and sleep, which but they picture be,
> Much pleasure; then from thee much more must flow,
>
> —*John Donne*

WHEN INDICATING INTERRUPTION OR HESITATION IN SPEECH OR THOUGHT

Often in story dialog or narration, an ellipsis is used to indicate hesitation or interruption in speech or thought:

> "Well . . . I couldn't make it. I didn't get to the exam."

5.6.8 SLASHES

USING SLASHES TO INDICATE LINES OF POETRY

The slash is used most often in academic writing to mark off lines of poetry when these have been incorporated into the essay text. Up to two or three lines from a poem can be quoted:

> Atwood's "Death of a Young Son by Drowning" opens with the haunting lines, 'He, who navigated with success / the dangerous river of his own birth / once more set forth'."

Leave one space before and one space after the slash. For quoted passages of poetry that are four or more lines in length, start each line of the poem on its own line, indented in the style of block quotations.

USING THE SLASH TO INDICATE OPTIONS OR PAIRED ITEMS
Sometimes the slash is used between options or paired items. Examples include *actor/producer, life/death, pass/fail*. In these cases, do not leave a space before and after the slash:

> Since the orchestra was short of funds, he served as artistic director/conductor.

AVOIDING AWKWARD AND AMBIGUOUS USE OF SLASHES
Avoid the use of constructions like *and/or* as they are informal and awkward in writing. Avoid the ambiguous use of the slash in which the reader is not sure whether the slash means *and, or*, both *and* and *or*, or either *and* or *or* but not both:

> Currently, their research focuses on tether space elevators/space fountains.

ANALYSIS Which? These are dramatically different approaches to space-elevator design.

6

SENTENCE STRUCTURE AND STYLE

6

Sentence Structure and Style

6   Sentence Structure and Style

Chapter 4, "Grammatical Sentences," is about the nuts and bolts of putting a sentence together. It provides you with the terminology you need to understand this chapter. This chapter reviews common grammar and usage problems. Think of **grammar** as the fundamental physics of a language. Grammar errors are primarily errors that no reasonably experienced user of the language would make—for example, *Book I read the*. Think of **usage** as the main area where the traditional errors occur. As its name implies, usage has to do with how people use language, how certain patterns—however arbitrary—have become rules. The dividing line between grammar and usage is not absolute; some usage rules indeed prevent confusion and increase clarity and precision of expression.

6.1 PROBLEMS WITH MODIFIERS

A **modifier** is a word, phrase, or clause that describes or limits another word, phrase, or clause within a sentence. If modifiers are not placed carefully and correctly, they can cloud and in some instances destroy sentence meaning. Keep modifying words close to the words they modify.

6.1.1 MISPLACED PHRASES AND CLAUSES

A **misplaced modifier** is a descriptive word, phrase, or clause that is incorrectly positioned within a sentence so that the meaning is illogical or unclear. The misplaced modifier relates to, or modifies, the wrong word or words in the sentence. When a modifier is misplaced, unusual misreadings can result.

Limiting words such as *just, even, only, almost, hardly, nearly*, and *merely* can easily be misplaced in a sentence. Place them directly before the verb they modify:

Misplaced modifier: <u>Only</u> capacitors and inductors become good conductors when the frequency reaches a certain level.

Revision: Capacitors and inductors become good conductors <u>only</u> when the frequency reaches a certain level.

ANALYSIS *Only* at the beginning of the sentence implies that only capacitors and inductors can become good conductors—nothing else.

Misplaced modifier: When a sound system is working properly, the speaker <u>nearly</u> produces the same vibrations that the microphone originally recorded and encoded on a tape or CD.

Revision: When a sound system is working properly, the speaker produces <u>nearly</u> the same vibrations that the microphone originally recorded and encoded on a tape or CD.

ANALYSIS In the first sentence, the speaker may not produce any vibrations at all!

The modifier *not* is often misplaced, a situation that can create confusing or unintended meaning:

Misplaced modifier: All speaker systems do <u>not</u> have midrange drivers.

Revision: <u>Not</u> all speaker systems have midrange drivers.

ANALYSIS In the incorrect version, a possible meaning is that no speaker system anywhere has a midrange driver, which could seriously reduce the quality of sound reproduction.

Phrases and clauses can also be misplaced:

Misplaced modifier: Charles F. Brush used his 12-kW horizontal-axis windmill to charge batteries <u>for over 20 years</u> to light his estate in Cleveland, Ohio.

Revision: <u>For over 20 years</u>, Charles F. Brush used his 12-kW horizontal-axis windmill to charge batteries to light his estate in Cleveland, Ohio.

ANALYSIS The batteries did not take 20 years to charge, nor did they stay charged for 20 years!

Misplaced modifier: <u>Sparked by the energy crisis and government tax incentives</u>, the last part of the twentieth century saw a resurgence of interest in wind power.

Revision: The last part of the twentieth century saw a resurgence of interest in wind power, <u>sparked by the energy crisis and government tax incentives</u>.

ANALYSIS The last part of the twentieth century was not sparked by the energy crisis or government tax incentives—interest in wind power was.

Misplaced modifier: Optimum size for wind turbine size is site-dependent—for example, the hills of California will require different sizes of turbines <u>with their complex wind patterns</u>, compared to offshore turbines in northwestern Europe.

Revision: Optimum size for wind turbine size is site-dependent—for example, the hills of California <u>with their complex wind patterns</u> will require different sizes of turbines, compared to offshore turbines in northwestern Europe.

ANALYSIS The complex wind patterns belong to the California hills—not wind turbines.

6.1.2 DANGLING MODIFIERS

A **dangling modifier** is a word, phrase, or clause that does not relate to any word within the sentence and, as a result, confuses the reader. Dangling modifiers usually appear at the start of the sentence, and the person who performs the action is not mentioned:

Dangling modifier: <u>Dividing the power output by the available wind power</u>, $\rho V^3_w A/2$, the efficiency or power coefficient, C_p, for the wind turbine can be determined.

Revision: The efficiency or power coefficient, C_p, for the wind turbine can be determined by <u>dividing the power output by the available wind power</u>, $\rho V^3_w A/2$.

ANALYSIS Neither the efficiency nor the power coefficient of the wind turbine can divide anything. One way to correct problems like these is to move the dangling modifier to the end of the sentence.

Dangling modifier: <u>After connecting one end of the wire to the terminal and moving the free end of the wire up and down the file</u>, small 9-volt sparks can be seen running along the file as the tip of the wire connects and disconnects with the file's ridges.

Revision: <u>After connecting one end of the wire to the terminal and moving the free end of the wire up and down the file</u>, *you* will see small 9-volt sparks running along the file as the tip of the wire connects and disconnects with the file's ridges.

ANALYSIS The first version implies that the sparks are doing the connecting and the moving. Another way to correct problems like these is to provide a real subject for the modifier—in this case, *you*.

Dangling modifier: <u>To ensure adequate generation and transmission stability</u>, bulk power supply reliability is greatly improved by strong system interconnections but requires special automatic controls.

Revision: <u>To ensure adequate generation and transmission stability</u>, strong system interconnections greatly improve bulk power supply reliability but require special automatic controls.

ANALYSIS Bulk power supply reliability does not ensure adequate generation and transmission stability—strong system interconnections do.

Dangling modifier: The operator controls the incoming generator while observing the synchronizing lamps or meters and a synchroscope <u>until synchronized</u>.

Revision: The operator controls the incoming generator while observing the synchronizing lamps or meters and a synchroscope <u>until the ac generators are synchronized</u>.

ANALYSIS "until synchronized" is an **elliptical clause**, a clause that can still be understood even though a word or phrase has been omitted. Here, the elliptical clause is unclear because the thing to be synchronized (ac generators) is nowhere in the sentence.

6.1.3 SPLIT INFINITIVES

An **infinitive** consists of *to* and the verb, as in *to love, to leave,* and *to forget.* In a **split infinitive**, a modifier is placed between *to* and the verb. Split infinitives can make sentences awkward:

Split infinitive: Financial analysts expected the stock prices *to,* after a period of sharp decline, dramatically *rise.*

Revision: Financial analysts expected the stock prices *to rise* dramatically after a period of sharp decline.

Split infinitive: Windmills and wind turbines have been used *to,* over the centuries, *grind* grain, pump water, and generate electricity.

Revision: Over the centuries, windmills and wind turbines have been used *to grind* grain, pump water, and generate electricity.

Split infinitive: In downwind configurations, the rotor yaws freely *to,* without active control, *face* into the wind.

Revision: In downwind configurations, the rotor yaws freely *to face* into the wind without active control.

However, if you find that using a split infinitive is the clearest, least awkward, and most succinct way to write a sentence, keep it.

6.2 SHIFTS

A **shift** is a sudden and unnecessary change in point of view, verb tense, mood, or voice, or a change from indirect to direct questions or quotations. Shifts can occur within and between sentences. They often blur meaning and confuse readers.

6.2.1 POINT OF VIEW

In writing, **point of view** is the perspective from which the work is written. Often, this is indicated by the pronouns the writer uses:

- First person: *I, we*: The first-person point of view often appears in more informal types of writing, such as journals, diaries, and personal letters.
- Second person: *you*: The second-person point of view is often found in directions or instructional types of writing, such as this handbook.
- Imperative: understood *you*: Although the imperative is not a point of view in grammatical terms, it is often involved in shifts in point of view. An **imperative** is a direct command: *take out the trash*.
- Third person: *he, she, it, one*, or *they*: The third-person point of view emphasizes the subject. It is used in informative writing, including the writing you do in many professional and academic contexts.

Shifts in point of view occur when writers begin their documents in one point of view, then shift carelessly back and forth between other points of view. To prevent needless shifts, determine which point of view is the most appropriate for your document and stick with that point of view throughout the document. Shifts in point of view are particularly noticeable and confusing in instructions. The best point of view for instructions is usually the imperative backed up with second person. Notice how potentially confusing the shifts are in Figure 6.1a.

Preparing the Pressurized Barrier Fluid System for Operation

To prepare a pressurized barrier fluid system for use:

1. Personnel must ensure that the seal barrier addition pump and addition container are free of all debris.
2. A 50%/50% mixture of propylene glycol/water must be mixed in the seal barrier addition container.
3. Connect the seal barrier addition pump hose to the quick-disconnect hydraulic coupling on the bottom of the seal pot.
4. Close the hand valve between the seal pot and nitrogen supply.
5. Personnel must vent the seal pot nitrogen to atmospheric pressure by opening the vent valve slowly.

FIGURE 6.1A
Shifts in point of view: This confusing example shifts between third person, passive voice, and imperative.

ANALYSIS The example in Figure 6.1a shifts between third-person singular, passive voice, and imperative. Although in some cultures, the imperative is too abrupt and thus rude, instructions in Western cultures typically use the imperative, supplemented by second person, as shown in the revised version in Figure 6.1b.

6.2.2 VERB TENSE

The **verb tense** tells the reader when the action in the document is taking place. Shifting from one verb tense to another, without a sound reason, can distract or confuse readers:

Tense shift: In Opticks (1704), Newton <u>argued</u> that light is composed of particles, but he <u>has</u> to associate them with waves to explain the diffraction of light.

Revision: In Opticks (1704), Newton <u>argued</u> that light is composed of particles, but he <u>had</u> to associate them with waves to explain the diffraction of light.

ANALYSIS The sentence begins in the past tense (*argued*) then shifts to the present tense (*has*). Although the sentence could be rewritten in the present tense, which is called the literary present tense, the tense of both verbs must be the same.

Preparing the Pressurized Barrier Fluid System for Operation

To prepare a pressurized barrier fluid system for use:

1. Ensure that the seal barrier addition pump and addition container are free of all debris.
2. Mix a 50%/50% mixture of propylene glycol/water in the seal barrier addition container.
3. Connect the seal barrier addition pump hose to the quick-disconnect hydraulic coupling on the bottom of the seal pot.
4. Close the hand valve between the seal pot and nitrogen supply.
5. Vent the seal pot nitrogen to atmospheric pressure by opening the vent valve slowly.

FIGURE 6.1B
Unified point of view: In this revision, the point of view is consistently imperative, which is the best choice for instruction writing.

In literature courses, you may have been instructed to write about the fictional events in the **literary present tense**. The same is true for any published work:

Shift from literary present: As an egocentric, Gabriel <u>has</u> "restless eyes" early in "The Dead." However, when he <u>displays</u> empathy near the end of the story, he <u>possessed</u> "curious eyes."

Revision: As an egocentric, Gabriel <u>has</u> "restless eyes" early in "The Dead." However, when he <u>displays</u> empathy near the end of the story, he <u>possesses</u> "curious eyes."

ANALYSIS The writer begins using the literary present tense convention, then abruptly shifts into the past tense with *possessed*.

Shift from literary present in a nonliterary work: In Opticks (1704), Newton <u>presents</u> his corpuscular theory of light. He <u>argued</u> that light <u>was</u> made up of extremely subtle corpuscles and that ordinary matter <u>was made</u> of grosser corpuscles

Revision: In Opticks (1704), Newton <u>presents</u> his corpuscular theory of light. He <u>argues</u> that light <u>is made</u> up of extremely subtle corpuscles and that ordinary matter <u>is made</u> of grosser corpuscles.

ANALYSIS If your focus is historical, you could use past tense for all four instances in this example. However, if your focus is scientific, the present tense is better. Conceptually, Newton's theory *still holds* that light and matter are made up of different kinds of corpuscles.

Be careful not to use past tense when the information in the sentence is still current:

Inaccurate past tense: The very high maximum speed potential of maglevs <u>made</u> them competitive with airline routes of 1,000 kilometers (600 miles) or less. The world's first commercial application of a high-speed maglev line <u>was</u> the initial operating segment demonstration line in Shanghai that <u>transported</u> people to the airport at a top speed of 431 km/h (268 mph) and at an average speed 250 km/h (150 mph).

Revision: The very high maximum speed potential of maglevs <u>makes</u> them competitive with airline routes of 1,000 kilometers (600 miles) or less. The world's first commercial application of a high-speed maglev line <u>is</u> the initial operating segment demonstration line in Shanghai that <u>transports</u> people to the airport at a top speed of 431 km/h (268 mph) and at an average speed 250 km/h (150 mph).

ANALYSIS These facts are still true at the time of this writing; therefore, they must be presented in the present tense.

6.2.3 VERB MOOD AND VOICE

MOOD

Shifts can also occur in the mood of verbs. The **mood** of the verb indicates the manner of action. There are three moods in English:

- The **indicative mood** is used to state facts or opinions, or to ask questions.

  Speech recognition <u>is</u> the process of electronically converting a speech waveform into words.

  In the typical speech-recognition process, a microphone <u>picks</u> up the acoustic signal of the speech to be recognized and converts it into an electrical signal.

  How accurate <u>are</u> speech-recognition systems?

- The **imperative mood** is used to give a command or advice, or make a request:

  <u>Tell</u> the bank's speech-recognition system that you want to apply for a loan.

  <u>Rewind</u> the videotape before returning it.

- The **subjunctive mood** is used to express doubt, wishes, or possibility:

  If I <u>were</u> a medical transcriptionist, I would use a speech-recognition system to streamline my work.

  Avoid shifts in mood:

  Confusing shift: Include more foreground than background by focusing closer to your main subject. <u>The reverse is also true</u>.

  Revision: Include more foreground than background by focusing closer to your main subject. <u>Include more background by focusing farther from your main subject</u>.

ANALYSIS The writer's purpose is to give advice on photography. In the problem version, he or she appropriately begins in the imperative mood, but abruptly shifts into the indicative mood.

VOICE

Voice refers to whether a verb is active or passive. A verb is **active** when the subject is the doer of the action. A verb is **passive** when the subject of the verb receives the action. Sudden shifts between voices can be distracting and confusing to the reader:

Shift in voice: Spend the necessary time training the system to your speaking style, and then <u>dictation of messages can be performed</u>.

Revision: Spend the necessary time training the system to your speaking style, and then <u>dictate your messages</u>.

ANALYSIS In the problem version, the second clause uses the passive voice, while the main clause uses the active voice.

6.2.4 INDIRECT AND DIRECT QUESTIONS OR QUOTATIONS

DIRECT AND INDIRECT QUOTATIONS

In a **direct quotation**, the writer repeats a speaker's words exactly, placing those words within quotation marks. In an **indirect quotation**, the writer summarizes or paraphrases what the speaker has said and no quotation marks are needed:

Direct quotation: U.S. General William C. Westmoreland said, "We'll blow them back into the Stone Age."

Indirect quotation: U.S. General William C. Westmoreland said his forces would bomb the enemy so relentlessly that they would be blown back into the Stone Age.

If certain words are particularly vivid and you want to associate them with the original writer or speaker, you can put them in quotation marks in a sentence that otherwise uses primarily indirect quotation:

Partial direct quotation: U.S. General William C. Westmoreland said his forces would bomb the enemy so relentlessly that they would be "blown back into the Stone Age."

However, be careful about shifting between direct and indirect quotations:

Shift from indirect to direct: The trainer explained that I should read samples of my writing out loud into the system for two hours, but <u>don't expect</u> perfect recognition right away.

Revision: The trainer explained that I should read samples of my writing out loud into the system for two hours but <u>that I should not expect</u> perfect recognition right away.

ANALYSIS The author shifts from indirect quotation to direct quotation with *don't expect perfect recognition right away.* The revision makes the sentence consistently indirect. An alternative direct quotation is *The trainer explained, "Read samples of your writing out loud into the system for two hours, but don't expect perfect recognition right away."*

DIRECT AND INDIRECT QUESTIONS

A direct question is one that is asked directly—for example:

Is the Hidden Markov Model still the underlying technology in most speech-recognition systems?

An indirect question reports that a question was asked, but does not ask it:

She asked whether the Hidden Markov Model is still the underlying technology in most speech-recognition systems.

Shifting from indirect to direct questions can make writing awkward and confusing:

Shift from indirect to direct: I asked whether the Hidden Markov Model is still the underlying technology in most speech-recognition systems, and, if so, <u>do you know</u> a good resource for learning it.

Revision: I asked whether the Hidden Markov Model is still the underlying technology in most speech-recognition systems, and, if so, <u>what</u> a good resource for learning it <u>would be</u>.

ANALYSIS The revision presents both questions indirectly. An alternate revision would be to pose both questions directly: *Is the Hidden Markov Model still the underlying technology in most speech-recognition systems, and if so, do you know a good resource for learning it?*

6.3 MIXED CONSTRUCTIONS

A sentence with a **mixed construction** incorrectly changes from one grammatical construction to another one that is incompatible with the first, thereby confusing the sentence's meaning.

6.3.1 MIXED GRAMMAR

When you draft a sentence, your options for structuring that sentence are limited by the grammatical patterns of English. You must consistently follow the pattern you choose within the sentence. You cannot start the sentence using one grammatical pattern and then abruptly change to another:

Mixed: By multiplying the number of specialty stations available to viewers via digital television increases the chance that cultural communities within the diverse cultural mosaic of the United States will be better served.

Revision: Multiplying the number of specialty stations available to viewers via digital television increases the chance that cultural communities within Canada's diverse cultural mosaic will be better served.

OR

Revision: By multiplying the number of specialty stations available to viewers via digital television, satellite and cable companies increase the chance that cultural communities within Canada's diverse cultural mosaic will be better served.

ANALYSIS The mixed construction, which begins with *by*, cannot serve as the subject of the sentence. In the first revision, *by* is dropped, so that the opening is no longer a participial phrase but is now a gerund and hence the subject. In the second revision, the participial phrase is retained, but a subject is added so that it is not a dangling modifier.

Another mixed construction problem involves incorrectly combining clauses.

Mixed: Although satellite dishes have become popular in many northern Canadian communities, but many viewers still prefer local stations.

Revision: Satellite dishes have become popular in many northern Canadian communities, <u>but</u> many viewers still prefer local stations.

ANALYSIS Here the clause beginning *although* is a subordinate clause. Hence, it cannot be linked to an independent clause with the coordinating conjunction *but*.

From time to time, you may encounter a sentence that you can't seem to fix. In instances such as these, try moving the words at the end of the sentence to the front and rephrasing—or else just start over.

Mixed: In the indirect method, students' errors are corrected only <u>when</u> they interfere with comprehension rather than <u>by</u> the direct method in which students' errors are corrected immediately to avoid habit formation.

Revision: In the indirect method, students' errors are corrected only when they interfere with comprehension; <u>in the direct method, students' errors are corrected immediately</u> to avoid habit formation.

ANALYSIS The mixed grammar occurs with the words *when* and *by*. A dependent clause, introduced by *when*, is being mixed with a prepositional phrase (which has a dependent clause embedded in it).

6.3.2 ILLOGICAL CONNECTIONS

Problems can occur when elements of the sentence do not logically fit together. Faulty predication is one example. In **faulty predication**, the subject and predicate do not make sense together. To remedy this problem, either the subject or the predicate must be changed.

Faulty predication: The original <u>function</u> of the Internet <u>was created</u> to exchange academic and military information.

Revision: Originally, the <u>Internet was created</u> to exchange academic and military information.

ANALYSIS The *function* was not created to exchange that information; the *Internet* was. In this instance, the subject is refined so that it connects logically with the predicate.

Faulty predication: The <u>decision</u> concerning routine mission scenarios and combat situations <u>was chosen</u> by the M5 computer installed in the USS *Enterprise* (NCC-1701).

Revision: The <u>decision</u> concerning routine mission scenarios and combat situations <u>was made</u> by the M5 computer installed in the USS *Enterprise* (NCC-1701).

ANALYSIS The *decision* didn't do the choosing. The *M5 computer* made the decision.

Another way that faulty predication occurs involves appositives. An **appositive** is a useful device: it is a noun or noun phrase that renames or explains a noun or noun phrase immediately before it. For example:

Daystrom, <u>the inventor of duotronic technology in 2243</u>, designed the M5, <u>the fifth prototype multitonic computer</u> . . .

The appositive must logically relate to the noun or noun phrase that precedes it; otherwise faulty predication occurs:

Faulty predication: Stock speculators, <u>a very risky business</u>, demands nerves of steel and a healthy bank account.

Revision: Stock speculation, <u>a very risky business</u>, demands nerves of steel and a healthy bank account.

ANALYSIS Stock speculation, not stock speculators, is a very risky business.

6.3.3 AVOIDING IS WHEN, IS WHERE, REASON . . . IS BECAUSE

In formal writing, avoid the following constructions.

• is *when* or is *where*:

Mixed construction: Computer dating <u>is when</u> the computer is used to match potential romantic partners according to their compatibility, interests, and desirability.

Revision: In computer dating, the computer is used to match potential romantic partners according to their compatibility, interests, and desirability.

ANALYSIS *Computer dating* is not a time but a service.

• *The reason is . . . because*:

Mixed construction: The <u>reason</u> that parts of the Romulan and Klingon empires are referred to as Alpha Quadrant powers <u>is because</u> they both spill into the Alpha Quadrant.

Revision: Parts of the Romulan and Klingon empires are referred to as Alpha Quadrant powers because they both spill into the Alpha Quadrant.

ANALYSIS Notice that the revised sentence is four unnecessary words shorter. Another way to avoid the awkward *reason is . . . because* construction is to replace *because* with *that.* Another option is to reverse the two parts of the sentence: *Because parts of the Romulan and Klingon empires spill into the Alpha Quadrant, they are referred to as Alpha Quadrant powers.*

6.4 COORDINATION AND SUBORDINATION

Coordination and subordination allow you to communicate the relationships between ideas in sentences. Coordination gives equal emphasis to ideas; subordination gives emphasis to one idea over another within a sentence.

COORDINATION

Coordination balances two or more equal ideas in a sentence, giving equal emphasis to each idea. You can coordinate ideas at the level of words, phrases, or clauses by using the following coordinating conjunctions: *and, but, for, nor, or, so, yet.*

> Nelson Mandela spent 28 years in a South African jail, <u>but</u> the great dignity with which he endured imprisonment made him a symbol of the struggle against apartheid.

When coordinating independent clauses (groups of words that can stand on their own as complete sentences), join the clauses with a comma or a semicolon and then the coordinating conjunction. In some cases, the semicolon is followed by a conjunctive adverb, such as these: *however, therefore, moreover, hence, indeed.*

> Nelson Mandela spent 28 years in a South African jail; <u>moreover,</u> the great dignity with which he endured imprisonment made him a symbol of the struggle against apartheid.

SUBORDINATION

Subordination allows you to communicate the relative importance of ideas within sentences. You give important ideas more emphasis by making them independent clauses; you give less important ideas less emphasis by making them subordinate clauses. Subordinate clauses begin with one of the following subordinating conjunctions or relative pronouns: *after, although, as, because, before, if, since, than, that, though, unless, until, when, whenever, where, wherever, whether, which, while, who, whom, whose.*

When drafting a sentence, you must decide which idea you would like to emphasize. Subordinating a clause within the Nelson Mandela sentence can give it a very different meaning:

> Although Nelson Mandela spent 28 years in a South African jail, the great dignity with which he endured imprisonment made him a symbol of the struggle against apartheid.

If you want to emphasize that the great personal price of spending 28 years in prison caused Mandela to become a symbol against apartheid, you might use the following subordinate pattern:

> While Nelson Mandela spent 28 years in a South African jail, the great dignity with which he endured imprisonment made him a symbol of the struggle against apartheid.

No version is right or wrong. However, one of the versions given here might be closer to the meaning that you, as a writer, intend.

6.4.1 CHOPPY SENTENCES

Too many consecutive short sentences in a passage can make your writing seem mechanically repetitive and choppy. In choppy text, some of the short sentences are not of equal importance compared to others. Some sentences should be emphasized; others should be deemphasized; still others should be balanced and given equal importance. Subordination and coordination enable you to accomplish these things and thus eliminate the problem of choppy writing:

> *Choppy:* Tennyson called lightning a "flying flame." Lightning travels at between 100,000 and 300,000 kilometers per second. Lightning reaches temperatures of 24,000 to 28,000 degrees C. It kills 20 people each day.

> *Revision:* Lightning, which Tennyson called a "flying flame," travels at between 100,000 and 300,000 kilometers per second and reaches temperatures of 24,000 to 28,000 degrees C, killing 20 people each day.

ANALYSIS The four very short sentences in the choppy version have become a single sentence in the revised version through the use of the subordinator *which* and the coordinator *and*.

Choppy: Intracloud lightning is the most common type of lightning. It occurs completely inside one cumulonimbus cloud. It is commonly called an anvil crawler.

Revision: <u>Commonly called an anvil crawler</u>, intracloud lightning is the most common type of lightning, which occurs completely inside one cumulonimbus cloud.

ANALYSIS Less significant detail about *intracloud lightning* is placed in the participial phrase using *called*.

Choppy: Franklin's experiments showed that lightning is a discharge of static electricity. This occurred more than 150 years ago. There has been little improvement in theoretical understanding of lightning since then.

Revision: Although more than 150 years ago Franklin's experiments showed that lightning is a discharge of static electricity, there has been little improvement in theoretical understanding of lightning since then.

ANALYSIS The subordinating word *Although* has been used to turn the first sentence into an adverbial clause.

Choppy: Positive lightning is a rarer form of lightning, and it occurs when the leader forms at the tops of positively charged clouds.

Revision: Positive lightning, <u>a rarer form of lightning</u>, occurs when the leader forms at the tops of positively charged clouds.

ANALYSIS A less important idea has been turned into an appositive describing positive lightning.

Choppy: Ball lightning is described as a floating, illuminated ball. It can occur during thunderstorms. It is rare. It can be fast moving. It can be slow moving. Or it can be nearly stationary. Some ball lightning makes hissing or crackling noises. Some ball lightning makes no noise at all. Some has been reported to pass through windows. Some even dissipates with a bang. Ball lightning has been described by eyewitnesses. It has rarely, if ever, been recorded by meteorologists.

Revision: Ball lightning is described as a floating, illuminated ball <u>that rarely occurs during thunderstorms</u>. It can be <u>fast moving, slow moving,</u>

or nearly stationary. Some ball lightning makes hissing or crackling noises or <u>no noise at all</u>. Some has been reported to pass through windows <u>and even dissipate with a bang</u>. Ball lightning has been described by eyewitnesses <u>but rarely, if ever, recorded by meteorologists</u>.

ANALYSIS The second sentence in the original has been turned into an adjective clause and joined to the first sentence in the revision. The third sentence in the original has been turned into an adverb and inserted into the first sentence in the revision. The remaining combinations use *and* and *or*.

Choppy: Research on positive lightning in the 1970s showed that positive lightning is typically six to ten times more powerful than negative lightning. It lasts around ten times longer. It can strike many kilometers from the clouds.

Revision: Research on positive lightning in the 1970s showed that positive lightning is typically six to ten times more powerful than negative lightning, lasts around ten times longer, and can strike many kilometers from the clouds.

ANALYSIS The first version uses three separate sentences to say three things about positive lightning. Using compound predicates, the revision says those three things in just one sentence.

6.4.2 INEFFECTIVE COORDINATION

Coordination is effective when you want to point out to the reader that two ideas are of equal importance:

<u>Speak softly</u> and <u>carry a big stick</u>.

ANALYSIS The words were stated by U.S. president Theodore Roosevelt. Balance is central to Roosevelt's meaning. He thought American foreign policy should employ diplomacy (*speaking softly*), but at the same time be backed up by military might (*a big stick*). Coordination is extremely effective here.

However, coordination problems arise when you attempt to coordinate ideas that are unrelated or unequal:

Weak coordination: Cell phones became very popular in the 1990s, <u>and</u> many users have experienced car accidents.

Revision: <u>Since</u> cell phones became very popular in the 1990s, many users have experienced car accidents.

ANALYSIS In the coordinated version of the sentence, the two ideas seem unrelated. By adding the subordinating word *Since* and turning the opening independent clause into a subordinate one, you foreground the cause–effect relationship between the ideas.

Weak coordination: With the new technology of refrigeration, many families now concentrated the entire week's food shopping into one trip to the market, <u>and</u> they stocked their refrigerators with perishables that would last a week or more.

Revision: With the new technology of refrigeration, many families now concentrated the entire week's food shopping into one trip to the market, <u>stocking</u> the refrigerator with perishables that would last a week or more.

ANALYSIS In the revision, the final independent clause is converted to a participial phrase, tightening up the sentence as a whole.

Weak coordination: Perhaps the most significant result of air conditioning was that people started moving south, <u>and</u> this reversed a northward demographic trend that had continued through the first half of the century.

Revision: Perhaps the most significant result of air conditioning was that people started moving south, <u>reversing</u> a northward demographic trend that had continued through the first half of the century.

ANALYSIS The revision is another example of converting an independent clause into a participial phrase. But notice how the revision gets rid of the awkward and potentially puzzling *this* in the original.

Weak coordination: Since 1940, the nation's fastest-growing states have been in the Southeast and the Southwest, <u>and</u> these regions that could not have supported large metropolitan communities before air conditioning made the summers tolerable.

Revision: Since 1940, the nation's fastest-growing states have been in the Southeast and the Southwest, <u>regions that</u> could not have supported large metropolitan communities before air conditioning made the summers tolerable.

ANALYSIS In this revision, the independent clause is converted to an appositive, a change that tightens up the sentence nicely.

Weak coordination: One foggy night, Carrier was waiting on a train platform in Pittsburgh, <u>and</u> he had a sudden insight into a problem he had been puzzling over for a while, <u>and</u> this was the complex relationship between air temperature, humidity, and dew point.

Revision: <u>While</u> he was waiting one foggy night on a train platform in Pittsburgh, Carrier had a sudden insight into a problem he had been puzzling over for a while—the complex relationship between air temperature, humidity, and dew point.

ANALYSIS Notice how the original sounds like the speech of a young child who strings idea after idea together with *and*. In the revision, the first independent clause is converted into an adverb clause using *while*. The final independent clause is converted into an appositive.

Weak coordination: Carrier's system had the cooling power of 108,000 pounds of ice a day, <u>and</u> this solved a printing problem that required controlled temperature for good color printing.

Revision: Carrier's system, <u>which</u> had the cooling power of 108,000 pounds of ice a day, solved a printing problem that required controlled temperature for good color printing.

ANALYSIS In the revision, the first independent clause is converted into an adjective clause beginning with *which*.

Weak coordination: In the 1930s, freon seemed to be the perfect refrigerant, <u>and</u> in the late 1980s, however, chlorofluorocarbons were found to be contributing to the destruction of Earth's protective ozone layer.

Revision: <u>Although</u> in the 1930s freon seemed to be the perfect refrigerant, in the late 1980s chlorofluorocarbons were found to be contributing to the destruction of Earth's protective ozone layer.

ANALYSIS Once again the revision uses a subordinate clause—specifically, an adverbial clause beginning with *although*—to bind the idea in the first clause tightly into the sentence as a whole.

Weak coordination: Carrier's main business was still commercial in 1928 when it introduced the "Weathermaker," <u>and</u> this was the first practical home air conditioner, <u>but</u> the company was slow to transform the "Weathermaker" into the smaller-scale designs that residential applications required.

Revision: <u>Because</u> the company's main business was still commercial in 1928 when it introduced the "Weathermaker," <u>the first practical home air conditioner</u>, Carrier was slow to transform this new product into the smaller-scale designs that residential applications required.

ANALYSIS The original illustrates how independent clauses can be heaped upon each other ineffectively. The revision converts the first independent clause into an adverb clause beginning with *because*. The second independent clause becomes an appositive (the first practical home air conditioner), enabling us to delete *but* from the final clause.

6.4.3 INEFFECTIVE SUBORDINATION

Avoid relegating the main idea of a sentence to subordinate status. Structure your sentences so that important ideas appear in independent clauses:

Ineffective subordination: Before the invention of the internal combustion engine, gasoline, which was considered a useless byproduct of the crude oil refining process, <u>had been known for many years</u>.

Revision: Before the invention of the internal combustion engine, gasoline, <u>which had been known for many years</u>, was considered a useless byproduct of the crude oil refining process.

ANALYSIS In the original, the really interesting information is relegated to a subordinate clause, which guarantees that it receives less emphasis. In the revision, that idea is moved into the main clause, promoting its importance in the sentence as a whole.

Ineffective subordination: Kerosene, which has more carbon atoms per molecule and a higher boiling point and which is easily separated from lighter gasoline, which has fewer atoms and a lower boiling point, as well as from other hydrocarbon compounds and impurities in the crude oil mix, was standard fuel for lamps through much of the nineteenth century.

Revision: Kerosene, the standard fuel for lamps through much of the nineteenth century, has more carbon atoms per molecule and a higher boiling point, making it easily separated from lighter gasoline, which has fewer atoms and a lower boiling point, as well as from other hydrocarbon compounds and impurities in the crude oil mix.

ANALYSIS Again, the revision promotes information from secondary status in subordinate clauses to primary status in the main clause. Also, the three adjective clauses (each introduced by *which*) are converted to the main verb of the sentence (*has more carbon atoms per . . .*), a participial phrase (*making it easily separated . . .*), and a prepositional phrase (*with fewer atoms and a lower boiling point*). Still, this is a long and complicated sentence that might be better as two or even three separate sentences:

Further revision: Kerosene, the standard fuel for lamps through much of the nineteenth century, has more carbon atoms per molecule and a higher boiling point. This chemical structure makes it easily separated from lighter gasoline, which has fewer atoms and a lower boiling point. This structure also makes kerosene more easily separated from other hydrocarbon compounds and impurities in the crude oil mix.

6.4.4 EXCESSIVE SUBORDINATION

In the preceding, you saw that when *and* is used ineffectively to string together ideas, the logic connecting those ideas is not stated; readers have to figure it out. As you saw in the preceding example, subordinate clauses (typically beginning with *that* and *which*) can pile up, blurring the focus and logic of a sentence. Usually there are two ways to address the excessive subordination problem:

- Recast the sentence.
- Divide the sentence into two.

Unclear: While jogging is good aerobic exercise, as you get older, since running is high impact on deteriorating knees, it is advisable to do it in moderation.

Revision: <u>Jogging is good aerobic exercise. However</u>, as you get older, since running is high impact on deteriorating knees, it is advisable to do it in moderation.

ANALYSIS Too many secondary, subordinate ideas were packed into one sentence.

6.5 PARALLELISM

Parallelism in writing means using the same style of phrasing for items in a series. Those series items can be individual nouns, pronouns, adjective, adverbs, prepositional phrases, subordinate clauses, predicates, and independent clauses. Errors in parallelism, known as **faulty parallelism**, occur when different styles of phrasing are used for items in a series. Writers and speakers have long used parallelism for rhetorical effect, balancing single words with single words, phrases with phrases, and clauses with clauses:

Parallel, Balanced Elements

Words: There are three sides to every story—yours, mine, and all that lie between.

–Jody Kern

Phrases: Do what you can, with what you have, where you are.

—Theodore Roosevelt

Clauses: Think like a man of action, act like a man of thought.

—Henri Bergson

When readers encounter items in a series, they expect each item to use the same style of phrasing—in other words, to be parallel in phrasing. However, when one or more of the items do not use that same style of phrasing, the sentence seems distracting and awkward to readers:

Faulty parallelism: The first functioning laser built in 1960 by Theodore Maiman began a rapid evolution of lasers, <u>taking</u> forms as big as a house and <u>some</u> were as small as a grain of sand.

Revision: The first functioning laser built in 1960 by Theodore Maiman began a rapid evolution of lasers, taking forms <u>as big as</u> a house and <u>as small as</u> a grain of sand.

ANALYSIS In the original, the phrase *some were* ruins the parallelism of the two phrases

Faulty parallelism: Laser light is typically near-monochromatic, that is, <u>composed</u> of a single wavelength or color and <u>which is emitted</u> in a narrow beam.

Revision: Laser light is typically near-monochromatic, that is, <u>composed</u> of a single wavelength or color and <u>emitted</u> in a narrow beam.

ANALYSIS The faulty version mixes two different types of phrasing: a participial phrase and a subordinate clause. The revision uses two present participial phrases: one beginning with *composed* and the other with *emitted*.

Parallel ideas can be connected in one of three ways:

- With a coordinating conjunction, such as *or, and*, or *but*
- With a pair of correlative conjunctions, such as *not only . . . but also* or *either . . . or*
- With comparative constructions using *than* or *as*

Whenever you relate ideas using one of these methods, always emphasize the connection between or among ideas by using just one parallel style of phrasing.

6.5.1 PARALLELISM WITH COORDINATING CONJUNCTIONS

Coordinating conjunctions are words such as *and, but, or, nor, for, yet*, and *so* that connect items in a series. Ensure that all items joined by coordinating conjunctions are parallel in phrasing.

Faulty parallelism: Townes was searching for a way to generate extremely short-wavelength radio waves or long-wavelength infrared waves that could be used <u>to probe</u> the structure of molecules and <u>how they behave</u>.

Revision: Townes was searching for a way to generate extremely short-wavelength radio waves or long-wavelength infrared waves that could be used to probe the <u>structure</u> and <u>behavior</u> of molecules.

ANALYSIS In the original, the *structure of molecules* and *how they behave* are not parallel. In the revision, they are both nouns.

Faulty parallelism: Einstein observed that a "stimulated" atom or molecule creates an identical twin—a second photon <u>that</u> perfectly matches the triggering photon in wavelength, <u>how</u> its wave crests and troughs align, and <u>in</u> the direction of travel.

Revision: Einstein observed the "stimulated" atom or molecule creates an identical twin—a second photon that perfectly matches the triggering photon <u>in</u> wavelength, <u>in</u> the alignment of wave crests and troughs, and <u>in</u> the direction of travel.

ANALYSIS The revision uses three prepositional phrases, each beginning with *in*.

Faulty parallelism: Laser light is essentially <u>monochromatic</u> (of essentially the same wavelength), <u>being</u> coherent (the crests and troughs of the waves perfectly in step, thus combining their energy), and <u>providing</u> highly directional function.

Revision: Laser light is essentially <u>monochromatic</u> (of essentially the same wavelength), <u>coherent</u> (the crests and troughs of the waves perfectly in step, thus combining their energy), and highly <u>directional</u>.

ANALYSIS The original version is complicated because of the parenthetical material. Strictly speaking, it does not contain faulty parallelism. However, if you revise for parallelism, it becomes more readable. If you ignore the parenthetical material, you see that the sentence makes three descriptive statements about laser light. All three of those items in the corrected version are adjectives.

Faulty parallelism: In manufacturing, infrared carbon dioxide lasers <u>cut</u> metal, <u>trim</u> computer chips, <u>drilling</u> holes in hard ceramics, <u>they</u> noiselessly slice through textiles, and <u>pierce</u> the openings in squeeze bottles.

Revision: In manufacturing, infrared carbon dioxide lasers <u>cut</u> metal, <u>trim</u> computer chips, <u>drill</u> holes in hard ceramics, noiselessly <u>slice</u> through textiles, and <u>pierce</u> the openings in squeeze bottles.

ANALYSIS In the revision, each of the underlined verbs reads grammatically with the subject *lasers*.

6.5.2 PARALLELISM WITH CORRELATIVE CONJUNCTIONS

Correlative conjunctions are pairs of words that join items in a series. Examples include *not only . . . but also, either . . . or*, and *both . . . and*. Ensure that each item linked by correlative conjunctions is parallel in phrasing:

Faulty parallelism: Laser excitation was generated <u>not only with</u> the synthetic ruby crystal, but <u>researchers</u> also discovered numerous other solids.

Revision: Laser excitation was generated <u>not only with</u> the synthetic ruby crystal <u>but also with</u> numerous other solids.

ANALYSIS A prepositional phrase follows *not only*, and so a prepositional phrase must follow *but also*.

Faulty parallelism: Either viewers criticized the television station for its inflammatory views, or it was criticized for its political stance.

Revision: Viewers *criticized* the television station <u>either</u> for its inflammatory views <u>or</u> for its political stance.

ANALYSIS The verb used with *either* must match the verb used with *or*. In the original sentence, one is in the active voice; the other, in the passive voice. The revision is more economical; the connection between the related ideas is clearer.

6.5.3 COMPARISONS LINKED WITH THAN OR AS

When you use *than* or *as* to make comparisons, use parallel phrasing for the items being compared.

Faulty parallelism: <u>Having</u> great wealth is not as satisfying as the <u>completion</u> of charitable works.

Revision: <u>Having</u> great wealth is not as satisfying as <u>completing</u> charitable works.

ANALYSIS Use the matching -*ing* form on both sides of the comparison.

Faulty parallelism: It is better to give than do the receiving.

Revision: It is better <u>to give</u> than <u>to receive</u>.

ANALYSIS Use the matching form—in this case, the infinitive form of the verb (with *to*)—on both sides of the comparison.

6.6 NEEDED WORDS

In your efforts to write concisely by deleting words, be careful not to cut essential words. This is especially true with the following:

- When using compound structures
- When *that* is required, if there is danger of misreading a sentence
- When using comparisons

Provide all words needed to make such sentences grammatically and logically complete.

6.6.1 IN COMPOUND STRUCTURES

In compound constructions, two or more elements (e.g., words, phrases, clauses) have equal importance and function as a unit:

> Code-division multiple-access (CDMA), the cellular technology in the United States, <u>uses</u> a bandwidth of 1.5%; Global System for Mobile Communications, the cellular technology in Europe based on time-division multiple access (TDMA), 0.3%.

> Code-division multiple access (CDMA), the cellular technology in the United States, <u>uses</u> a fractional bandwidth of 1.5%; Global System for Mobile Communications, the cellular technology in Europe based on time-division multiple access (TDMA), <u>uses</u> a <u>bandwidth of</u> 0.3%.

ANALYSIS If these two independent clauses were simpler, *uses* could be omitted in the second since the omitted words are common to both parts of the compound structure.

When the parts in a compound structure differ in any way, all words must be included in each part of the compound construction:

> *Incorrect:* I have and will continue to support your right to remain silent.

> *Correct:* I <u>have supported</u> and will continue to support your right to remain silent.

ANALYSIS In the incorrect version *have . . . support* is not grammatically acceptable.

> *Incorrect:* Many people are terrified and understand lasers in terms popularized by movies such as the James Bond film *Goldfinger*.

> *Correct:* Many people are terrified <u>by</u> and understand lasers <u>in terms</u> popularized by movies such as the James Bond film *Goldfinger*.

ANALYSIS People may indeed be terrified in general, but in this sentence they are terrified *by* popularized notions of lasers.

6.6.2 IN *THAT* CLAUSES

Sometimes when the word *that* introduces a subordinate clause, it can be omitted, but only if the omission does not present a danger of misreading the sentence:

No potential misreading: The laser finder can reveal the precise location in the sky [that] the telescope is pointed.

Potential misreading: At these wavelengths, a laser beam is so completely attenuated as it passes through the eye's cornea no light remains to be focused by the lens onto the retina.

Revision: At these wavelengths, a laser beam is so completely attenuated <u>that</u> as it passes through the eye's cornea no light remains to be focused by the lens onto the retina.

OR

Revision: A laser beam at these wavelengths is so completely attenuated as it passes through the eye's cornea <u>that</u> no light remains to be focused by the lens onto the retina.

ANALYSIS Notice how the placement of *that* creates a potential difference in meaning. In the first, the laser beam is "so completely attenuated" *before* it passes through the cornea. In the second, laser beam's is attenuated *after* it passes through the cornea.

6.6.3 IN COMPARISONS

Make comparisons only between like items. Comparisons between unlike items are illogical and jarring to the reader:

Illogical comparison: Ultrawideband's low susceptibility to multipath fading improves performance compared to the wireless local-area network standard (WiFi).

Revision: Ultrawideband's low susceptibility to multipath fading improves performance compared to <u>that of</u> the wireless local-area network standard (WiFi).

ANALYSIS The incorrect version compares *performance* with WiFi—two distinctly different items. The sentence requires that *performance* be compared with *performance*.

Illogical comparison: Channel capacity increases as the function of bandwidth much faster than the function of signal power.

Revision: Channel capacity increases <u>as the function of</u> bandwidth much faster than <u>as the function of</u> signal power.

ANALYSIS The original illogically compares the speed of channel capacity to signal power.

Illogical comparison: The FCC limit for power spectral density emission is the same as unintentional emitters in the UWB band, but is significantly lower in certain segments of the spectrum.

Revision: The FCC limit <u>for</u> power spectral density emission is the same as <u>for</u> unintentional emitters in the UWB band, but is significantly lower in certain segments of the spectrum.

ANALYSIS The original illogically compares the *FCC limit* to *unintentional emitters*.

6.6.4 IN COMPARISONS WITH *ANY* AND *ANY OTHER*

Comparisons using *any* and *any other* can be confusing. Writers sometimes omit *other*, making comparisons illogical. Follow these guidelines:

- Use *any other* when comparing an item with other items in the same group.

  Incorrect: Toronto has a larger population than any city in Ontario.

  Revision: Toronto has a larger population than any <u>other</u> city in Ontario.

  ANALYSIS The incorrect sentence suggests that Toronto has a larger population than any city in Ontario, a province that includes Toronto. In other words, Toronto is bigger than any city in Ontario, including itself.

- Use *any* when comparing an item with other items in a different group.

   Incorrect: Chicago has a larger population than any <u>other</u> city in Iowa.

   Revision: Chicago has a larger population than any city in Iowa.

   ANALYSIS The incorrect sentence suggests that Chicago is in Iowa, which is untrue.

6.6.5 IN COMPARISONS WITH *THAN*

Comparisons must be complete so that it is clear to the reader just what is being compared:

   Incomplete comparison: Because a laser can maintain light intensity over larger distances and because it is more precise, it can be used for detecting problems in automated production.

   Revision: Because a laser can maintain light intensity over larger distances <u>than normal light</u> and because it is more precise <u>than normal light</u>, it can be used for detecting problems in automated production.

Leave no chance for ambiguity in the comparisons you make. For example, the following sentence can be interpreted in a number of ways:

   Ambiguous: Bell Labs pumped more money into laser research than Columbia University.

   One possibility: Bell Labs pumped more money into laser research than <u>did</u> Columbia University.

   OR

   Another possibility: Bell Labs pumped more money into laser research than Columbia University <u>did</u>.

   ANALYSIS The one other possibility here is that Bell Labs pumped more money into Columbia University than it did into laser research—a less likely though possible alternative.

6.7 SENTENCE VARIETY

After you complete an engineering document, you may be relieved just to have the ideas on paper. When you carefully review the draft, you may find that many sentences are very similar in structure and length. Traditionally, this writing style is considered dull and monotonous. The traditional remedy is to revise by creating sentences of varied structures and lengths, thus making your draft "more lively." However, in an engineering context—or, for that matter, in any business or scientific context—monotony and liveliness should not be the main issues. Clarity, succinctness, and readability should be. If repetitive sentence structures and lack of sentence variety call attention to themselves and distract readers, revise. Otherwise, don't bother.

The version in Figure 6.2a is not just monotonous but comical in its repetitiveness. The repetition of "maglev" at the beginning of every sentence distracts readers from getting the information—the repetition calls attention to itself and away from the message. More importantly, the repetition destroys the continuity of the text, making the flow of ideas difficult to follow. In the revision, notice how each successive sentence echoes some key word in the sentence, described as the "old-to-new" pattern in Chapter 1. Notice how transition words like *however* and *consequently* guide readers through the discussion.

Maglevs are trains that use a form of transportation called magnetic levitation to suspend, guide, and propel them by means of electromagnetic force. Maglevs can be faster and more comfortable than wheeled mass transit systems. Maglevs can potentially reach velocities comparable to turboprop and jet aircraft (500 to 580 km/h). Maglevs have been hindered by scientific and economic limitations, however. Maglevs must be designed as separate, complete transportation systems because they cannot share existing transportation infrastructure. Maglevs do not make any physical contact with the track over which they are suspended, meaning that the only friction occurs between the vehicles and the air. Maglevs can potentially travel at very high speeds with reasonable energy consumption and noise levels. Maglevs have been proposed that operate at up to 650 km/h (404 mph), far faster than any conventional rail transport. Maglevs with this high maximum speed potential are thus competitors to airline routes of 1,000 kilometers (600 miles) or less. Maglevs had their first commercial application in a demonstration line in Shanghai that transports people 30 km (18.6 miles) to the airport in just 7 minutes 20 seconds (top speed of 431 km/h or 268 mph, average speed 250 km/h or 150 mph). Maglevs are being studied worldwide for their feasibility.

FIGURE 6.2A
Draft lacking sentence variety. Maglev begins so many sentences that the repetition calls attention to itself and becomes distracting.

If you use the ideas for achieving coherence discussed in Chapter 1, you should not have any problems with sentence variety.

6.7.1 SIMPLE, COMPOUND, AND COMPLEX SENTENCES

Too many of any one type of sentence structure can make your writing repetitive and monotonous. More importantly, too much repetition of the same sentence structure may indicate problems with continuity and coherence in a rough draft. Recall that sentences come in three basic structures—simple, compound, and complex—and that sentence structures can be combined to create, for example, compound-complex sentences. Guard against overuse of one sentence type: too many simple sentences can make your writing difficult to follow; too many unnecessarily complex sentences can make your writing seem difficult, tedious, or even pretentious.

> Magnetic levitation transport, or maglev, is a transportation technology that suspends, guides and propels vehicles by means of electromagnetic force. This technology can be used to achieve faster and more comfortable transportation than wheeled mass transit systems. Trains that use this technology could potentially reach velocities comparable to turboprop and jet aircraft (500 to 580 km/h). However, scientific and economic limitations have hindered the proliferation of the technology. One problem is that maglev technology is not compatible with conventional railroad systems; maglevs require separate, complete transportation systems. But the essential technological advantage of maglevs outweighs this problem: because there is no physical contact between the track and the vehicle, the only friction exerted is that between the vehicles and the air. Consequently, maglevs can potentially travel at very high speeds with acceptable energy consumption and noise levels. Systems have been proposed that can operate at up to 650 km/h (404 mph), far faster than is practical with conventional rail transport. This high-speed potential makes maglevs competitors with airline routes of 1,000 kilometers (600 miles) or less. The world's first commercial application of this technology is the demonstration line in Shanghai that transports people to the airport at a top speed of 431 km/h (268 mph) and at an average speed 250 km/h (150 mph). Other such projects are being studied worldwide for feasibility.

FIGURE 6.2B
Revision with sentence variety. Maglev begins so many sentences that the repetition calls attention to itself and becomes distracting. Numerous independent clauses in the original version are converted to dependent clauses and phrases and then combined with remaining independent clauses.

6.7.2 SENTENCE OPENINGS

In most English text, sentences begin with the subject first, followed by the verb, and then the object. Too many sentences using the subject-verb-object pattern can create a monotonous effect. More importantly, such writing may lack coherence and needed emphasis. To improve coherence and add emphasis to your writing as you revise, consider beginning some sentences differently.

ADVERBIAL MODIFIERS

You can easily place an adverbial modifier in front of the sentence subject, if such a construction is needed to achieve emphasis and coherence. These modifiers can be single words, phrases, or clauses:

Original: Ultrawideband and Bell Labs Layered Space Time technology are two wireless technologies that will make significant advances in the communications <u>in the next 10 years</u>.

Initial phrase modifier for variety: <u>In the next ten years</u>, Ultrawideband and Bell Labs Layered Space Time technology are two wireless technologies that will make significant advances in the communications.

Original: The students in residence were in their rooms <u>for seven days</u>, where they were deprived of sleep, overwhelmed by assignments, and driven to distraction with worry.

Initial phrase modifier for variety: <u>For seven days</u>, the students in residence were deprived of sleep, overwhelmed by assignments, and driven to distraction with worry.

ANALYSIS Remember nothing is wrong with the original versions of these sentences.

ADJECTIVE AND PARTICIPLE PHRASES

Often adjective phrases and participle phrases can be moved to the start of the sentence without affecting meaning—and sometimes improving the readability of the sentence and the coherence of that area of the document.

Original: The state government employees, expecting to be terminated, dusted off their résumés.

Initial phrase modifier for clarity: Expecting to be terminated, the state government employees dusted off their résumés.

Poor readability: UWB fractional bandwidth, defined as $(F_u - F_l)/F_c$, where F_c is the center frequency of the utilized band, F_u is the upper frequency, and F_l is the lower frequency, is equal to or greater than 25%.

Initial phrase modifier for clarity: Defined as $(F_u - F_l)/F_c$, where F_c is the center frequency of the utilized band, F_u is the upper frequency, and F_l is the lower frequency, UWB fractional bandwidth is equal to or greater than 25%.

When using adjective or participle phrases, ensure that the subject is clearly identified, or else you may be creating a dangling modifier.

6.7.3 INVERTED ORDER

Changing the common subject-verb-object sentence pattern may not only create sentence variety but also—in some instances—improve coherence and add emphasis. In an inverted sentence, the subject appears toward the end of the sentence:

Subject-verb-object: A new product line for observational astronomers—the laser finder—came with the advent of higher power green laser pointers.

Inverted order: With the advent of higher power green laser pointers came a new product line for observational astronomers—the laser finder.

ANALYSIS The inverted order is dramatic and emphatic because the keyword in the sentence, *the laser finder*, occurs at the end of the sentence.

Be careful with inverted sentences; they often sound awkward and artificial:

Awkward inverted order: Into the laser beam, the gain medium transfers external energy.

Revision: The gain medium transfers external energy into the laser beam.

Awkward inverted order: By an external energy source, the gain medium is energized, or pumped.

Revision: The gain medium is energized, or pumped, by an external energy source.

Awkward inverted order: Placed on a handgun or rifle aligned to emit a beam parallel to the barrel, a laser sight is a small, usually visible-light laser.

Revision: A laser sight is a small, usually visible-light laser placed on a handgun or rifle aligned to emit a beam parallel to the barrel.

7

USAGE

7

Usage

7   Usage

This chapter is about words: similar-sounding or similar-looking words that are commonly confused; words that are non-standard; words that cause redundancy; words that are pompous, inflated, and empty; words that are stuffy and overly formal; words that are too informal for formal writing situations; words that have unwanted connotations; words that are vague; words that have become clichés; words that are sexist.

7.1 GLOSSARY OF USAGE

Skim the items in the following glossary, paying attention to the ones that you are unfamiliar with. After that, use this glossary as a quick-reference tool for usage items you are uncertain about.

a, an. Use *a* before a word that begins with a consonant sound, even if the word begins with a vowel: *a computer, a desk, a unique individual, a university*. Use *an* before a word that begins with a vowel sound, even if the word begins with a consonant: *an iguana, an oak, an hour, an honor*. Words beginning with the letter *h* often present problems. Generally, if the initial *h* sound is hard, use *a*: *a hotdog, a heart attack*. However, if the initial *h* is silent, use *an*: *an honest mistake*.

accept, except. *Accept* is a verb meaning "to receive" or "take to (oneself)." *He accepted the lottery prize.* Although *except* can be used as verb in rare instances; it is most commonly a preposition meaning "to exclude." *A maser is a device similar to a laser, except that it works at microwave frequencies.*

adapt, adopt. *Adapt* means to "adjust oneself to" or "make suitable," and it is followed by the preposition *to. The lizard will adapt to its surroundings.* The word *adapt* can also mean "revise," in which case it is used with the preposition *for* or *from. They will adapt the novel for the silver screen. Adopt* means "to take or use as one's own." *They plan to adopt the idea for their computer game.*

adverse, averse. *Adverse* means "unfavorable." *Smoking has an adverse effect on your health. Averse* means "opposed" or "having an active dislike"; or, it means "reluctant," in which case it is followed by the preposition *to. She was averse to fighting of any kind.*

advice, advise. *Advice* is a noun that means "an opinion about what should be done." *Take my advice and sell while you can. Advise* is a verb that means "to offer advice." *The high-priced lawyer will advise us on what course of action to take.*

affect, effect. *Affect* is a verb that most commonly means "to influence." *Pollution severely affects the fish habitat. Effect* most commonly means "result." *Even very small impurities can have large effects on semiconductor devices. Effect* can also be used as a verb meaning "to bring about or execute." *The cost-cutting moves will effect a turnaround for the business.*

aggravate, irritate. *Aggravate* is a verb that means "to make worse or more severe." *Suburban sprawl aggravates two groups of economic and social problems. Irritate,* a verb, means "to make impatient or angry." Although *aggravate* is often used colloquially to mean *irritate,* do not substitute *aggravate* for *irritate* in formal writing. *His constant complaining irritated* [not *aggravated*] *me.*

agree to, agree with. *Agree to* means "to consent to." *The two sides will agree to the proposal. Agree with* means "to be in accord with." *The witness's version of events agrees with theirs.*

ain't. Nonstandard usage for "am not," "are not," or "is not." It is nonstandard English and should not be used in formal writing.

all ready, already. *All ready* means "completely prepared." *The sprinter is all ready for the starter's gun. Already* is an adverb that means "before this time; previously; even now." *By the 1940s, low-cost, small-sized refrigerators were already a fixture in most American homes.*

all right, alright. *All right* is always written as two words. *Alright* is nonstandard English for *all right* and should not be used in formal writing. *It's all right [not alright] to eat dinner if Desmond is late.*

all together, altogether. See *altogether, all together.*

allude, elude. *Allude* means "to refer to indirectly or casually." Do not use it to mean "to refer to directly." *In his presentation, Freud specifically referred to [not alluded to] the importance of the subconscious. Elude* means "to evade or escape from, usually with some daring or skill," or "to escape the understanding or grasp of." *However, the goal of unifying the laws of physics eluded Einstein.*

allusion, illusion, delusion. *Allusion* is an "implied or indirect reference." *The prosecuting attorney made an allusion to her criminal past.* The word *illusion* means "an appearance or feeling that misleads because it is not real." *When the bus beside ours backed up, it created the illusion that we were moving. Delusion* means "a false and often harmful belief about something that does not exist." *The paranoid reporter had the delusion that every e-mail contained a virus.*

alot, a lot. *A lot* is always written as two words. *We have not had a lot [not alot] of snow this winter.* Avoid using *alot* in formal writing.

altogether, all together. *Altogether* means "completely, entirely." *The new piston design enabled Joseph Day to do away with valves altogether, giving rise to the classic two-stroke piston.* The phrase *all together* means "together in a group." *For Robert Kahn, the problem was how to put the ground packet radio net, the packet satellite network, and the Arpanet all together into one network of networks.*

a.m., p.m., A.M., P.M. Use these abbreviations only with specific times, when numerals are provided: *10 a.m.* or *1 p.m.* Do not use the abbreviations as substitutes for *morning, afternoon,* or *evening. The mother had to wake up early in the morning [not a.m.] to take her daughter to the hockey game.*

among, between. See *between, among.*

amoral, immoral. *Amoral* means "not having any morals; neither moral or immoral." *The cabinet adopted an amoral perspective when it considered tax cuts.* The word *immoral* means "morally wrong or wicked." *It is immoral to steal food from the food bank.*

amount, number. *Amount* is used to refer to things in bulk or mass, things that cannot be counted. *A large amount of litter can be found along the highway. Number* is used to refer to things that can be counted. *He gobbled down a number of bedtime snacks every evening.*

an, a.. See *a, an.*

and . . . etc. *Etc.* (*et cetera*) means "and so forth." Do not use *and . . . etc.* because it is redundant. See also *etc.*

and/or. Sometimes used to indicate three possibilities: one, or the other, or both. It is acceptable in some business, technical, or legal writing.

ante-, anti-. *Ante-* is a prefix that means "before; earlier; in front of." *The reporter waited in an anteroom until the politician could see her.* The prefix *anti-* means "against" or "opposed to." *The pacifists strongly support the antigun law.* Use *anti-* with a hyphen when it is followed by a capital letter (*Anti-American*) or a word beginning with i (*anti-intellectual*). Otherwise, consult a dictionary.

anxious, eager. *Anxious* means "nervous," "troubled," or "worried." *The dark looming clouds made Tim anxious. Eager* means "looking for-ward" and is often followed by the preposition *to. Stella was eager to receive the Christmas parcel.* Do not use *anxious* to mean "eager." *Faraday was eager* [not *anxious*] *to leave his bookbinding job and work for Sir Humphrey Davy, president of the Royal Society.*

anyone, any one. *Anyone* is an indefinite pronoun that means "any person at all." *Anyone* is singular. *Can anyone tell me what to do?* In *any one,* the pronoun *one* is preceded by the adjective *any.* Here the two words refer to any person or thing in a group. *Once the last of the patrons has left, you can jump into any one of the bumper cars.*

anyways, anywheres. *Anyways* and *anywheres* are nonstandard for *anyway* and *anywhere,* respectively. Always use *anyway* and *anywhere* in formal speaking and writing.

as. Substituting *as* for *because, since,* and *while* may make a sentence vague or ambiguous. *CT scan time must be kept to a minimum, because* [not *as*] *patients must hold their breath during the examination.* If *as* were used in this sentence, the cause–effect relationship would be unclear.

as, like. See *like, as*.

as for, as far as. The phrases *as far as* and *as for* are commonly confused. An object follows *as for. As for the laser light itself, it could be made available in a broad range of wavelengths, from infrared to ultraviolet.* The elements of a clause—subject and verb—must follow *as far as. As far as anyone knew, ordinary glass could not be made transparent enough to carry laser light at any useful distance.*

averse, adverse. See *adverse, averse*.

awful, awfully. Originally, the adjective *awful* meant "filled with awe," although the word now contains such negative connotations that it cannot be used in its original sense. *The Northern Lights are an awe-inspiring* [or *awesome*, but not *awful*] *sight. At the 1939 First World Science Fiction Convention, Asimov saw Fritz Lang's movie* Metropolis, *which he thought was awful.* The adverb *awfully* is sometimes used in informal speech as an intensifier to mean "extremely" or "very." *He was awfully upset when he opened the bill.* Avoid such colloquial usage in formal writing.

awhile, a while. *Awhile* is an adverb. *Stay awhile, if you wish.* Use the article and noun, *a while*, as the object of a preposition. *In 1902, Carrier had been puzzling over for a while about the complex relationship between air temperature, humidity, and dew point.*

bad, badly. *Bad* is an adjective. *They felt bad* [not *badly*] *about resigning from the project.* The word *badly* is an adverb. *His infected hand hurt badly.*

being as, being that. Both *being as* and *being that* are nonstandard expressions used in place of the subordinate conjunctions *because* or *since. Since* [not *Being that*] *vandals had written on the walls, tough security measures were put in place.*

beside, besides. *Beside* is a preposition meaning "by the side of" or "near." *Grass grows beside the stream. Besides* is a preposition meaning "also," "moreover," or "further." *In a typical large transformer, the price difference between copper and aluminum wire has a relatively small effect on total cost; besides, copper offers other savings. Besides* can also be an adverb meaning "in addition." *Besides space exploration, materials science has been a driving force in technologies such as plastics, semiconductors, and biomaterials.*

between, among. Use *among* when referring to relationships involving more than two people or things. *Kuwait's Burgan Field and Mexico's Cantarell Field are among the largest oil fields in the world. The World Trade Center's 15 million square feet of floor space was distributed among seven buildings.* Use *between* when referring to relationships involving two people or things. *Neural networks are nonlinear statistical data modeling tools that can be used to model complex relationships between inputs and outputs or to find patterns in data.*

bring, take. Use *bring* when something is being moved toward the speaker or object. *The new POWER single-chip solution enables Apple and Motorola to bring machines to market much faster.* Use *take* when something is being moved away. *At a starting estimate of US$5 billion, it could take at least a decade to recoup the expenses for even a very primitive space elevator.*

burst, bursted; bust, busted. *Burst* is an irregular verb meaning to "fly apart suddenly with force; explode; break open." *The water-filled balloon burst when it hit the pavement. Bursted* is the nonstandard past-tense form of *burst.* Avoid the nonstandard *bust* and its past-tense form *busted.*

can, may. *Can* means "know how to" or "be able to." *With a space elevator, vehicles can climb the tether and escape the planet's gravity without the use of rockets. May* means "be allowed to" or "have permission to." *Given the catastrophic potential if a space elevator were cut at 25,000 km, governments may not allow such projects to be implemented.* The distinction in meaning between *can* and *may* is important. *May* also indicates doubt or uncertainty. *The current market may not be large enough for space elevators to be economically feasible.*

capital, capitol. *Capital* refers to a city where the government of a country, province, or state is located. *Edmonton is the capital of Alberta. Capital* can also mean "the amount of money a company or person uses in carrying on a business." A *capitol* is a building in which American lawmakers meet. When referring to the building in which the U.S. Congress meets, capitalize the first letter, as in *Capitol.*

censor, censure. The verb *censor* means "to edit or remove from public view on moral or other grounds." *Some government officials have attempted to censor the disturbing projections about global warming.* The verb *censure* means "to express strong disapproval." *The House will censure the minister for giving misleading information.*

cite, site, sight. The verb *cite* means "to quote, especially as an authority." *In Terry Pratchett's* The Dark Side of the Sun, *a robot cites the "Eleventh Law of Robotics, Clause C, As Amended."* The noun *site* often means "a particular place." *The prototype of IBM's Blue Gene, the world's fastest supercomputer in 2005, was developed at IBM's Rochester site.* Of course, the Internet has also made use of the word: *The Cal Poly Robotics Club is a web site that includes robot project descriptions, robot tutorials, and robot development tools.* The word *sight* can be either a verb or noun, meaning *to view* or *spot visually* or *something seen.*

climactic, climatic. *Climactic* is an adjective derived from *climax; climax* means "the highest point; point of highest interest; the most exciting part." *In the 1996 feature film,* Star Trek: First Contact, First Contact *refers to the climactic moment in human history in which humanity achieves warp drive and encounters extraterrestrial life for the first time.* The adjective *climatic* means "of or having to do with climate." *The Ferengi language has over 160 words for "rain" because of the planet Ferenginar's well-known climatic conditions.*

coarse, course. *Coarse* usually means "heavy and rough in texture" or "crude." *Shelley used a coarse sandpaper to finish the table. Course* means "a line of movement," "a direction taken," "a way, path, or track," or "a playing field." *Seeking help for your drinking problem is the right course of action.*

compare to, compare with. *Compare to* means "to represent as similar." *A CoM, a collection of different neural networks that together "vote" on a situation, generally provides better results compared to other neural network models. Compare with* means "to point out how two persons or things are alike and how they differ." *From 1995 to 2002, China lost 15 million manufacturing jobs of its own, compared with 2 million lost in the U.S.*

complement, compliment. The verb *complement* means "to reinforce, add to, or complete something." *The International Linear Collider will complement the Large Hadron Collider by allowing more precise measurements of the properties of newly found particles.* As a noun, *complement* is something that completes. *Compliment* as a verb means "to say

something in praise." *He should compliment your work on this project.*
As a noun, *compliment* means "a remark of praise."

compose, comprise. *Compose* means "make up" or "make." *The solar system is composed of a sun and eight planets. Comprise* means "is made up of" or "consists of." *The solar system comprises a sun and eight planets.* It is technically incorrect to say *The solar system is comprised of a sun and eight planets.* To simplify matters, you can say *The solar system is made up of a sun and eight planets* or *A sun and eight planets make up the solar system.*

conscience, conscious. *Conscience* is a noun meaning "the sense of moral right and wrong." *Sisko explained that a guilty conscience was a small price to pay for the safety of the Alpha Quadrant. Conscious* is an adjective that means "aware; knowing." *Writer Rene Echevarria made a conscious effort not to glamorize the same-sex kiss between Jadzia Dax and Lenara Khan in* Star Trek: Deep Space Nine.

consensus of opinion. *Consensus* means "general agreement." As a result, the phrase *consensus of opinion* is redundant. *A consensus* [not *consensus of opinion*] *is required before the motion will be passed.*

continual, continuous. *Continual* means "repeated many times; very frequent." *When the roofers were here, there was continual hammering. Continuous* means "without a stop or a break." *The advantage of circular accelerators over linear accelerators is that the ring topology allows continuous acceleration.*

could of. *Could of* is nonstandard for *could have. If not for his injury, Mr. Martin could have* [not *could of*] *become a professional basketball player.*

council, counsel. *Council* is a noun used to describe "a group of people called together to talk things over, or give advice"; it also applies to "a group of people elected by citizens to make up laws." *A tribal council will decide the appropriate punishment. A councillor is a member of the council. Counsel* as a noun means "advice." *The chief gives wise counsel. Counsel* can act as a noun meaning "lawyer." *A counsellor is someone who gives advice or guidance.*

course, coarse. See *coarse, course.*

criteria, criterion. *Criteria* are rules for making judgments. *Criteria* is the plural form of *criterion. The major criteria for the job are a background in multimedia and a willingness to work overtime.*

data, datum. *Data* are "facts or concepts presented in a form suitable for processing in order to draw conclusions." *Data* is the plural form of *datum*, which is rarely used. Increasingly *data* is used as a singular noun; however, careful writers use it as a plural. *The new data reveal* [or, increasingly, *reveals] that the economy is rebounding.*

delusion, illusion. See *allusion, illusion, delusion.*

differ from, differ with. *Differ from* means "to be unlike." *More powerful computers differ from conventional computers by dividing work between more than one main CPU. Differ with* means "to disagree with." *Many analysts differ with the peak oil theory, citing the abundance of low-grade forms of energy.*

different from, different than. In standard English the preferred form is *different from. What makes a cell phone different from a regular phone?* However, use *different than* when a full clause follows or is implied. *The IBM Corporation is different now than it was twenty years ago.*

discreet, discrete. *Discreet* means "prudent and tactful in speech and behavior." *The mayor was very discreet when talking about the manager's personal life. Discrete* means "separate; distinct." *There are discrete parts of the cell that perform specialized functions.*

disinterested, uninterested. *Disinterested* means "impartial." *The chairperson appointed a disinterested third party to mediate the dispute. Uninterested* means "lacking in interest," or "bored." *Businesses are notoriously uninterested in fundamental health and safety standards.*

don't. A contraction for *do not. Don't slam the door.* Do not use *don't* as a contraction for *does not*; the correct contraction is *doesn't. Selma doesn't* [not *don't] want to shovel the walk.*

due to. A phrase meaning "caused by" or "owing to." It should be used as an adjective phrase following a form of the verb *to be. The inquest ruled that the death was due to driver negligence.* In formal writing, *due to* should not be used as a preposition meaning "because of." *Work was canceled because of* [not *due to] the heavy snowstorm.*

each. *Each* is singular.

eager, anxious. See *anxious, eager*.

effect, affect. See *affect, effect*.

e.g. The Latin abbreviation for *exempli gratia*, which means "for example." Governmental and corporate style guides commonly recommend avoiding Latin phrases and abbreviations such as *e.g.* and instead to use phrases such as *for example* or *for instance* instead. *Many fish, for example salmon and trout, will be affected.*

either. *Either* is singular.

elicit, illicit. *Elicit* is a verb meaning "to draw forth" or "bring out." *The research project was designed to elicit information about students' opinions concerning written communications in engineering.* The adjective *illicit* means "unlawful." *The neighbors had an illicit growing operation in their basement.*

elude, allude. See *allude, elude*.

emigrate from, immigrate to. *Emigrate* means "to leave one's own country or region and settle in another" and requires the preposition *from. The Bhuttos emigrated from Pakistan. Immigrate* means "to enter and permanently settle in another country" and requires the preposition *to. Mr. Bhutto's cousin now plans to immigrate to Canada.*

eminent, immanent, imminent. *Eminent* means "distinguished" or "exalted." *The eminent scientist delivered the lecture. Immanent* is an adjective that means "inherent" or "remaining within." *The Transcendentalists believed that divinity was immanent within all of nature. Imminent* is an adjective meaning "likely to happen soon." *Given the troop movements, the general felt that an attack was imminent.*

enthused, enthusiastic. *Enthused* is an informal term meaning "showing enthusiasm." Use *enthusiastic* instead. *He becomes enthusiastic [not enthused] about the Oilers chances to make the playoffs.*

-ess. Many readers find the *-ess* suffix demeaning. Write *actor*, not *actress; singer*, not *songstress*.

etc. An abbreviation that means "and other things." Governmental and corporate style guides commonly recommend avoiding Latin phrases and abbreviations *etc.* and instead to use expressions such as *and so on.* See also *and . . . etc.*

eventually, ultimately. *Eventually* often means "an undefined time in the future." *Ultimately* commonly means "the greatest extreme or furthest extent." *Eventually* and *ultimately* are frequently used interchangeably. Use *eventually* when referring to time and *ultimately* when referring to greatest extent. *After many frustrations, the Bell team eventually developed the first transistor in 1947. Many names were suggested for the newly invented device, but ultimately "transistor" was chosen.*

everybody, everyone. *Everybody* and *everyone* are both singular.

everyone, every one. *Everyone* is an indefinite pronoun meaning "every person." *Everyone wanted to purchase a ticket. Every one* is a pronoun, *one*, modified by an adjective, *every*; the two words mean "each person or thing in a group." *Every one* is frequently followed by *of. Every one of the roads in Oakville experiences gridlock.*

except, accept. See *accept, except.*

except for the fact that. Avoid this wordy, awkward construction; use *except that.*

explicit, implicit. *Explicit* means "clearly expressed; directly stated." *Laws about the use of the title "engineer" are enforced when an unlicensed individual uses that title and is explicitly offering engineering services to the public. Implicit* means "meant but not clearly expressed or directly stated." *Implicit in the idea of superconductivity is the underlying concept of zero electrical resistance and the rich technological possibilities that concept implies.*

farther, further. In formal English *farther* is used for physical distance. *The nearest stellar system, Alpha Centauri, is much farther away from Earth than the Klingon Homeworld, which is a four-day journey at warp 5. Further* is used to mean "more" or "to a greater extent." *In 2374, the crew of the USS Defiant hoped that further study of the rare subspace compression phenomenon in Federation space would lead to the creation of working transwarp drives.*

female, male. *Female* and *male* are considered jargon if substituted for "woman" and "man." *Sixteen men [not males] and seventeen women [not females] made the team.*

fewer, less. Use *fewer* only to refer to numbers and things that can be counted. *There are fewer houses up for sale than there were last year at this time.* Use *less* to refer to collective nouns or things that cannot be counted. *Generally, there is less traffic congestion at midday.*

finalize. A verb meaning "to bring to a conclusion." The word, though often used, is considered jargon by many people. Use a clear, acceptable alternative. *The theory of high-temperature superconductivity has not yet been completed* [not *finalized*].

flunk. Colloquial for *fail*, and it should be avoided in formal writing.

folks. *Folks* is informal for "one's family; one's relatives." In formal writing, use a more formal expression than *folks*. *My mother and father* [not *folks*] *are organizing the family reunion.* In business contexts, *folks* is also used to refer informally to any group of people. *The folks in HR are hosting a benefits kickoff.* In any formal situation, avoid this usage.

fun. When used as an adjective, *fun* is colloquial; it should be avoided in formal writing. *The Jawbreaker was an exciting* [not *fun*] *ride.*

further, farther. See *farther, further*.

get. A common verb with many slang and colloquial uses. Avoid the following uses of *get*: in *He got cold*, use *became*; in *Gillian got back at Ted for the rumors he spread*, use "got revenge"; in *His constant complaining really got to me*, use "annoyed"; in *The final scene in the movie really got to me*, use "elicited an emotional response".

good, well. *Good* is an adjective. *Michael is a good skier. Well* is nearly always an adverb. *The racing team skis well.*

hanged, hung. *Hanged* is the past tense and past participle of *hang*, which means "to execute." *The man was convicted of treason and hanged. Hung* is the past tense and past participle of *hang*, which means "to fasten or be fastened to something." *Decorations for the dance hung from the ceiling.*

hardly. Avoid double negative expressions such as *not hardly* or *can't hardly. I can* [not *can't*] *hardly find words to express myself.*

has got, have got. Avoid using *have got* or *has got* when *have* or *has* alone communicate intended meaning. *I have* [not *have got*] *two more books to finish reading to complete the course requirements.*

he. Do not use only *he* when the complete meaning is "he or she." In modern usage, this is not inclusive.

he/she, his/her. Use *he or she*, or *his or her* in formal writing.

hisself. Nonstandard for *himself*.

hopefully. An adverb meaning "in a hopeful manner." *Hopefully* can modify a verb, an adjective, or another adverb. *They waited hopefully for news from the surgeon on how the operation had gone.* In formal writing, do not use *hopefully* as a sentence modifier with the meaning "I hope." *I hope* [not *Hopefully*] *the operation will be a success.*

hung, hanged. See *hanged, hung*.

i.e. An abbreviation for the Latin *id est*, which in English means "that is." Governmental and corporate style guides commonly recommend avoiding Latin phrases and abbreviations such as *i.e.* and instead using the English equivalent, *that is*.

if, whether. *If* is used to express conditions. *If the MOSFET is an N-Channel or nMOS FET, then the source and drain are N+ regions and the body is a P region.* Use *whether* to express alternatives. *In the MOSFET, the voltage applied to the gate determines whether the switch is on or off.*

illicit, elicit. See *elicit, illicit*.

illusion, allusion. See *allusion, illusion*.

immanent, imminent, eminent. See *eminent, immanent, imminent*.

immigrate to, emigrate from. See *emigrate from, immigrate to*.

immoral, amoral. See *amoral, immoral*.

implement. A word that means "to carry out" and can seem pretentious. *The president carried out* [not *implemented*] *the board's recommendations.*

implicit, explicit. See *explicit, implicit*.

imply, infer. A word that means to "express indirectly." *A business cannot use a name that implies that it offers professional engineering services unless it employs at least one licensed professional engineer.* Infer means "to conclude by reasoning." *Soil Inference System is a proposed knowledge base that could be used to infer soil properties.*

in, into. A word that generally indicates a location or condition. *The terms engineer and scientist are often confused in the minds of the general public. Into* indicates a direction, a movement, or a change in condition. *Chemical engineering involves the process of converting raw materials into more useful forms.*

individual. Sometimes used as a pretentious substitute for *person*. *A few persons* [not *individuals*] *were invited to apply for the position.*

ingenious, ingenuous. *Ingenious* means "clever" or "skillful." *The criminal devised an ingenious plan to rob the bank. Ingenuous* means "frank" and "simple." *His country manner was quite ingenuous.*

in regards to. Instead of *in regards to*, use *in regard to. Talk to your supervisor in regard to the application.* Or just use the simple word *about. Talk to your supervisor about the application.*

irregardless. Nonstandard English. Use *regardless* instead.

irritate, aggravate. See *aggravate, irritate.*

is when, is where Do not use *when* or *where* following *is* in definitions. *Photosynthesis is the process by which* [not *is when*] *plant cells make sugar from carbon dioxide and water in the presence of chlorophyll and light.*

its, it's. *Its* is a possessive pronoun. *Because of his achievements in electricity and radio, Nikola Tesla was selected as a fellow of the IEEE (at the time the AIEE) and was awarded its most prestigious prize, the Edison Medal.* [*It's* is a contraction for *it is.*] *In 1975, IEEE created a Nikola Tesla Award. It's given to those who make outstanding contributions to the generation or utilization of electric power.*

kind, kinds. *Kind* is singular and should not be treated as a plural. *This* [not *These*] *kind of uncertainty is the focus of fuzzy logic and fuzzy probability. Kinds* is plural. *These kinds of buildings could not be built without employing twentieth-century air conditioning.*

kind of, sort of. *Kind of* and *sort of* are colloquial expressions meaning "rather" or "somewhat." Do not use these colloquialisms in formal writing. *The first automobile air conditioner introduced in the 1939 Packard used a rather* [not *kind of* or *sort of*] *awkward system.*

lay, lie. See *lie, lay.*

lead, led. *Lead* is a verb meaning to conduct or guide and a noun referring to a soft heavy metal. *Led* is the past tense of the verb *lead*. *His accurate directions led me to the correct address.*

learn, teach. *Learn* means "to gain knowledge of or a skill by instruction, study, or experience." *Early engineers learned much from Fourier's ideas about heat transfer and dimensional homogeneity. Teach means "to impart knowledge or a skill." In his early years, Fourier, often considered the father of modern engineering, taught [not learned] students mathematics at French universities.*

leave, let. *Leave* means "to go away." *Let* means "to allow or permit." Do not use *leave* with the nonstandard meaning "to permit." *Let [not leave] me help you trim the fruit trees.*

led, lead. See *lead, led.*

less, fewer. See *fewer, less.*

liable. A word that means "legally responsible." Avoid using it to mean "likely." *Jeff is not likely [not liable] to catch any fish on this trip.*

lie, lay. *Lie* means "to recline." It is an intransitive verb, which means it does not take a direct object. The principal forms of the verb are *lie, lay,* and *lain. Lie down now.* Lay means "to put" or "to place." It is a transitive verb, which means it always requires a direct object. The principal parts of the verb are *lay, laid,* and *lain. He lays the book on the table.*

like, as. *Like* is a preposition and should be followed by a noun or a noun phrase. *Daniel looks like a million dollars.* As is a subordinating conjunction and should be used to introduce a dependent clause. *As always happens, he is late.*

loose, lose. *Loose* is an adjective meaning "not firmly fastened." *On October 9, 1963, a landslide near the Vaiont Dam broke loose and caused a huge wave of water to surge over the top the dam, flooding towns downstream. Lose is a verb meaning "to misplace" or "to be defeated." Smaller primordial black holes would emit more cosmic microwave background radiation than they absorb and thereby lose mass.*

lots, lots of. *Lots* and *lots of* are colloquial substitutes for *many, much,* and *a great deal.* They should not be used in formal writing.

male, female. See *female, male.*

mankind. An exclusive term because it excludes women. Use terms such as *humans, humanity, the human race,* or *humankind.*

may, can. See *can, may.*

may of, might of. *May of* and *might of* are nonstandard English for *may have* and *might have. Without Einstein, the general relativistic laws of gravity might not have [not might not of] been discovered until decades later.*

maybe, may be. *Maybe* is an adverb meaning "perhaps." *Maybe we should build the outdoor rink tomorrow. May be* is a verb phrase. *Since the temperature will be lower on Tuesday, that may be a better day.*

media, medium. *Media* is the plural of *medium. The media are offering too much coverage of sensational stories.*

moral, morale. *Moral* is a noun meaning "an ethical conclusion." *Norbert Weiner believed that "cybernetics," the science of information feedback systems, would have enormous moral implications. Morale* means "the attitude as regards courage, confidence, and enthusiasm." *According to the author, Erhard Seminars Training (est) would ultimately destroy the morale and social cohesion of the community in the Augmentation Research Center, which Douglas Engelbart had founded.*

most. When used to mean "almost," *most* is colloquial. This usage should be avoided in formal writing. *Almost [not Most] any algorithm will work well with the correct hyperparameters for training on a particular fixed dataset.*

must of. See *may of, might of.*

myriad. Commonly used to refer to a large number of an unspecified size. It is incorrect to say "a myriad of." *Parameters such as electrical resistance, heat transfer coefficient, and material modulus, and the myriad others like them, are the engineering tools now used to analyze natural phenomena.*

myself. A reflexive pronoun. *I hurt myself. Myself* can also be an intensive pronoun. *I will go myself.* Do not use *myself* in place of *I* or *me. Jeremy and I [not myself] are going on a trip.*

neither. A word that takes a singular verb when it is followed by a singular noun, when it is followed by a prepositional phrase containing a singular noun, or when a singular noun is implied.

none. A word that takes a singular verb when it is followed by a prepositional phrase containing a singular noun, or when a singular noun is implied.

nowheres. Nonstandard English for *nowhere*.

number, amount. See *amount, number*.

of. A preposition. Do not use it in place of the verb *have* after *could, should, would, may, must,* and *might. The Johnsons might have* [not *of*] *left their garage door open.*

off of. Nonstandard for *off. The young boy fell off* [not *off of*] *the table.*

OK, O.K., okay. In formal writing and speech, avoid these colloquial expressions.

parameters. A mathematical term that means "a quantity that is constant in a particular calculation or case but varies in other cases." It is sometimes used as jargon to mean any limiting or defining element or feature. Avoid such jargon. *The whole project had very vague guidelines* [not *parameters*].

passed, past. *Passed* is the past tense of the verb *pass*, which means "to go by." *In his research on the 1919 solar eclipse, Arthur Eddington confirmed Einstein's prediction that star light would bend as it passed close to the Sun. Past* commonly means "gone by; ended." Never use *past* as a verb. *Einstein believed that any distinction between past, present, and future was a "stubbornly persistent illusion."*

people, persons. Use *people* to refer to a group of individuals who are anonymous and uncounted. *The people of South Africa have a long history of apartheid.* Generally, use *persons* or *people* when referring to a countable number of individuals. *Only five persons* [or *people*] *attended the town meeting.*

percent, per cent, percentage. Always use *percent* (also spelled *per cent*) with specific numbers. *The problem facing fiber-optics research in the 1960s was that light traveling along a fiber lost about 99 percent of its energy by the time it had gone just 30 feet. Percentage* means "part of" or "portion," and it is used when no number is provided. *From 2002 to 2005, the percentage of U.S. workers employed in industrial jobs decreased from about 14 percent to 10 percent.*

phenomenon, phenomena. *Phenomenon* means "a fact, event or circumstance that can be observed." *A wormhole is a hypothetical phenomenon of space-time that might enable shortcuts through space and time. Phenomena is the plural of phenomenon. Quantum mechanics can provide accurate and precise descriptions for many phenomena that "classical" Newtonian mechanics and classical physics cannot.*

plus. A nonstandard substitute for *and*. Do not use *plus* to join independent clauses. *He has a driver's license; and [not plus], it is expired.*

p.m. See *a.m., p.m., a.m., p.m.*

practice. A noun meaning "an action done several times over to gain a skill" as well as a verb meaning "to do something again and again in order to learn it." *An engineer is someone who practices the profession of engineering.*

precede, proceed. *Precede* means "to go or come before." *In product development, reviews of the manufacturability and producibility of the product precede full-scale production in order to prevent costly redesign. Proceed means "to go on after having stopped" or "to move forward." Statistical process control in manufacturing usually proceeds by randomly sampling and testing a fraction of the output.*

principal, principle. The noun *principal* means "a chief person" or "a sum of money that has been borrowed." *After Mr. Toutant's retirement from J.P. Morgan Elementary School, a new principal was appointed.* The noun *principle* means "a fact or belief on which other ideas are based." *The engineering principle on which both refrigeration and air conditioning are based is called mechanical refrigeration.* The adjective form, *principal*, means "main." *The principal problem with integrated circuits used in nanometer-scale devices is leakage current.*

proceed, precede. See *precede, proceed.*

quote, quotation. *Quote* is a verb meaning "to repeat the exact words of." *Tesla was quoted as saying that Edison "had a veritable contempt for book learning and mathematical knowledge, trusting himself entirely to his inventor's instinct and practical American sense." Quotation is a noun* meaning "a passage quoted." Do not use *quote* as a shortened form of *quotation*. *Often using a relevant quotation* [not *quote*] *is a good way to begin a speech.*

raise, rise. *Raise* means "to move to a higher level; to elevate." It is a transitive verb, which means it requires a direct object. *The stage manager raised the curtain. Rise* means "to go up." It is an intransitive verb, which means it does not require a direct object. *The smoke rises.*

real, really. *Real* is an adjective; *really* is an adverb. *Don was really* [not *real*] *excited.* In formal writing, avoid *real* and *really* as intensifiers to mean "extremely" or "very."

reason is because. A redundant expression. Use *reason is that* instead. *The reason that Thomas Edison continued trying to perfect his own phonograph design in the 1880s was that* [not *was because*] *Alexander Graham Bell and his associates had redesigned a phonograph using wax-coated cardboard cylinders.*

reason why. A redundant expression. In its place use either *reason* or *why. I still do not know why* [not *the reason why*] *she rejected my project design.*

regretfully, regrettably. *Regretfully* means "full of regret." It describes a person's attitude of regret. *Regretfully, he wrote to apologize. Regrettably* means that circumstances are regrettable. *Regrettably, the site visit was rained out today.*

relation, relationship. Traditionally, *relation* has been used to describe the association between two or more things, and *relationship* to describe the association or connection between people. While this usage remains true for *relation*, in practice *relationship* is used for both things and people. *The scientist studied the relation between lung cancer and smog. Peter and Olga had a professional relationship that soon blossomed into a personal one.*

respectfully, respectively. *Respectfully* is an adverb meaning "showing or marked by proper respect." *She respectfully presented her counterargument in the debate. Respectively* is an adverb meaning "singly in the order designated or mentioned." *Chand, Doug, and Lenore are a surgeon, bus driver, and company vice-president, respectively.* Notice that *respectively* must be punctuated with a comma.

rise, raise. See *raise, rise.*

sensual, sensuous. *Sensual* is an adjective meaning "relating to gratification of the physical senses." *An ascetic seeks to reject all sensual pleasures. Sensuous* is an adjective meaning "pleasing to the senses." *Sensuous* is always favorable and often applies to the appreciation of nature, art, or music. *An ascetic avoids sensuous delights such as food, drink, sex, and music.*

set, sit. *Set* means "to set in place, position, or put down." It is a transitive verb, requiring a direct object and its principal parts are *set, set, set. Ali set the book on the ledge. Sit* means "to be seated." It is an intransitive verb, not requiring a direct object; its principal parts are *sit, sat, sat. Set* is sometimes used as a non-standard substitute for *sit.* Avoid this usage in formal writing. *The dog sat* [not *set*] *down.*

shall, will. *Shall* was once used with the first-person singular and plural as the helping verb with future-tense verbs. *I shall visit grandfather on Wednesday. We shall deliver the results on Thursday.* In modern usage *will* has replaced *shall. I will see you on Friday.* The word *shall* is still often used in polite questions. *Shall I bring the newspaper to your door?* However, *shall* is used to mean *must* in legalistic contexts, such as in specifications. *2.2.A All structural steel sections and welded plate members shall be designed in accordance with applicable AISC specifications.*

she/he, her/his. See *he/she, his/her.*

should of. Nonstandard for *should have. He should have* [not *should of*] *submitted his status report on time.*

since. Mainly used in situations describing time. *We have been waiting for the meteor shower since midnight.* Do not use *since* as a substitute for *because* if there is any chance of confusion. *Since we lost the division, we have been playing our second-string players.* Here *since* could mean "from that point in time" or "because." See also *as.*

sit, set. See *set, sit.*

site, cite. See *cite, site.*

somebody, someone. *Somebody* and *someone* are singular.

something. *Something* is singular.

sometime, some time, sometimes. *Sometime* is an adverb meaning "at an indefinite or unstated time." *Let's meet sometime on Thursday.* In the phrase *some time*, the adjective *some* modifies the noun *time*. *We haven't seen the Jebsons for some time. Sometimes* is an adverb meaning "at times; now and then." *Sometimes, I'm not sure which major to pursue.*

sort of, kind of. See *kind of, sort of.*

stationary, stationery. *Stationary* means "not moving." *At the club, he rode on a stationary bike. Stationery* refers to paper and other writing products. *I will need to buy the stationery at the business supply store.*

suppose to, use to. See *use to, suppose to.*

sure and. Nonstandard for *sure to. Please be sure to* [not *sure and*] *edit your work carefully.*

take, bring. See *bring, take.*

teach, learn. See *learn, teach.*

than, then. *Than* is a conjunction used to make comparisons. *Then* is an adverb used to indicate past or future time. *An ambulance's siren starts out higher than its stationary pitch, slides down as it passes, and then continues lower than its stationary pitch as it moves away from you.*

that, who. See *who, which, that.*

that, which. Most writers use *that* for restrictive clauses and *which* for nonrestrictive clauses.

theirselves. *Theirselves* is nonstandard English for *themselves. They amused themselves* [not *theirselves*] *by going to the drive-in.*

them. Nonstandard when it is used in place of *those. Margaret, please place those* [not *them*] *flowers on the kitchen table.*

then, than. See *than, then.*

there, their, they're. *There* is an adverb meaning "at or in that place." *I'll call home when I get there. There* can also be used in an expletive, a phrase at the beginning of a clause. *There are two beautiful dogs in the garage. Their* is a possessive pronoun. *It was their first house. They're* is a contraction for *they are. They're first in line for tickets.*

this kind. See *kind, kind(s).*

thru. Colloquial spelling of *through*. Do not use *thru* in formal writing.

to, too, two. *To* is part of a preposition. *We need to talk. Too* is an adverb. *There are too many people in the city. Two* is a number. *I have two red pens.*

toward, towards. Both versions are acceptable.

try and. Nonstandard English. Instead use *try to. Try to* [not *try and*] *be polite.*

ultimately, eventually. See *eventually, ultimately.*

uninterested, disinterested. See *disinterested, uninterested.*

unique. Like *straight, round*, and *complete, unique* is an absolute. There are no degrees of uniqueness. Avoid expressions such as *more unique* and *most unique.*

usage, use. *Usage* refers to conventions, most often of language. *Placing "ain't" in a sentence is incorrect usage. Use* means "to employ." Do not substitute *usage* when *use* is required. *I do not think surfing the Internet is the proper use* [not *usage*] *of your study time.*

use to, suppose to. *Use to* and *suppose to* are nonstandard for *used to* and *supposed to. Engineers used to* [not *use to*] *use slide rules for complex calculations.*

utilize. A word that means "to put to use." Often *use* can be substituted; *utilize* can make writing sound pretentious. *Control engineers often use* [not *utilize*] *feedback when designing control systems.*

wait for, wait on. *Wait for* means "to await." *The girls are waiting for the commuter train. Wait on* means "to serve." It should not be used as substitute for *wait for. Jeff will wait on our table. We will wait for* [not *wait on*] *the morning bus.*

ways. Colloquial in usage when designating distance. *Denver is quite a way* [not *ways*] *from Buffalo.*

weather, whether. *Weather* is a noun describing "the state of the atmosphere at a given time and place." *The weather in central Canada has been unseasonably warm. Whether* is a conjunction that signals a choice between or among alternatives. *Grif did not know whether to stay or go.*

well, good. See *good, well.*

where. Nonstandard in usage when it is substituted for *that* as a subordinate conjunction. *I read in Wikipedia that* [not *where*] *Clifford Milburn Holland, for whom the Holland Tunnel is named, died before the project was completed.*

whether, if. See *if, whether.*

which. See *that, which* and *who, which, that.*

while. Do not use *while* as a substitute for "although" or "whereas" if such usage risks ambiguity. *The wafer is then cut into small rectangles called dice (die is the singular form of dice, although* [not *while*] *dies is also used as the plural).* If *while* were used, it could mean "although" or "at the same time." *Software engineering promotes the development of high quality software systems, while minimizing and estimating costs and timelines.*

who, which, that. Use *who* not *which* to refer to persons. Use *that* to refer to things. *Alan Arnold Griffith was a British engineer who* [not *that*] *is best known for his work on stress and fracture in metals, concepts that are now known as metal fatigue and fracture mechanics.* However, *that* may be used to refer to a class or group of people. *The design team that wins the first fully autonomous humanoid robot soccer will be famous.*

who, whom. *Who* is used for subjects and subject complements. *Who is coming to dinner? Whom* is used for objects. *He did not know whom to ask.*

who's, whose. *Who's* is a contraction for *who is. Who's going to the dinner? Whose* is a possessive pronoun. *Whose life is it anyway?*

will, shall. See *shall, will.*

would of. Nonstandard English for *would have. He would have* [not *would of*] *achieved a perfect score if he had obtained one more strike.*

you. Avoid using *you* in an indefinite sentence to mean "anyone." *Any collector* [not *You*] *could identify it as a fake.*

your, you're. *Your* is a possessive pronoun. *Your bicycle is in the garage. You're* is a contraction for *you are. You're the first person I contacted about the job.*

7.2 WORDINESS

Effective writing is concise, clear, and direct. Concise writing does not necessarily mean fewer words or shorter sentences. It means words that function clearly and sentences that express their point succinctly. A longer sentence may be succinct if it is required to express a complex idea. When you revise, review each sentence with an eye to eliminating any phrase or word that is not absolutely necessary to your intended meaning.

7.2.1 REDUNDANCIES

Redundancy is the use of unnecessary words in a sentence, the expression of the same idea two or more times:

> It is 6:30 a.m. ~~in the morning.~~

Other common examples of redundancy include *final completion, important essentials, close proximity, consensus of opinion,* and *actual fact.* Here are some additional examples:

> ~~The reason~~ Nebuchadnezzar stopped his conquest ~~was~~ because he heard of his father's death and his own succession to the throne.

> Because the board members did not want to repeat the debate ~~again,~~ they had a frank ~~and honest~~ discussion during which they identified some basic ~~essential~~ ideas.

> When people are in ~~situations of~~ conflict at a meeting, they should circle ~~around~~ the speaker and try to ~~attempt to~~ form a consensus of opinion.

> The bridge ~~that people cross to get~~ to Burlington is ~~sort of~~ rectangular ~~in shape,~~ and ~~it~~ is made of strong materials such as reinforced steel, and concrete~~, and etc~~.

7.2.2 UNNECESSARY REPETITION OF WORDS

Sometimes you may want to repeat words or phrases to create an effect or for emphasis, as in parallel constructions. However, when words are repeated for no apparent reason, they make writing sloppy and awkward. As you revise, eliminate unnecessary repeated words:

> *Unnecessary repetition:* A robot doctor is equipped with microphones, cameras, and databases that allow a doctor and a patient to communicate almost as if <u>the doctor and the patient</u> were in the same room.

Revision A robot doctor is equipped with microphones, cameras, and databases that allow a doctor and a patient to communicate almost as if <u>they</u> were in the same room.

Unnecessary repetition: Proponents of telemedicine argue that the technology will enhance care as long as <u>the technology</u> is used intelligently.

Revision: Proponents of telemedicine argue that the technology will enhance care as long as <u>it</u> is used intelligently.

7.2.3 EMPTY OR INFLATED PHRASES

Some writers, either consciously or unconsciously, attempt to make their writing sound more important by using a style of phrasing that turns out to be empty and inflated. It's easy to fall into this trap because we commonly hear this style of phrasing. However, these phrases are just padding that only increase word count and contain little meaning. Effective writers state what they mean as simply and directly as possible. As you revise your work, trim sentences of any wordy, empty, or inflated phrases:

Empty, inflated phrasing: <u>By virtue of the fact that</u> the 1–1.5% error rate for manual installment of pistons could be reduced to 0.00001% with automation, for example, the automobile and truck industries <u>at the current time</u> are rapidly moving to automated machine installation.

Revision: Because the 1–1.5% error rate for manual installment of pistons could be reduced to 0.00001% with automation, for example, the automobile and truck industries currently are rapidly moving to automated machine installation.

You can use concise words or phrases without affecting your meaning.

ELIMINATING WORDY OR INFLATED PHRASES	
WORDY/INFLATED	**CONCISE**
along the lines of	like
as a matter of fact	in fact
at all times	always

at the present time	now, currently, presently
at this point in time	now, currently, presently
because of the fact	because
being that	because
by means of	by
by virtue of the fact that	because
due to the fact that	because
for the purpose of	for
for the simple reason that	because
have a tendency to	tend to
have the ability to	be able to
in the nature of	like
in order to	to
in spite of the fact that	although, even though
in the event that	if
in the final analysis	finally
in the neighborhood of	about
in the world of today	today
it is necessary that	must
on the occasion of	when
prior to	before
until such a time as	until
with regard to	about

7.2.4 SIMPLIFYING STRUCTURE

The following word-trimming strategies will help you make your sentences simple, clear, and direct.

CONVERT NOMINALIZATIONS; STRENGTHEN VERBS

A **nominalization** is a verb that has been turned into noun. Often these nominalizations can be turned back into verbs to make the sentence more direct and active:

Nominalized verb: During the strike, the <u>accumulation</u> of garbage carried on for fifteen days.

Revision: During the strike, garbage <u>accumulated</u> for fifteen days.

ANALYSIS The noun phrase *the accumulation of garbage* has been turned into the verb *accumulated*.

AVOID COLORLESS VERBS

The verbs *is, are, was, were,* and *have* are weak and often create wordy sentence constructions:

Weak verb: The budget proposal before the legislature <u>is to do with</u> tax cuts and massive reductions in public sector spending.

Revision: The budget proposal before the legislature recommends tax cuts and massive reductions in public sector spending.

ANALYSIS The weak verb *is* has been replaced with the more active verb *recommends*.

REVISE EXPLETIVE CONSTRUCTIONS

An **expletive construction** uses *there* or *it* and a form of the verb *be* in front of the sentence subject. Often these constructions create excess words. You might remove the expletive and revise the sentence to make it more concise and direct:

Wordy expletive: <u>There is</u> a picture of Pierre Trudeau playing baseball <u>that</u> shows the energy he brought to the prime minister's office.

Revision: A picture of Pierre Trudeau playing baseball shows the energy he brought to the prime minister's office.

Wordy expletive: <u>It is</u> remote surgery, minimally invasive surgery, and unmanned surgery <u>that</u> are the three major advances of robot-aided surgery.

Revision: Remote surgery, minimally invasive surgery, and unmanned surgery are the three major advances of robot-aided surgery.

Wordy expletive: <u>It is important that</u> doctors use telemedicine technology to supplement, not replace, in-person visits to patients.

Revision: Most importantly, doctors use telemedicine technology to supplement, not replace, in-person visits to patients.

WHERE POSSIBLE, USE THE ACTIVE VOICE

The active voice is generally more concise and direct than the passive voice. Use the active voice when you want to be direct and to focus on the action of a sentence:

Passive: The research was conducted by members of the robot-soccer development team who were planning to enter the RoboCup.

Active: Members of the robot-soccer development team who plan to enter the RoboCup conducted the research.

7.2.5 REDUCING CLAUSES AND PHRASES

In many instances, modifying clauses and phrases can be tightened. Where possible, reduce clauses to phrases and phrases to single words:

Unnecessary clauses and phrases: RoboCup, <u>which is</u> an international robotics competition founded in 1993, promotes the development of autonomous robots as well as research and education in <u>the field of</u> artificial intelligence.

Revision: RoboCup, an international robotics competition founded in 1993, promotes the development of autonomous robots as well as research and education in artificial intelligence.

Unnecessary clauses and phrases: <u>The ultimate goal of the</u> RoboCup project <u>is</u> to develop a team of humanoid robots that are fully autonomous and that can win against the human world-champion team <u>in soccer</u> by 2050.

Revision: The RoboCup project to develop a team of humanoid robots that are fully autonomous and that can win against the human world-champion soccer team by 2050.

7.3 DICTION AND AUDIENCE

The effectiveness of your writing will in large measure depend on the appropriateness of the language you decide to use for your audience. **Diction** is the business of choosing the wording that best suits the context and the audience of your writing. Consider these elements as you choose your words:

- Subject
- Audience (their needs, expectations, and feelings)
- Purpose
- Voice (as reflected in your unique writing style)

The following section guides you in selecting appropriate language for your writing assignments.

7.3.1 JARGON

Jargon is the specialized language of a particular group or occupation. In this sense, jargon is useful: it often enables specialists to communicate in an abbreviated way. For example, it is much easier to use the term SQUIDS, as opposed to superconducting quantum interference devices. In the context of computer databases, it's easier to talk about "normalizing" data as opposed to making all similar data identical in spelling, capitalization, and spacing. When your audience is familiar with the jargon, using the jargon can make your communications more concise. However, if your audience is not, avoid jargon and use plain English in its place. Moreover, jargon can be used so heavily that it becomes incomprehensible even to those who are familiar with it:

Jargon: Positive input into the infrastructure impacts systematically on the functional base of the organization in that it stimulates meaningful objectives from a strategic standpoint.

Revised: Positive feedback to the organization helps it formulate concrete, strategic objectives.

ANALYSIS Notice that the jargon makes the meaning of the original sentence virtually incomprehensible; the writer needed to rethink these ideas completely and then recast the sentence.

As you can see from these examples, jargon can be misused in an effort to impress readers rather than to communicate information and ideas effectively. Jargon-filled language is often found in business, government, education, and military documents. Sentences containing jargon terms are difficult to read and extremely unclear:

Jargon: The Director of Instruction implemented the optimal plan to ameliorate poor test scores among reading-at-risk students.

Clear: The Director of Instruction carried out the best plan to improve poor test scores among students having trouble reading.

Jargon: We will endeavour to facilitate a viable trash-recovery initiative for all residences in the neighborhood.

Clear: We will try to create a workable garbage pickup plan for all neighborhood homes.

Watch for inflated words or phrases like the following in your documents, and consider alternative words that are simple, clear, and precise in meaning.

ELIMINATING JARGON

WORDS DESIGNED TO IMPRESS	SIMPLE ALTERNATIVE(S)
ameliorate	fix, improve
commence	begin, start
components	parts
endeavor	attempt, try
exit	go, leave
facilitate	help
factor	cause, consideration
finalize	complete, finish
impact on	effect
implement	carry out
indicator	sign

initiate	start, begin
optimal	best
parameters	boundaries, limits
prior to	before
prioritize	order, rank
utilize	use
viable	workable

7.3.2 PRETENTIOUS LANGUAGE AND EUPHEMISMS

AVOID PRETENTIOUS LANGUAGE

When writing for business and technical audiences and purposes, it is tempting to use elevated language. However, using uncommon or unnecessarily long words can highlight rather than obscure deficiencies in content—and make writing seem pretentious. Business, technical, and even academic writing does not require that you use longer, difficult words for their own sake. State your ideas in words that *you* and your audience understand:

Pretentious: Most students who have matriculated from civil engineering school commence their careers with employment positions of minimal responsibility.

Plain language: Most civil-engineering graduates start with jobs of low responsibility.

Pretentious: In the standard lexicon, the term civil engineer has reference to any individual who engages in the practice of civil engineering.

Plain language: The term civil engineer refers to any individual who practices civil engineering.

AVOID EUPHEMISMS

A **euphemism** is a word or expression intended to lessen the impact of harsh or uncomfortable words or phrases. An example of a euphemism in a military context is *collateral damage*, a term sometimes used to describe civilian casualties. While in some writing situations using euphemisms is acceptable, avoid euphemisms because they blur meaning.

AVOIDING EUPHEMISMS	
EUPHEMISM	**PLAIN ENGLISH**
chemical dependency	drug addiction
correctional facility	jail
declared redundant	laid off
developing nations	poor countries
downsizing	laying off or firing employees
economically deprived	poor
incendiary device	bomb
laid to rest	buried
leather-like	vinyl
military solution	war
misleading phrase	lie
pre-owned automobile	used car
starter home	small house
strategic withdrawal	defeat or retreat

7.3.3 SLANG, REGIONALISMS, NONSTANDARD ENGLISH

SLANG

Slang is the informal, colorful vocabulary that is often unique to and coined by subgroups such as teenagers, college students, musicians, skateboarders, computer programmers, street gangs, rap artists, and soldiers. Outsiders are not likely to understand these insiders' slang. Slang can of course become well known, current and trendy, but it quickly becomes overused and cliched. Think of how tiresome phrases like *cool dude* and *check it out* have become as of

2005. While slang can often make story dialog lively and authentic, it is inappropriate in the formal writing of an engineer:

Slang: It's like civil engineering is the profession that makes the world a cooler place in which to hang out.

Revision: Essentially, civil engineering is the profession that makes the world a more agreeable place in which to live.

Slang: Construction engineering is all about doing the up-front skullwork on structures such as highways, bridges, airports, railroads, buildings, dams, and reservoirs and then riding herd on the management of their construction.

Revision: Construction engineering concerns planning and managing the construction of structures such as highways, bridges, airports, railroads, buildings, dams, and reservoirs.

REGIONAL EXPRESSIONS

A **regional expression** is common to a particular area of the country. For instance, in Atlantic Canada, a *barachois* is "a tidal pond partly obstructed by a bar" (*Nelson Canadian Dictionary*, p.108):

Murray could see the skiff beyond the barachois.

Regional expressions, like slang, can add color and authenticity; however, they may not be familiar to a general audience and should be avoided in formal business and academic writing:

Regional expression: After he caught the winning salmon, they threw the fisherman in the salt chuck.

Revision: After he caught the winning salmon, they threw the fisherman in the ocean.

ANALYSIS *Salt chuck* is a regional expression used in British Columbia and the U.S. Pacific Northwest. It might not be known to most people.

Many dictionaries have labels indicating whether a word or expression is regional.

NONSTANDARD ENGLISH

Nonstandard English is acceptable in informal social and regional contexts, but it should be avoided in any formal writing. Examples of nonstandard English include the following words and phrases from the Glossary of Usage:

ain't, anyways, bursted, nowheres, hisself, theirselves

Standard English, on the other hand, is the written English commonly expected and used in businesses, government, educational institutions, and other contexts in which people must formally communicate with one another. Use standard English in all of your business and academic writing. If you are in doubt about whether a word or phrase is standard or nonstandard English, check in the Glossary of Usage in this handbook or in a good dictionary.

Nonstandard: The guy was nowheres in sight, and he could of left town, but she didn't care anyways.

Standard: The man was nowhere in sight. He could have left town, but she did not care anyway.

If you speak a nonstandard dialect, identify how your dialect differs from standard English. In Chapter 8, you can read about language areas that often present writing problems for speakers of nonstandard dialects.

7.3.4 LEVELS OF FORMALITY

Informal writing is casual in language and tone and is appropriate for communication in such forms as friendly letters, e-mails, journal entries, and brief memorandums to people you know very well. **Formal writing** is formal in tone and language and is appropriate for business or academic writing such as essays, research reports, job application letters, and business letters and reports.

In any business or academic writing you do, use a formal level of writing and assume a serious tone. The following opening line of a career application letter is too informal:

Too informal: I'm just dropping you a few lines to put my name in for that engineering job I saw somewhere in the engineerjobs.com a few weeks back.

More formal: I am writing to apply for the Project Engineer-Mechanical/Projects/Modifications (34667) position on engineerjobs.com, dated June 6, 2006.

LANGUAGE THAT IS TOO FORMAL CAN ALSO BE A PROBLEM:

Too formal: When the illustrious Maple Leafs exited from the frozen playing surface trailing their less renowned opponents, the Wild, by the modest score of 1–0, the assembled spectators vigorously voiced their disapproval. The officials in charge of the National Hockey League were authentic demons for having the audacity to schedule these mismatched contests between the annual All-Star Game and the hockey tournament that is part of Olympic competition.

More appropriate: When the Leafs left the ice trailing the Wild 1–0, a smattering of boos rained down from the crowd. The NHL was the real culprit for slipping lopsided games like these between the All-Star Game and the Olympics.

7.3.5 NONSEXIST LANGUAGE

Sexist language is biased in attributing characteristics and roles to people exclusively on the basis of gender. Sexist language can be explicit, as in calling an attractive young woman a *hot chick*. It can be patronizing by referring to a mature woman as a *girl*. It can reflect stereotypical thinking by unnecessarily drawing attention to a person's gender, as in *a female university president*. And sexist language can be subtle, yet still highly biased, by including only male pronouns when more inclusive language is needed—for example, *an athlete always needs to maintain his composure*.

Sexist language can apply to men as well as women; for instance, if a writer describes *a male kindergarten teacher* or a *male nurse*. You can use several strategies to avoid sexist language:

- Treat all people equally in your descriptions of them:

  Unequal treatment: Mr. Delmonico, Mr. Habib, Mr. Dawson, and Tillie, the secretary, arrived for the meeting.

  Acceptable: Mr. Delmonico, Mr. Habib, Mr. Dawson, and Ms. Lord arrived for the meeting.

- Avoid stereotypes:

  Stereotyping: Like all men, he hates to cook.

- Use pairs of pronouns to indicate inclusive gender references:

  Exclusive: A professor is motivated by his students.

  Acceptable: A professor is motivated by his or her students.

- Rewrite the sentence in the plural:

  Acceptable: Professors are motivated by their students.

- Rewrite the sentence so there is no gender problem:

  Acceptable: A professor is motivated by students.

In addition, you can make gender-neutral word choices, such as the following:

AVOIDING SEXIST LANGUAGE

INAPPROPRIATE	GENDER-NEUTRAL
alderman	city council member, councillor
anchorman	anchor
businessman	businessperson, entrepreneur
chairman	chairperson, chair
clergyman	member of the clergy, minister
coed	student
craftsman	artisan, craftsperson
fireman	firefighter
forefather	ancestor
foreman	supervisor
freshman	first-year student
housewife	homemaker
mailman	mail carrier, letter carrier, postal worker
male nurse	nurse
mankind	people, humankind, human

---

manpower	personnel, human resources
newsman	journalist, reporter
policeman	police officer
salesman	salesperson, sales clerk
stewardess	flight attendant
to man	to staff, to operate
weatherman	weather forecaster
waitress	server
workman	worker, laborer, employee

7.4 PRECISION IN LANGUAGE

When trying to choose the most precise word to communicate your meaning, you may find a number of language reference books helpful. Among the most useful will be a good dictionary and a thesaurus.

7.4.1 CONNOTATIONS

Many words have two kinds of meaning: a denotative meaning and a connotative meaning. The **denotative meaning** of a word is its common, literal, dictionary meaning. The **connotative meaning** is the emotional meaning of the word, which includes experiences and associations you make when you see a word in print or hear it spoken. For example, the dictionary meaning of *eagle* is "a large bird of prey." However, the word *eagle* also carries additional emotional and associative meanings such as "power," "pride," "majesty," and "fierceness."

Consider both the denotative and the connotative meanings of the words you use. Some words are loaded with connotations that can imply a meaning you do not intend. For example, "crusade" has extremely negative connotations in the Middle East. Review all listed meanings in a dictionary entry to get a sense of a word's connotations.

The computer industry has a history of using words with unfavorable connotations:

Press Cancel to <u>abort</u> the program.

To <u>invoke</u> the program, press Start.

To <u>execute</u> the command, click Complete.

Entries with spaces in them are <u>invalid</u>.

ANALYSIS The word *abort* obviously connotes abortion; *invoke*, magical incantations and spells; *execute*, execution of prisoners and criminals; and *invalid*, people with disabilities.

7.4.2 CONCRETE NOUNS

Be aware of the types of nouns and use them appropriately.

GENERAL AND SPECIFIC NOUNS
Nouns can be very general or very specific. Suppose someone asks, *What is your project about?* You respond: *I am doing something with artificial intelligence. Artificial intelligence* is a very broad, general area. It would be far less vague and abstract to say that the project involves writing a computer program to enable a vehicle to successfully navigate an obstacle course.

ABSTRACT AND CONCRETE NOUNS
Nouns can be abstract or concrete. Abstract nouns refer to concepts, ideas, qualities, and conditions. They are not concrete; for instance, *love, charity, kindness, humanism, youth*, and *integrity*. Concrete nouns name things that are detectable by your senses; for instance, *snake, dill, sunset, coffee, caramel*, and *harp*.

While creative writers search for precise words to communicate an idea or feeling, in your engineering writing you need precise words to describe, explain, and evaluate general and abstract content. While general and abstract language is sometimes useful, specific and concrete language is more so:

General, abstract words: Hazy city air made things uncomfortable as we put the boat in the water.

Specific, concrete words: Chicago's smog made it difficult to breathe as we launched the sailboat into Lake Michigan.

General abstract nouns, such as *things, considerations,* and *aspects,* are extremely vague and lacking in color.

7.4.3 ACTIVE VERBS

Where possible, choose precise verbs that give your writing impact and power.

WHICH VERBS ARE WEAK?
Weak verbs are forms of the verb *to be* (*be, am, is, are, was, were, being, been*). None of these verb forms communicates a specific action. Also, verbs in the passive voice tend to be lacking in power. Combine the two—the *be* verb and the passive voice—and you have lifeless, uninspiring writing: *An acceptable job was done by her.*

HOW CAN I USE VERBS TO MAKE MY WRITING DIRECT AND EMPHATIC?
Choose precise, vigorous, emphatic, direct, or descriptive verbs in the active voice. In the following examples, the sentence shift from using the verb *be* in the passive voice to a direct, emphatic verb in the active voice:

Passive voice: Written technical proposals for recommended process modifications will be created by the mechanical process engineer.

Active voice: The mechanical process engineer will write technical proposals for recommended process modifications.

*WHEN SHOULD THE **BE** VERB BE REPLACED?*
Change the *be* verb form when it creates a wordy construction. Look for a **nominalization**, word that can be turned into a verb:

Nominalization: Keeping the prisoners in cages would <u>be an infringement of</u> their human rights.

Revision: Keeping the prisoners in cages would infringe on their human rights.

ANALYSIS Using the verb *infringe* is more direct and economical than using an *infringement.*

WHEN SHOULD THE *BE* VERB NOT BE REPLACED?

- You should keep forms of *be* (*be, am, is, are, was, were, being, been*) when you want to link the subject of a sentence with a noun that renames the subject or an adjective that describes it:

    Another reason for loss of prestress <u>is</u> the elastic shortening of concrete and steel caused by compression of the concrete.

    Prestressed concrete <u>is</u> particularly advantageous for beams.

- Keep *be* verbs when they function as helping verbs before present participles:

    The elk <u>are</u> vanishing.

    In this example, the draped arrangement of the tendons <u>is</u> counteracting the diagonal tension near the beam ends.

- The *be* verb forms are acceptable when expressing ongoing action:

    This design <u>is using</u> post-tensioning to fabricate the tendons.

WHEN SHOULD A PASSIVE VERB *BE* CHANGED TO ACTIVE?

With sentences in the active voice, the subject performs the action:

Active voice: José hammered the nail.

With sentences in the passive voice, the subject receives the action:

Passive voice type 1: The nail was hammered by José.

In some passive sentences the subject is not mentioned:

Passive voice type 2: The nail was hammered.

Strong writing clearly states who or what performs actions. Use the active voice by making the person or thing that performs the action the subject of the sentence:

Passive voice: In post-tensioning, the tendons are prevented from bonding to the concrete by encasement in sheaths.

Active voice: In post-tensioning, encasement in sheaths prevents the tendons from bonding to the concrete

ANALYSIS The sentence is more direct and vigorous in the active voice. Right away, we know that encasement prevents bonding.

WHEN SHOULD A PASSIVE VERB BE USED?
Use the passive voice in the following writing situations:

- You want to emphasize who or what receives the action.
- You want to de-emphasize the person or thing that performs the action.
- The person or thing that performs the action is not known.

For example, in the preceding communication situation involving José and the nail, you would select the active voice if you wished to emphasize José. If you wanted to emphasize the importance of the nail being hammered, you would use the passive voice. And if hammering the nail was of central importance and José of no importance whatsoever, or you didn't know who did the hammering, you would use *The nail was hammered.*

7.4.4 MISUSED WORDS

Watch out for misuse of words—it can confuse your overall meaning and create unintentional humor. If you have any doubt or hesitation about the meaning or spelling of a word, check for it in a good dictionary.

Misused word: Burns is <u>conscience</u> of his own powers of destruction.

Revision: Burns is <u>conscious</u> of his own powers of destruction.

Misused word: While producing a product that works every time it comes off the assembly line requires <u>conscious</u> engineering effort, that effort can considerably reduce the costs.

Revision: While producing a product that works every time it comes off the assembly line requires <u>conscientious</u> engineering effort, that effort can considerably reduce the costs.

Misused word: The committee found that she was <u>illegible</u> for the chief administrative position.

Revision: The committee found that she was <u>ineligible</u> for the chief administrative position.

Misused word: Scientists are still trying to <u>dissolve</u> the mystery of the unified theory.

Revision: Scientists are still trying to <u>solve</u> [or <u>resolve</u>] the mystery of the unified theory.

7.4.5 STANDARD IDIOMS

An **idiom** is an expression whose meaning can't be determined by simply knowing the definition of each word within the idiom. An idiom always appears in one particular form, one that may not necessarily be taken literally. An example of an idiom is *beside himself* [or *herself*]. *She was beside herself* means "She was in a state of extreme excitement or agitation."

Using idiomatic expressions with prepositions can be tricky. An unidiomatic expression may make better literal sense, but the idiomatic expression is used because it is accepted English usage. If you are in doubt, check a good dictionary by looking up the word before the preposition.

AVOIDING UNIDIOMATIC EXPRESSIONS	
UNIDIOMATIC	**IDIOMATIC**
according with	according to
angry at	angry with
capable to	capable of
comply to	comply with
desirous to	desirous of
different than	different from
go by	go to
intend on doing	intend to do
off of	off
plan on doing	plan to do
preferable than	preferable to
prior than	prior to
recommend her to do	recommend that she do

superior than	superior to
sure and	sure to
try and	try to
type of a	type of
wait on a person	wait for a person
wait on line	wait in line
with reference in	with reference to

7.4.6 CLICHÉS

A **cliché** is an overused phrase or expression that has become trite and predictable and, hence, is ineffective for freshly communicating writing ideas. Here are some clichés to avoid in your writing:

add insult to injury	at long last	a word to the wise
cool as a cucumber	cold as ice	easier said than done
few and far between	first and foremost	for all intents and purposes
finishing touches	good as gold	hit the nail on the head
in the long run	it stands to reason	narrow escape
red-letter day	this day and age	

If you listen to a sportscaster describing a sporting event on TV, you can quickly construct a sizeable list of sports clichés. Not only are clichés obnoxious, but they also make a reader think that the writers of those clichés cannot think originally or independently. Of course, clichés used self-consciously can inject some fun and freshness into less formal writing:

He is as strong as an ox; unfortunately, that describes his odor, too.

7.4.7 FIGURES OF SPEECH

In figurative language, words carry more than their literal meaning. **Figures of speech** are particular types of figurative language. Common examples of figures of speech are similes, metaphors, and personification. In a **simile**, a comparison is made between two different ideas or objects using *like* or *as*. In a **metaphor**, a comparison is made between two otherwise dissimilar ideas or objects; here, the comparison does not use *like* or *as*. And in **personification**, human traits are assigned to something that is not human.

Used effectively, figures of speech can add color and emphasis to your writing. However, used without care, they can make writing clumsy and too informal. A common writing problem is mixing metaphors. In a **mixed metaphor**, two or more incongruous images are mingled:

Mixed metaphor: She was able to take a firm foothold in the eye of public opinion.

Revision: She was able to keep her focus while in the eye of public opinion.

Mixed metaphor: The Grand Canyon of Harry's depression reached the pinnacle when his pet died.

Revision: Harry reached the depths of depression when his pet died.

7.5 DICTIONARY AND THESAURUS

7.5.1 DICTIONARY

You need at least one good dictionary in your personal reference library. The standards are

- *Merriam-Webster's New Collegiate Dictionary*
- *American Heritage Dictionary of the English Language*

A sample entry from the *American Heritage Dictionary* appears on the next page. The labels indicate the range of information you can obtain from a typical dictionary entry. Watch out for cheap dictionaries. You can spot them by looking at how many entries are associated with words; cheap dictionaries have few entries. Stress-test dictionaries: check to see how detailed they are with problem words like *criteria, dialog, likely* and *apt, between* and *among,* license vs. licence.

Dictionaries also vary in their content and features. It's a good idea to explore the front matter of any dictionary you own for information on what it has to offer and how to use this invaluable resource. In the following excerpt from the *American Heritage Dictionary*, notice how contemporary dictionaries indicate nonstandard usage: the "or" quietly and nonjudgmentally tells you that *criterions* is not standard usage. Notice too that the usage note warns you against using *criteria* as a singular word:

pronunciation guide   nonstandard usage indicator

entry word

cri·te·ri·on (krī-tîr′ē-ən) *n., pl.* -te·ri·a (-tîr′ē-ə) or -te·ri·ons, A standard, rule, or test on which a judgment or decision can be based. [Gk. *kritērion* < *krites, judge* < *krinein*, to separate.] —cri′te′ri·al (-əl) *adj.*

Usage: criteria is a plural form only. It should not be substituted for the singular *criterion*.

part of speech
v = verb
n = noun   usage note   word origins
adj = adjective
prep = preposition
conj = conjunction
pron = pronoun
interj = interjection

SPELLING, WORD DIVISION, PRONUNCIATION

The main entry (*criterion* in the sample entry) shows the correct spelling of the word. When there are two spellings of a word (*pickaxe* or *pickax*, for example) both spellings are given, with the preferred spelling provided first.

If the word is a multisyllabic word (as in *cri-te-ri-on*), the entry shows how to divide the word into syllables. The dot between *cri-* and *te-*separates the word's first two syllables. If a word is a compound word, the main entry shows how to write it: as one word (*poolroom*), as a hyphenated word (*pooper-scooper*), or as two words (*poop deck*).

The pronunciation of the word is given just after the main entry. If the word is a multisyllabic word, accents indicate which syllables are stressed. Other marks help the reader pronounce the word. These marks are explained in a pronunciation key. In this dictionary, the pronunciation key is in the lower far right-hand corner of the two-page spread. The placement of this feature varies from dictionary to dictionary.

WORD ENDINGS AND GRAMMATICAL LABELS

When a word takes endings to indicate grammatical functions, which are called **inflections**, the endings are listed in boldface. In a dictionary, you will see three inflections listed for **poor: poor·er, poor·est**, and **poor·ness**.

The labels for the parts of speech and for other grammatical terms used in entries are abbreviated. The most commonly used dictionary abbreviations are

n.	noun
pl.	plural
sing.	singular
v.	verb
tr.	transitive (as in transitive verb)
intr.	intransitive (as in intransitive verb)
adj.	adjective
adv.	adverb
pron.	pronoun
prep.	preposition
conj.	conjunction
interj.	interjection

MEANINGS, WORD ORIGIN, SYNONYMS, AND ANTONYMS

Each word meaning is given with a separate number, and the most common meanings are listed first.

Some words can function as more than one part of speech (*positive*, for example, can be a noun or an adjective). In such cases, all meanings for one part of speech are given, then all meanings for another part of speech, and so on. After the final meaning, any idioms containing the word are given. For example, with the entry for *pop*, the idiom *pop the question* is listed; as well, the idiom's meaning is provided. In square brackets at the end of the entry appears the *etymology*, or information about the origins of the word. According to the etymology for *poor*, the word originated from the Middle English word *poure*, as well as from Old French and Latin.

Synonyms, or words with similar meaning to the main entry word, are listed for some dictionary entries. *Indigent* is listed as one of seven synonyms for *poor*.

Antonyms are words that have a meaning opposite to that of the main entry word. For example, *poor* is an antonym of *rich*.

USAGE

Usage notes follow many entries in this dictionary. These notes present important information and guidance on matters of grammar, diction, pronunciation, and nuances of usage.

In the *American Heritage Dictionary*, status labels indicate that an entry word or an entry is limited to a particular level or style of usage. These labels include *archaic, slang*, and *informal*. Usage labels can vary among dictionaries. Seeking to show respect for cultural diversity, dictionaries do not mark the words and phrases of different regional, folk, or ethnic groups as incorrect or nonstandard. Instead, the nonstandard word is preceded by *var, variant, or*, or *also*.

7.5.2 THESAURUS

You may find yourself in a writing situation in which you know a word but want to find a more precise or colorful word with the same or a similar meaning. Or, you have repeatedly used a word within a paragraph or even a sentence and don't want your writing to sound mechanical. *Roget's International Thesaurus* is an excellent resource for finding synonyms, or words with a similar meaning.

HOW DO I USE THE THESAURUS?

Suppose you want to find a synonym for *abundance*. Your first step is to find *abundance* in the extensive index at the back of the thesaurus. Of the possibilities listed in the entry, the closest to the meaning and part of speech you want is *plenty*. Turn to the Abstract Relations section at the front of the thesaurus to the number listed in the index beside *plenty*, under *abundance*.

A possible word you may want to use as an alternative to *abundance* is *cornucopia*. But read about *cornucopia* in your dictionary: it may have meaning you don't want. As you explore other word choice options in the thesaurus, always double-check their meaning in your dictionary.

Use the thesaurus to locate the best word. Some writers use a thesaurus to find the most difficult or exotic synonym possible. Always strive for simplicity and clarity in your writing. If you use the thesaurus to find inflated vocabulary, often you'll risk misusing words, and you may also make your writing seem pretentious. For example:

Pompous word choice: Our department is audited every *lustrum*.

Revision: Our department is audited every *five years*.

8

ESL

8

ESL

8   ESL

This section is written especially for readers whose first language is not English. Covered here are some specific features of English that are typically problematic and may cause errors in usage.

8.1 ARTICLES

The articles in English are *a*, *an*, and *the*. They are **determiners**, which signal that a noun will follow and that any modifiers appearing before that noun refer to that noun. *A* and *an* are the indefinite articles; *the* is the definite article.

Definite article: the television, the portable television

Indefinite article: a cat, a Siamese cat

Other determiners that may mark nouns include:

- Possessive nouns

  David's car, Mom's birthday

- Numbers

  two cats, 100 balloons

- Pronouns

my	your	his	her
its	our	their	whose
this	that	these	those
a lot of	a great deal of	a number of	all
any	both	each	either
enough	every	few	many
more	most	much	neither
no	several	some	

8.1.1 WHEN TO USE *A* (OR *AN*)

If the noun is a singular count noun, use *a* or *an* before the noun. To decide whether the noun is a singular count noun, use the following checklist:

- Is it a common noun (rather than a proper name)?
- Does the noun refer to a person, place, or thing that can be counted, such as *one woman, one city, two bowls?*
- Does the noun name something unknown to the reader because it is being used for the first time or because its specific identity is not known even to the writer?

If the answer to all three of these questions is yes, then use *a* or *an* before the noun. To choose between *a* and *an*, use this checklist:

- Use *a* if a consonant sound follows the article.
- Use *an* if a vowel sound follows the article, remembering that *h* or *u* may make either vowel or consonant sounds:

  h, u as consonants: a Himalayan cat, a horrible incident, a unicycle, a union member

  h, u as vowels: an honorable act, an honest man, an unlikely story, an unfamiliar face

Note: A or *an* usually means "one among many" but may simply mean "any."

Note also: Some **collective nouns** (nouns that name a group of things) are always treated as plural. These include *clergy, military, people, police.* To refer to one member of the group, use a singular noun with these collective nouns, such as *a member of the clergy, a military officer, a man of the people, a police officer.*

8.1.2 WHEN NOT TO USE *A* (OR *AN*)

If the noun is a noncount noun, do not use *a* or *an* before the noun. To decide whether a noun is a noncount noun, determine whether the noun refers to an entity that cannot be counted, such as *philosophy, ice,* or *fatigue.*

To express a particular amount of a noncount noun, you can modify it by using another determiner, such as *some, any,* or *more.* Remember:

- A *few* or *a little* means "some," whereas *few* and *little* mean "almost none."
- Use *much* with noncount nouns and *many* with count nouns.
- Use *an amount of* with noncount nouns and *a number of* with count nouns.
- Use *less* with noncount nouns (*less money*) and *fewer* with count nouns (*fewer hours*).

NONCOUNT NOUNS

GROUPS OF ITEMS THAT MAKE UP A WHOLE
baggage, money, silver, research, furniture, mail, clothing, real estate

ABSTRACT NOUNS
wealth, awareness, joy, discord, esteem

LIQUIDS
tea, milk, water, beer

GASES
smoke, steam, oxygen, air, fog

MATERIALS
wood, steel, wool, gold

FOOD
pork, pasta, butter, salmon

GRAINS OR PARTICLES
wheat, dust, rice, dirt

SPORTS OR GAMES
rugby, hockey, chess, poker

LANGUAGES
French, English, Mandarin, Farsi

FIELDS OF STUDY
chemistry, engineering, nursing

NATURAL EVENTS
lightning, cold, sunlight, darkness

You can also add a count noun in front of a noncount noun to make it more specific: a jug of wine, a piece of jewelry, a game of bridge.

> The amount of seawater being converted to potable water had increased to 22,700,000 m^3/day (6,000,000,000 gal/day) by 1998.

Note: Some noncount nouns can also be used as count nouns, depending on their meaning. This is usually the case when the noun can be used individually or as part of a larger whole made up of individual parts.

> Wind-energy <u>technology</u> is primarily focused on the generation of electric power.

> However, it is <u>a technology</u> that can be used in water desalination.

Invest in a dictionary that tells you whether a noun is noncount or count.

8.1.3 WHEN TO USE *THE*

Use *the* in the following situations:

• When or if the noun has been previously mentioned:

> When a salt solution is separated from pure water by a semi permeable membrane, water tends to diffuse through the membrane into the salt solution.

ANALYSIS The noun phrase *salt solution* is preceded by *A* when first named. When named again, it is preceded by *the* since you now know its specific identity.

Note: Use a after the first mention of a noun if a descriptive adjective comes between the article and the noun.

Distillation is a process that turns seawater into vapor by boiling and then condenses that vapor to produce potable water. It is an energy-intensive process, requiring about 2261.2 kilojoules of energy to convert water to vapor.

• When a modifying word, phrase, or clause after the noun restricts its meaning:

The program *you are about to see* has been edited for television.

ANALYSIS The clause *you are about to see* restricts the meaning of the word *program*. In other words, it identifies the specific program.

• When a superlative makes *the* specific:

The multistage flash distillation process is currently the most popular desalination process.

ANALYSIS The superlative *most popular* restricts the identity of the noun *process*.

• When the noun is a unique person, place, or thing:

Some people still maintain that the Earth is flat.

ANALYSIS *The Earth* refers to this specific planet.

• When the context makes the noun's specific identity clear:

One major problem with the multistage flash distillation process is scale formation on the equipment surface.

ANALYSIS The context makes it clear that this particular desalination process has this problem.

Note: In phrases beginning with *one of the*, the noun that follows must be in the plural.

The multistage flash distillation process is currently one of the most popular desalination processes (not *process*).

8.1.4 WHEN NOT TO USE *THE*

Do not use *the*:

- When the reference is to a general group of things:

  Elephants are mammals.

  ANALYSIS *The* is not needed because the statement refers to all elephants,—that is, to elephants in general.

- Don't use *the* with most proper nouns. **Proper nouns** name specific persons, places, and things and begin with a capital letter. *The* is not used with proper names, except in the following cases.

8.1.5 WHEN TO USE *THE* WITH PROPER NAMES

Although the general rule is not to use *the* with proper names, these are the exceptions:

- With nouns that follow the pattern of *the . . .of . . .*

  the President of the United States, the state of Iowa

- With plural proper nouns

  the Barrets, the Chicago Cubs, the United States

- With collective proper nouns

  the Santa Fe Opera Company, the Federal Bureau of Investigation, the Supreme Court

- With some proper names of geographical features, such as large areas, deserts, mountain ranges, peninsulas, oceans, seas, gulfs, canals, and rivers, and with names of ships:

  the West Coast, the Sahara Desert, the Pyrenees, the Iberian Peninsula, the Pacific Ocean, the St. Lawrence Seaway, the Gulf of Mexico, the Suez Canal, the Ganges, the *Titanic*

8.2 VERBS

All writers in English encounter problems with verbs. In addition to the material below, helpful information will be found in this chapter.

The verb in a sentence may be one verb or a phrase made up of a main verb and a helping verb. Every main verb (with the exception of *be*) has five forms that are used to create tenses:

Base (simple) form: work, eat

Past tense: worked, ate

Past participle: worked, eaten

Present participle: working, eating

-s form: works, eats

Note: The -s form is used with the third-person singular in the present tense.

Work is an example of a regular verb; *catch* is an example of an irregular verb.

8.2.1 HELPING VERBS

Helping (auxiliary) verbs always come before main verbs:

Japanese engineers *have developed* Wakamaru, a humanoid robot intended for elderly and disabled people. However, these robots *do* not display actual intelligence.

Some helping verbs—*be*, *do*, and *have*—are used to conjugate verbs into their various tenses. *Be*, *do*, and *have* thus change form.

FORMS OF *BE*, *DO*, and *HAVE*

be, am, is, are, was, were, being, been

do, does, did

have, has, had

MODALS

Other helping verbs, called **modals**, usually do not change form to indicate tense or to indicate singular or plural.

- One-word modals do not change form: can, could, may, might, must, shall, should, will, would, ought to.

- Most two-word modals do change form to indicate tense or to indicate singular or plural: need to, has to.

Note: Modals do not require the word *to* in front of the base verb, except in the case of *need to, has to,* and *ought to.*

USE OF BE *TO FORM THE PROGRESSIVE TENSE*
The **progressive tense** is used to indicate action that is in progress. Create the progressive tense by using *be, am, is, are, was, were, been,* or *will be* followed by the present participle of the main verb.

Present progressive: Robots are *being* used to defuse roadside bombs.

Past progressive: Engineers *have been attempting* to develop robots with natural gaits for years.

Future progressive: Some engineers believe that robots *will be demonstrating* human-like intelligence in this century.

Present perfect progressive: The automobile industry has been using robots for years.

Past perfect progressive: By 2005, over a hundred Wakamaru robots *had been reminding* elderly and disabled people in Japan to take their medicine and calling for help if something seemed wrong.

Future perfect progressive: Despite that reluctance, soon viewers *will have been watching* animated films for almost a century.

Always make sure to use the *-ing* form to create the progressive tense.

Incorrect: Robots *are compete* in a wide range of tasks including combat, games, maze solving, and navigational exercises.

Correct: Robots *are competing* in a wide range of tasks including combat, games, maze solving, and navigational exercises.

Note: *Been* and *be* need other helping verbs to form the progressive tense. *Been* requires either *have, has,* or *had* to form the tense. *Be* forms the progressive tense with one of the modal verbs.

Some verbs are not commonly used in the progressive tense in English. These are considered **static** verbs, which describe a state of

being or a mental activity: appear, believe, belong, contain, know, seem, think, understand, want.

> *Incorrect:* The word robot is referring to machines that are capable of movement and can perform physical tasks.

> *Correct:* The word robot refers to machines that are capable of movement and can perform physical tasks.

USE OF BE *TO FORM THE PASSIVE VOICE*

A verb is in the **passive voice** when its subject is the receiver of the action, rather than the performer of the action:

> The word cryptography *is derived* from Greek κρυπτός, meaning "hidden," and from γράφειν, meaning "to write."

The passive voice is formed with one of the forms of the verb *be* and the main verb's past participle.

> *Passive voice, present tense:* Cryptography *is being used* in everyday applications such as ATM card security, computer passwords, and electronic commerce.

> *Passive voice, past tense:* Until recently, cryptography *had been used* in the sense of encryption, the process of converting ordinary information into an unreadable ciphertext.

> *Passive voice, future tense:* For an attack against an application using the Data Encryption Standard, one known plaintext and 2^{55} decryptions will be required.

The passive voice can occur in any tense of a transitive verb (that is, a verb that takes an object to complete its meaning). However, if a verb is intransitive, it cannot be used to form the passive voice:

> The accident happened last night.

> The accident *was* last night.

Note: When *be, being,* or *been* is used to form the passive voice, it must be preceded by another helping verb. *Been* must be preceded by a form of *have; being* must be preceded by a form of *to be,* and *be* must be preceded by a modal verb.

USE OF DO *TO FORM QUESTIONS AND FOR EMPHASIS*

The **auxiliary verbs** *do, does*, and *did* are used to indicate questions, negatives, and emphasis. The verb *do* is always used with base forms of the main verb that do not change.

It *does* not look like rain.

Did you ever hear anything like that in your life?

I *do* believe that you will succeed in your endeavors.

In the preceding examples, *look, hear,* and *believe* are base forms of the main verb.

USE OF HAVE *TO FORM THE PERFECT ASPECT*

After the **helping verbs** *have, has,* or *had*, use the past participle to form the perfect forms. Note that past participles frequently end in *-d, -ed, -en, -n,* or *-t*.

Though Yuen *has* never *lived* in China, Cantonese is his native language.

Before 1997, few people knew that asymmetric cryptography *had* been *invented* by James H. Ellis at GCHQ, a British intelligence organization, in the early 1970s.

Have, has, and *had* may also be preceded by modal verbs, such as *will*, used in forming the future perfect:

In the process of decryption, a key *will be needed*.

MODAL VERBS BESIDES BE, DO, *AND* HAVE

Use **modal auxiliary verbs** before main verbs to indicate ability, necessity, advisability, and probability.

- *Can* is used to express ability in the present or future tense. *Could* is used to express ability in the past tense:

    Statistical analysis of ciphertexts produced by classical ciphers *can reveal* enough clues to enable an informed hacker to break the code.

    The Digital Millennium Copyright Act criminalized all cryptanalytic activities that could circumvent digital rights management methods.

- *Must, have to,* and *need to* express the necessity of doing something. *Must* is used in the immediate present and the future tense. The others are used in all verb tenses:

  In symmetric-key systems, the communicating parties *must share* a different key.

  In symmetric-key systems, the communicating parties *will have to share* a different key.

  In symmetric-key systems, the communicating parties *need to share* a different key.

- *Should* and *ought to* indicate that a certain action is advisable or expected in the present or the future. The past tenses are *should have* and *ought to have. Had better* implies advice with a warning:

  You *should use* a public-key cryptography system. It *should provide* better security for your operations in the future. I *ought to have advised* you earlier. You *had better make* this change right away.

- *May, might, could,* and *must* sometimes express probability or possibility. To form the past tense, add *have* and the past participle of the main verb after the modal. *Could* or *might* are also used for polite requests:

  You *may regret* continuing to use symmetric-key cryptography. You *could start* the process of switching to a public-key cryptography system today. You *might have* been hacked already.

- *Will* and *would* suggest promise or agreement. *Would* may also be used for polite requests:

  Any attempt to move a digital signature from one document to another *will be* detected. *Would* you please *monitor* for such attempts?

- *Would rather* is used to express preference. To form the past tense, add *have* and the past participle of the main verb after the modal:

  Modern cryptographers *would rather* hide the meaning of a message than its existence.

- *Be supposed to* indicates that the subject has a plan or an obligation. This modal can be used in the present and in the past tense:

  He *was supposed to* arrive at work by 8:30.

• *Used to* and *would* indicate that a repeated action or habit is now past. Only *used to* can express an action that lasted a length of time in the past:

> I *used to* smoke. I *would have* a cigarette as soon as I woke up in the morning.

> *Incorrect:* I would live in the Middle East.

> *Correct:* I *used to* live in the Middle East.

> ANALYSIS The "incorrect" version would be correct if the writer meant that he would live in the Middle East if certain conditions were right.

8.2.2 VERBS FOLLOWED BY GERUNDS OR INFINITIVES

Make a point of learning which verbs are followed by gerunds and which verbs by infinitives. A **gerund** is a form of the verb that ends in *-ing* and is used as a noun.

> The cost of wind energy technology will continue *declining* as larger multi-megawatt turbines are mass produced.

> One wind-power company is *attempting to commercialize* tethered aerial turbines suspended with helium to take advantage of high-speed winds at high altitude.

An **infinitive** is the base form of the verb preceded by *to*, which marks its use as an infinitive. *To* in front of the base form of a verb is not a preposition, but an indication of the infinitive use.

> The cost of wind energy technology will continue to decline as larger multi-megawatt turbines are mass produced.

> In wind farms that contain many turbines, adequate spacing enables the turbines to harvest more wind energy.

A few verbs, such as *like*, can be followed by either a gerund or an infinitive, although some of those change their meaning as a result. Other verbs take a gerund, but never an infinitive; others

take an infinitive (some with an intervening noun phrase), but not a gerund.

- Some verbs can take either gerunds or infinitives. These common verbs may take either a gerund or an infinitive, with negligible or no differences in meaning. Here are just a few examples:

begin	go	love
can't bear	hate	prefer
can't stand	like	start
continue		

Infinitive: The opponents of wind-generated electricity *can't bear <u>to think</u>* about the birds killed by wind turbines.

Gerund: The opponents of wind-generated electricity *can't bear <u>thinking</u>* about the birds killed by wind turbines.

- Some other verbs can take both gerunds and infinitives, but their meaning changes significantly in that case:

forget	remember	stop	try

Gerund: Opponents of wind-generated electricity want us to *stop building* wind farms.

Infinitive: But few have stopped *to think* about how birds are killed by automobiles.

- Some verbs take only gerunds:

acknowledge	discuss	object to
admit	dislike	postpone
adore	dream about	practice
advise	enjoy	put off

anticipate	escape	quit
appreciate	finish	recall
avoid	give up	recommend
can't help	have trouble	resent
complain about	imagine	resist
consider	include	risk
consist of	insist on	suggest
contemplate	keep	tolerate
delay	mention	understand
deny	mind	detest
miss		

Gerund: The World Wind Energy Association *recommends getting* the local population involved in the development of wind farms.

- Some verbs take only infinitives:

afford	demand	pretend
agree	deserve	promise
aim	expect	refuse
appear	fail	say
assent	give permission	seem
arrange	have	struggle
ask	hesitate	tend
attempt	hope	threaten
be able	intend	volunteer
be left	know how	wait
beg	learn	want

care	manage	wish
claim	mean	would like
consent	offer	
decide	plan	
decline	prepare	

Infinitive: The wind-energy development company *plans to invite* the local population to invest in the project.

- Some verbs require infinitives but need a noun or pronoun in between. Some verbs in the active voice usually require an infinitive, but a noun or pronoun must come between the verb and the infinitive. The noun or pronoun usually names a person affected by the action in the sentence.

admonish	forbid	persuade
advise	force	remind
allow	have	request
cause	instruct	teach
challenge	invite	tell
command	hire	urge
convince	oblige	warn
dare	order	
encourage	permit	

Infinitive: The World Wind Energy Association is attempting *to persuade* the public <u>to support</u> the growth of wind-generated power.

- A few verbs can take an infinitive *either* directly or with a noun or pronoun in between:

ask	need	would like
expect	want	

Direct infinitive: The World Wind Energy Association would like to see 120,000 MW of wind-generation capacity installed worldwide by 2010.

Infinitive preceded by pronoun: By 2010, the World Wind Energy Association *expects* 120,000 MW of wind-generation capacity *to be installed* worldwide.

- "Sense verbs" take a noun or pronoun with an unmarked infinitive or gerund. Some verbs, called "sense verbs," take a noun or pronoun with an **unmarked infinitive** (the base form of the verb without *to*) or a gerund:

feel	listen to	see
watch	have	look at
notice	hear	make (meaning "force")
let		

Here are some examples of unmarked infinitives:

Watch that horse *run*.

Watch that horse *running*.

Note: The verb *help* may be used either with a marked or unmarked infinitive:

Marked infinitive: Help the homeless to find shelter.

Unmarked infinitive: Help the homeless find shelter.

8.2.3 PHRASAL VERBS

Make a point of learning how to use phrasal verbs idiomatically. **Phrasal verbs** (also called two-word verbs) consist of verbs with prepositions or adverbs, known as **particles**. Phrasal verbs have distinctive idiomatic usage that requires careful study since they cannot be understood literally. Most phrasal verbs can be separated to allow a noun or pronoun object to come between them. When the verb can be separated from its particle, a pronoun object must always be placed between them. In the following sentences are the phrasal verbs that can be separated:

Turn on the lamp.

Turn the lamp on.

Turn in the application.

Turn the application in.

Turn down the offer.

Turn the offer down.

Other phrasal verbs cannot be separated:

Incorrect: The engineers went the design over with the client.

Correct: The engineers *went over* the design with the client.

COMMON PHRASAL VERBS

Here is a list of common phrasal verbs. If the particle (preposition or adverb) cannot be separated from the verb by a direct object, it is marked with an asterisk (*). Check meanings of phrasal verbs carefully in the dictionary if you have doubts about their usage.

ask out	burn down	call up
break down	burn up	clean up
bring about	call back	come across
bring up	call off	cut up
do over	help out	speak to*
drop in	keep on	speak up
drop off	keep up with*	speak with*
figure out	leave out	stay away from*

fill out

fill up

get along
   with*

get away
   with*

get back

get off

get up
   (intransitive)

give away

give back

give in
   (intransitive)

give up

go out

go over*

grow up
   (intransitive)

hand in

hand out

hang on
   (intransitive)

hang on to

hang up

look after*

look around*

look into*

look out for*

look over

look up

make up
   with*

pick out

pick up

play around

point out

put away

put back

put off

put on

put out

put together

put up with*

quiet down

run across*

run into

run out*

see off

shut off

stay up*

take care of*

take off

take out

take over

take up

talk about*

talk to*

talk with*

talk over

think about*

think over

think up

throw away

throw out

try on

try out

turn down

turn on

turn out*

turn up

wake up

wear out

wrap up

8.2.4 OMITTED VERBS

Verbs such as *to be* must be included in all English sentences. They cannot be omitted as they can in some languages:

Incorrect: Dolores very industrious.

Correct: Dolores *is* very industrious.

Incorrect: Wayne Gretzky in Manhattan.

Correct: Wayne Gretzky *lives* in Manhattan.

Incorrect: Roads in Montana wide.

Correct: Roads in Montana *are* wide.

8.2.5 CONDITIONAL SENTENCES

Conditional sentences state a relationship between one set of circumstances and another. Conditional sentences can be used to express a cause-and-effect relationship that is factual, to predict future possibilities, or to speculate about what might occur or might have occurred.

- Use the conditional to express factual relationships. Such sentences may express scientific truth, or they may simply describe something that usually or habitually happens:

   Present conditional: When a leap year occurs, *there are* 366 days in the year.

   Past conditional: Whenever a leap year occurred, *they celebrated* on the 29th day of February.

- Use the conditional to predict the future and to express plans or possibilities: Normally, use the present tense in the subordinate clause and the future tense in the main clause.

   If the signal strength of a transmitter is insufficient, noise *will corrupt* the information in the signals.

- Sometimes a modal verb, such as *may, might, can, could,* or *should* is used in the main clause instead of a future tense:

   If you become an electrical engineer, you *might want* to work in the telecommunications field.

319

- Use the conditional to speculate about unlikely events in the future. The verb in the main clause is usually a modal verb, such as *will*, *can*, *could*, *may*, *might*, *should*, followed by the base form of the main verb:

If you know your exact location on Earth, you *would be able* to pinpoint where a satellite is in its orbit by measuring the Doppler distortion.

If you knew your exact location on Earth, you *could pinpoint* where a satellite is in its orbit by measuring the Doppler distortion.

- Use the conditional to speculate about events that did not happen, that are hypothetical or that are contrary to fact:

If Allied forces had not broken the code messages of the Nazi Wehrmacht Enigma machine, World War II *might have* lasted another two years.

If Polish cryptologist, Marian Rejewski, had not developed the "cryptologic bomb" in 1938, the Allies *would not have been able* to break German Enigma machine ciphers.

According to British Prime Minister Winston Churchill, if the Allies had not had Ultra, they *would not have* won the war.

If she *were* a mathematician, she could work in the field of cryptanalysis.

Note: The subjunctive mood is used in the *if* clause; the subjunctive is used when you are describing a hypothetical case or something contrary to fact.

8.2.6 INDIRECT QUOTATIONS

An **indirect quotation** reports what someone said or wrote but with slight changes in verb tense and without quotation marks. Indirect quotations usually occur in subordinate clauses. When the present tense is used in the main clause, the verb in the subordinate clause is in the same tense as the original quotation:

Indirect quotation: In this web article, Tim Wilkinson of the University of Sydney *states* that, after the initial plane impacts, most observers *did* not initially *believe* that the damage to the World Trade Center was enough to cause collapse.

Original quotation: According to Tim Wilkinson of the University of Sydney, "Initially this [damage to the World Trade Center] was not enough to cause collapse."

When the past tense is used in the main clause, the verb in the subordinate clause changes tense from the original quotation. The past tense and the present perfect tense change to the past perfect tense; however, the past perfect tense does not change:

Indirect quotation: Historian Lewis Mumford denounced the World Trade Center as an example of the "purposeless giantism and technological exhibitionism" that was ruining every great city.

Original quotation: Historian Lewis Mumford denounced the World Trade Center as an "example of the purposeless giantism and technological exhibitionism that are now eviscerating the living tissue of every great city."

When the direct quotation states a general truth, or reports a situation that is still true, use the present tense in the indirect quotation regardless of the verb in the main clause:

Indirect quotation: Canada's Prime Minister Pierre Trudeau *said* that the state *has* no place in the nation's bedrooms.

Original quotation: "The state has no place in the nation's bedrooms."

Note: Indirect quotations can include direct quotations of distinctive words from the original and are usually introduced by *that*.

8.3 PROBLEMS WITH SUBJECTS AND OBJECTS

Special problems for non-native speakers and writers of English also occur with grammatical subjects and objects.

8.3.1 OMITTED SUBJECTS; OMITTED *THERE* OR *IT*

English requires a subject in all sentences except for commands, in which the subject *you* is understood. If your first language does omit subjects in some cases, pay particular attention to this usage in English:

Incorrect: Have a diploma in accounting.

Correct: I have a diploma in accounting.

Incorrect: Your daughter is accomplished; seems very gifted.

Correct: Your daughter is accomplished; she seems very gifted.

If a subject is placed in its normal position in front of a verb, it requires an expletive pronoun (*there* or *it*) at the beginning of the clause. *There* used in this way points to the location or existence of something:

Incorrect: Is a letter in the mailbox.

Correct: There is a letter in the mailbox.

Incorrect: As I have explained, are many reasons for my decision.

Correct: As I have explained, there are many reasons for my decision.

Note that the verb after *there* agrees with the subject that follows it: *letter is, reasons are.*

The word *it* may also function as an expletive, calling attention to something and introducing it in an impersonal way:

Incorrect: Is important to get a good education.

Correct: It is important to get a good education.

Incorrect: Is clear that he must arrive on time.

Correct: It is clear that he must arrive on time.

The word *it* is also used as the subject in sentences that

- Describe temperature or weather conditions:

  It snows less in North Dakota than people think.

  In the summer, *it* can be extremely warm.

- State the time:

  It is midnight.

- Indicate distance:

  It is a long way from Houston to El Paso.

- State a fact about the environment:

  It gets busy on the highways on the long weekend.

8.3.2 REPEATED SUBJECTS

Do not restate a subject as a pronoun before the verb. State the subject only once in the clause:

Incorrect: The temperature it reached 30 degrees C.

Correct: The temperature reached 30 degrees C.

Incorrect: The professor she gave a lecture on globalization and the new media.

Correct: The professor gave a lecture on globalization and the new media.

Note that there is no need for a pronoun, even if other words intervene between subject and verb:

Incorrect: The letter I received today it brought good news.

Correct: The letter I received today brought good news.

8.3.3 REPEATED OBJECTS AND ADVERBS IN ADJECTIVE CLAUSES

Adjective clauses begin with relative pronouns: *who, whom, whose, which, that, where(ever)*, or *when(ever)*. The first word of an adjective clause replaces another word, either the subject, an object, or a pronoun.

Make sure not to restate the word being replaced in the adjective clause. Such repetition occurs in other languages, but never in English:

Incorrect: He usually works at the desk that I am sitting at it.

Incorrect: He usually works at the desk I am sitting at it.

Correct: He usually works at the desk I am sitting at.

Correct: He usually works at the desk at which I am sitting.

ANALYSIS The pronoun *that* replaces *desk* in the adjective clause; hence it is not needed as the object after the preposition *at*. Even when the pronoun *that* is left out, it is still understood.

Adverbs in adjective clauses, like relative pronouns, do not need to be repeated:

Incorrect: The city where she lives there is accessible by bus or train.

ANALYSIS The adverb *there* is not needed in an adjective clause beginning with *where*.

8.4 GENDER IN PRONOUN–ANTECEDENT AGREEMENT

In English, the gender of a pronoun should match its antecedent (that is, the noun to which it refers) and not a noun that the pronoun may modify.

Incorrect: Kevin gave the diamond ring to her fiancée.

Correct: Kevin gave the diamond ring to his fiancée.

Note: Nouns in English are neuter unless they specifically refer to males or females. Hence, nouns such as *chair, newspaper, moon*, and *sun* take the pronoun *it*.

Incorrect: Kevin gave him to his fiancée.

Correct: Kevin gave it to his fiancée.

8.5 PROBLEMS WITH ADJECTIVES AND ADVERBS

Placement of adjectives and adverbs can be a problem for non-native speakers and writers of English, as well as the use of participles as adjectives.

8.5.1 PLACEMENT OF ADJECTIVES

In English, adjectives normally precede the noun, though they may also follow linking verbs. The list that follows shows the proper word order for cumulative adjectives (those not separated by commas); for example, *I ate another five beautiful round ripe black Spanish olives*. Note that some exceptions do occur.

Remember that long lists of adjectives in front of a noun may be awkward. Try to use no more than two or three of them between the determiner and the noun itself.

Word Order for Cumulative Adjectives

1. **Determiners, if there are any**
   a, an, the, my, your, Canada's, those

   2. **Expressions of order, including ordinal numbers, if any**
      first, second, next, final

      3. **Expressions of quantity, including cardinal numbers, if any**
         one, two, each, some, all

         4. **Adjectives of opinion, if any**
            beautiful, fascinating, ugly, dull

            5. **Adjectives of size or shape, if any**
               tiny, huge, tall, rotund, triangular

               6. **Adjectives of age and condition, if any**
                  brand-new, ancient, ripe, rotten

                  7. **Adjectives of color, if any**
                     yellow, purple, magenta, chartreuse

                     8. **Adjectives of nationality, if any**
                        Chinese, Portuguese, German

                        9. **Adjectives of religion, if any**
                           Muslim, Jewish, Protestant, Catholic

                           10. **Adjectives of material, if any**
                               gold, mahogany, silk, wood

                               11. **Adjectives that can also be used as nouns, if any**
                                   business, English, government

                                   12. **The noun being modified**

8.5.2 PLACEMENT OF ADVERBS

Adverbs and adverbial phrases are flexible in English, and they can appear at the beginning, middle, or end of a clause:

Correct: Because the planes had *only recently* taken off, the fire was *initially* fuelled by large volumes of jet fuel.

Correct: Because the planes had taken off *only recently,* the fire was fuelled initially by large volumes of jet fuel.

An adverb may not be placed between a verb and a direct object, however:

Incorrect: Carrying nearly 24,000 U.S. gallons of jet fuel, each aircraft became *effectively* a guided missile.

ANALYSIS   The adverb *effectively* must be placed either at the beginning or the end of the sentence or immediately before the verb. It cannot appear after the verb because the verb is followed by the direct object *a guided missile.*

Placement of adverbs in English sentences depends on the type of adverb:

- Adverbs of **manner** describe how something is done. They usually appear in the middle of the clause or at the end.

  On September 11, 2001, the Pentagon was *severely* damaged by fire, and one section of the building collapsed.

- Adverbs of **time** describe when an event takes place or how long it lasts. They usually go at the beginning or end of a clause:

  When the crew and passengers tried to seize control of the plane, the hijackers *then* rocked the plane, attempting to throw the passengers off balance.

- Adverbs of **place** describe where an event occurs. They usually go at the end of a clause:

  The fourth hijacked aircraft crashed *into a field* in rural Somerset County, Pennsylvania.

- Adverbs of **frequency** describe how often an event occurs. They usually appear in the middle of a clause or at the beginning of a clause to modify the whole sentence:

  According to one source, the al-Qaeda leaders *never* planned to include Moussaoui because they had doubts about his reliability.

- Adverbs of **degree** or **emphasis** describe how much or to what degree and are used with other modifiers. They come immediately before the word they modify:

  In a poll conducted two years after the attacks, 32 percent of Americans believed that Saddam Hussein's involvement in the September 11th terrorist attacks was *very* likely.

- Some adverbs modify an entire sentence, using transitional words and words like *however, therefore*, and *doubtless*. These usually appear at the beginning of a clause:

    As a result, thousands of tons of toxic debris were created by the collapse of the Twin Towers including asbestos, lead, and mercury, as well as high levels of dioxin and PAHs from the fires, which burned for three months afterward.

8.5.3 PRESENT AND PAST PARTICIPLES AS ADJECTIVES

Both present and past participles can be used as adjectives in English:

Present participles: debilitating, horrifying, inspiring

Past participles: debilitated, horrified, inspired

To decide whether to use the present participle or past participle, ask whether the noun modified is causing or experiencing what is being described.

Present participles

- Always end in *-ing* in English.
- Modify a noun or pronoun that is the cause of the action.

    The tons of toxic debris resulting from the collapse of the Twin Towers led to *debilitating* illnesses among rescue and recovery workers and the death of NYPD officer James Zadroga.

    ANALYSIS The *illnesses*, the noun modified, causes debilitation; hence, the present participle, which causes the illnesses, is correct.

    Some stories about 9/11 are *horrifying*; others are *inspiring*.

Past participles

- Usually end in *-ed, -d, -en*, or *-t*, though many other endings are possible.
- Modify a noun or pronoun that experiences what is being modified.

    Some rescue and recovery workers *were debilitated* by the toxic debris resulting from the collapse of the Twin Towers.

ANALYSIS The noun *workers* experienced the debilitation; hence, the past participle, which describes people experiencing the illness, is correct.

I am both *horrified* and *inspired* by 9/11 stories.

ANALYSIS The pronoun *I* experiences the horror and inspiration.

Note: In English, both the past-tense and past-participle forms of regular verbs (such as *talk, work, play*) are created by adding *-ed* or *-d* to the base (simple) form of the verb. But this pattern is not followed for all irregular verbs (for example, *ring, rang, rung*).

Although the present and past participles of a verb sound similar, they can mean very different things:

- *He is annoying* (present participle) means "He is causing an annoyance"—an active use of the verb.
- *He is annoyed* (past participle) means "He is feeling annoyed"—a passive use of the verb.

PARTICIPLES THAT MAY CAUSE CONFUSION

PRESENT PARTICIPLE	PAST PARTICIPLE
amazing	amazed
amusing	amused
annoying	annoyed
appalling	appalled
astonishing	astonished
boring	bored
confusing	confused
depressing	depressed
disgusting	disgusted
embarrassing	embarrassed
exciting	excited
exhausting	exhausted

fascinating	fascinated
frightening	frightened
frustrating	frustrated
insulting	insulted
interesting	interested
offending	offended
overwhelming	overwhelmed
pleasing	pleased
reassuring	reassured
satisfying	satisfied
shocking	shocked
surprising	surprised
tiring	tired
worrying	worried

8.6 PROBLEMS WITH PREPOSITIONS

Special problems also occur with prepositions *at*, *on*, and *in* when the context is time or place.

Time
At *a specific time:* at 9:00 a.m., at midnight, at breakfast

On *a specific day or date:* on Friday, on June 16

In *part of a 24-hour period:* in the morning, in the daytime, in the nighttime (*but* at night)

In *a year or month:* in 2003, in June

In *a period of time:* in two weeks

Place
At *a specific location:* at home, at school

At *the edge of something:* at the corner

At a target: aiming the dart at the board

On a surface: on the wall, on the floor, on the road (but in the newspaper)

On a street: the school on my street

In an enclosed place: in the room, in the camera, in the car (*but* on a plane)

In a geographic location surrounded by something else: in Winnipeg, in the Northwest Territories

Prepositions used with verbs vary widely in meaning. Check in an ESL dictionary for the meanings of verbs in combination with different prepositions. Keep track of their meanings in a notebook as you discover them in different contexts. Idiomatic uses of verbs with prepositions must be memorized because no particular rules apply.

9

SPELLING AND MECHANICS

9

Spelling and Mechanics

9   Spelling and Mechanics

Spelling and mechanics often make or break a document. Referring to "Robert Bondar" in a fundraising letter when you meant to refer to Canadian astronaut Roberta Bondar may have serious repercussions for your fundraising campaign and will certainly make the reader think twice about your credibility as a writer. Little things mean a lot.

9.1 SPELLING

Checking spelling should be one of the final steps in your writing process—and an extremely important step. Presenting error-free written work is vital to creating a good impression in academic and business contexts. Spelling errors distract readers and, at their worst, can cause them to lose your meaning. When misspellings suggest entirely different meanings, they can completely confuse readers.

Spell-checkers in word-processing programs can be useful tools for helping you to spot *some* potential spelling problems. However, spell-checkers have limitations that allow many spelling errors to be missed. These limitations include the following:

- Countless words, such as new and very specialized vocabulary, are not included.
- Spell-checkers cannot distinguish between commonly confused words that have entirely different meanings (e.g., *allusion* and *illusion*).

- They cannot indicate that you have made a simple typographical error, such as using *do* when you meant *to*.
- They cannot indicate that you have left out a word, such as *no* or *not*.
- The majority of proper nouns are not included.

Checking writing drafts for spelling errors demands your complete attention, and a computer spell-checker is just one of many tools and strategies at your disposal.

Good spelling is a challenge in English. Because it is a highly unphonetic language (sound and letter patterns frequently do not correspond), many spelling rules have exceptions. Still, you can greatly improve your spelling by

- Knowing basic spelling rules and their exceptions
- Recognizing words that sound alike but have entirely different meanings and spellings
- Identifying and remembering commonly misspelled words

Once you have mastered these spelling strategies, there are three more you can apply to help ensure that any manuscript you submit is free of spelling errors:

1. Proofread drafts meticulously to identify and fix any spelling problems.
2. Keep a dictionary nearby to check the spelling of words you do not know; have access to specialized dictionaries or other resources to check concerns you have about spelling specialized vocabulary.
3. Make and maintain a list of your own recurrent spelling problems; focus on list items as you proofread your draft so you don't repeat these errors.

9.1.1 SPELLING RULES

Know the major spelling rules:

Patterns with ie and ei: Use *i* before *e* except after *c* or when the *ei* sounds like *ay*, as in *weigh*:

- *i* before *e*:

  believe, chief, niece, yield, fierce, grieve

- *e* before *i* after *c*:

  ceiling, deceive, perceive, receive

- *ei* sounding like *ay*:

  eight, vein, neighbor, freight, reign

- Exceptions:

  either, leisure, weird, foreign, seize

  Silent e with suffixes: Drop the final silent *e* when adding a suffix that begins with a vowel, and keep the final *e* when the suffix begins with a consonant.

- Drop the final *e*:

  love + able = lovable

  race + ing = racing

  fame + ous = famous

- Exceptions:

  dyeing, changeable, hoeing

- Keep final *e*:

  achieve + ment = achievement

  hope + ful = hopeful

  love + ly = lovely

- Exceptions:

  judgment, truly, argument

  Plurals of words ending in y: When adding -ed or -s to words ending in y, you usually change the y to i when y is preceded by a consonant. However, do not change the y when it is preceded by a vowel.

  Preceded by a consonant:
  try + ed = tried

  melody + es = melodies

If you need to spell a proper name ending in y and preceded by a consonant, do not change the y to i when adding s. For example, when describing *the Dunwitty family*, you would use *the Dunwittys*.

Preceded by a vowel:
stay + ed = stayed

donkey + s = donkeys

DOUBLED CONSONANTS WITH SUFFIXES

- Double the final consonant when the word has only one syllable:

  bet + ing = betting

  fit + ed = fitted

- Double the final consonant when the word ends in an accented syllable:

  com<u>mit</u> + ing = committing

  oc<u>cur</u> + ed = occurred

- Do not double the final consonant if the word begins with an accented syllable:

  <u>ben</u>efit + ed = benefited

  cancel + ed = canceled

  Adding -s to form plurals of nouns: Add -s to make most nouns plural:

- Most nouns:

  plant + s = plants

  fork + s = forks

  satellite + s = satellites

- Nouns ending in -s, -sh, -ch, -x, or –z:

  stress + s = stresses

  bush + s = bushes

  peach + s = peaches

  fax + s = faxes

  buzz + s = buzzes

  Creating other plurals: Usually, add -s to nouns ending in -o when -o follows a vowel. However, add -es when -o follows a consonant.

- Following a vowel:

  studio + s = studios

  video + s = videos

- Following a consonant:

  echo + s = echoes

  hero + s = heroes

To make the plural of a hyphenated compound word, add -s to the main word of the compound, even if that word does not appear at the end of the compound:

father-in-law + s = <u>fathers</u>-in-law

Some English words come from other languages, such as French or Latin. These words are made plural as they would be made plural in the original language.

SINGULAR	PLURAL
criterion	criteria
datum	data
phenomenon	phenomena

9.1.2 SPELLING: WORDS THAT SOUND ALIKE

Words that sound the same but have different meanings and spellings are called **homophones**. Homophones are often the source of spelling problems. As you proofread your work, look carefully for homophones you may have used or spelled incorrectly.

accept a verb meaning "to receive"

except a preposition meaning "other" or verb meaning "to exclude"

affect a verb meaning "to cause change"

effect usually a noun meaning "the result of change"

cite a verb meaning "to quote"

site a noun meaning "location"

sight a noun meaning "vision"

desert a verb meaning "to abandon"

dessert a noun meaning "sweet course after the main course of a meal"

its possessive pronoun meaning "belonging to it"

it's contraction of *it is*

loose adjective meaning "not well attached"

lose verb meaning "to misplace" or "to part with"

principal noun meaning "the chief person," as in a school, or adjective meaning "main"

principle noun meaning "a basic truth"

their possessive pronoun meaning "belonging to them"

they're contraction of *they are*

there adverb meaning "in that place"

who's contraction of *who is* or *who has*

whose the possessive form of *who*

your possessive form of *you*

you're contraction of *you are*

The Glossary of Usage in Chapter 7 contains many homophones, as well as words that sound nearly the same and can cause spelling problems. Here you will also find definitions for each word in the following sets of words.

HOMOPHONES AND SIMILAR-SOUNDING SETS OF WORDS IN THE GLOSSARY OF USAGE

accept, except	adopt, adapt
adverse, averse	advice, advise
affect, effect	aggravate, irritate
all ready, already	all together, altogether
allusion, illusion	allude, elude
amoral, immoral	anyone, any one

awhile, a while	beside, besides
capital, capitol	censor, censure
cite, site	climactic, climatic
coarse, course	complement, compliment
conscience, conscious	continual, continuous
council, counsel	discreet, discrete
disinterested, uninterested	elicit, illicit
emigrate, immigrate	eminent, imminent
everyone, every one	explicit, implicit
farther, further	forth, fourth
ingenious, ingenuous	its, it's
lead, led	licence, license
loose, lose	maybe, may be
moral, morale	passed, past
practice, practise	precede, proceed
principal, principle	respectfully, respectively
sometime, some time, sometimes	stationary, stationery
than, then	their, there, they're
to, too, two	weather, whether
who's, whose	your, you're

9.1.3 COMMON SPELLING ERRORS

The following is a list of words commonly misspelled by students. Check that they are spelled correctly in your writing drafts.

COMMONLY MISSPELLED WORDS

abbreviate	abhor	absence	absorption
absurd	abysmal	acceptable	accidentally

accommodate	accomplish	accumulate	acquaintance
acquire	address	aggressive	all right
almost	amateur	analyze	annual
apology	apparently	appearance	appropriate
arctic	argument	arrangement	ascend
association	attendance	attorney	audience
awkward	bachelor	barbarous	basically
becoming	beginning	behavior	believe
beneficial	boundary	brilliant	Britain
bureau	burial	business	cafeteria
calendar	candidate	canister	carburetor
career	Caribbean	category	cemetery
changeable	characteristic	choose	chosen
column	commission	committee	comparative
competitive	compulsory	concede	conceivable
conference	conqueror	conscience	conscientious
conscious	consensus	courteous	criticism
criticize	curiosity	curriculum	cylindrical
dealt	decision	definitely	descend
description	despair	desperate	diarrhea
dictionary	dilemma	disagree	disappear
disappointment	disastrous	discipline	dissatisfied
dissipate	dominant	dormitory	ecstasy
efficient	eighth	eligible	elimination
embarrassment	eminent	enthusiastic	entirely
entrance	environment	equipped	equivalent
erroneous	especially	exaggerated	exceptionally

exercise	exhaust	exhilarate	existence
experience	explanation	extraordinary	extremely
fallacy	familiar	fascinate	February
fictitious	foreign	foreseen	forty
frantically	friend	fundamental	further
gauge	genealogy	generally	grammar
guarantee	guard	guerrilla	guidance
harass	height	hereditary	heroine
hindrance	humorous	hungrily	hypocrisy
hypothesis	illiterate	imaginary	imagination
imitation	immediately	impromptu	incidentally
incredible	indefinitely	independent	indispensable
inevitable	infinite	ingenious	initiation
inoculate	intelligence	interesting	interpretation
involve	iridescent	irrelevant	irresistible
jealousy	knowledge	laboratory	legitimate
liaison	license	lightning	literature
liveliest	loneliness	luxury	magazine
maintenance	maneuver	marriage	marshal
mathematics	medieval	miniature	mischievous
misspell	moccasin	mortgage	mysterious
necessary	negotiation	nevertheless	noticeable
obligation	obstacle	occasion	occasionally
occur	occurred	occurrence	omission
opinions	opportunity	optimistic	original
outrageous	pamphlet	parallel	paralyze
particularly	pastime	peer	perform

performance	permanent	permissible	perseverance
perspiration	Philippines	physically	picnicking
playwright	politics	practically	precedence
preference	preferred	prejudice	preparation
prevalent	primitive	privilege	probably
proceed	professor	prominent	pronunciation
propaganda	psychology	quantity	quiet
quite	quizzes	recede	receive
recommendation	reference	referred	regard
religious	reminiscent	repetition	resistance
restaurant	rhythm	ridiculous	roommate
sacrifice	sandwich	schedule	secretary
seize	separate	sergeant	several
siege	similar	simultaneous	sincerely
soliloquy	sophomore	specimen	strictly
subtly	succeed	supercede	surprise
syllable	tariff	temperament	temperature
tendency	thorough	threshold	tragedy
transferred	tries	truly	typical
tyranny	unanimous	unnecessarily	until
usually	vacuum	vengeance	villain
weird	whether	written	

9.2 HYPHENS

Hyphens are used for compound words, fractions, numbers, and prefixes. Hyphens also have an important role in preventing misreading.

9.2.1 HYPHENS FOR COMPOUND WORDS

A **compound word** is made up of two or more words that combine to express one concept. It may be written in one of three ways:

- As separate words, as in *half sister*
- As one word, as in *stepfather*
- As a hyphenated word, as in *mother-in-law*

Check in the dictionary to determine whether to write a compound word as separate words, one word, or a hyphenated compound. If a compound word does not appear in the dictionary, treat it as two words:

> The tractor-trailer swerved and almost hit us.

> He has an extensive resource library on the bookshelf of his dormitory room.

> Include a cross-reference to your source of information on robotics.

> Industrial by-products are commonly used as fuel.

9.2.2 HYPHENATED ADJECTIVES

When two or more words function as an adjective before a noun, they are hyphenated.

> Jack Kilby, the well-known electrical engineer at Texas Instruments, discovered that a crystalline wafer of silicon could be used to create a fully functioning circuit.

> Jack Kilby, a fifty-five-year-old electrical engineer at Texas Instruments, made this discovery in 1958.

> Manufacturers were soon able to use inexpensive, off-the-shelf microprocessors for practically any application.

> Engineers and scientists are currently exploring three-dimensional architectures for circuits.

In most cases, do not use a hyphen when the compound follows the noun:

> As a result of his discoveries, he became quite well known.

> At the time of his discoveries, he was fifty five years old.

Hyphens are suspended if the modifying words are in a series:

> The earliest internal combustion engine used both two- and four-stroke cycles.

9.2.3 HYPHENS WITH FRACTIONS AND COMPOUND NUMBERS

Use a hyphen with compound numbers from twenty-one through ninety-nine and with fractions:

> Gloria dreaded the thought that she would be fifty-one on her next birthday.

> I use one-third of my basement as an office.

9.2.4 HYPHENS WITH PREFIXES AND SUFFIXES

Use a hyphen with the prefixes *all-*, *ex-*, *great-*, *quasi-*, and *self-* and with the suffix *-elect*:

> By the mid-1970s, transistors were not much larger than bacteria, and they cost mere hundredths of a cent each.

> The ex-premier always has a difficult time because he is frequently asked his position on controversial issues.

> The mayor-elect was impatient to begin implementing her agenda.

9.2.5 HYPHENS TO AVOID AMBIGUITY

A hyphen is used in some words to eliminate awkward double or triple letters—for example, *co-opt*.

Some pairs of words are spelled the same but have entirely different meanings and could cause confusion. In such cases, a hyphen is traditionally used in one of the words to distinguish it from the other. For example, *recount* means "to tell a story," while *re-count* means "to count again":

> My uncle used to <u>recount</u> terrible stories about life in a concentration camp during the Second World War.

> The candidate for council demanded a <u>re-count</u> after her opponent received only marginally more votes.

9.2.6 HYPHENS FOR WORD DIVISION AT THE END OF A LINE

Most word-processing programs automatically hyphenate words at the end of a line. However, some older programs may not do so, or you may need to create a manuscript using a typewriter. When creating or proofreading a text, it is important to know end-of-line hyphenation rules:

- Divide words only between syllables:

  Incorrect: If you want to write well, you must follow a systematic proc-ess that includes more than one draft.

  Correct: If you want to write well, you must follow a systematic pro-cess that includes more than one draft.

- Do not divide one-syllable words:

  Incorrect: Wounded in the extremely heavy fighting, the officer knew dea-th was approaching, and he accepted it with great dignity.

  Correct: Wounded in the extremely heavy fighting, the officer knew death was approaching, and he accepted it with great dignity.

- Do not divide a word so that one or two letters remain at the end of the line:

  Incorrect: She found the stale air in the room oppressive, so she went to o-pen the window and then turned on the fan.

  Correct: She found the stale air in the room oppressive, so she went to open the window and then turned on the fan.

  Incorrect: Neighbors of the accused man told reporters that the man lived a-lone and was very quiet.

  Correct: Neighbors of the accused man told reporters that the man lived alone and was very quiet.

- Divide a hyphenated word at the hyphen, and divide a closed compound only between complete words:

  Incorrect: He is not naïve in any way; most people consider him a very sel-f-aware individual.

  Correct: He is not naïve in any way; most people consider him a very self-aware individual.

  Incorrect: The next step, after installing the new carpet, is to nail on the ba-seboards where they are required.

Correct: The next step, after installing the new carpet, is to nail on the baseboards where they are required.

9.3 CAPITALIZATION

Capitalize the first word of every sentence. In addition, you will need to capitalize specific types of words within sentences. Use the following rules as general guidelines for capitalization. Consult your dictionary to determine which words must be capitalized.

9.3.1 PROPER VS. COMMON NOUNS

Capitalize proper nouns, and words derived from them, but do not capitalize common nouns. Proper nouns are the names of specific people, places, and things. Common nouns include all other nouns. Usually, capitalize the following:

- Names of religions, religious practitioners, holy books, special religious days, and deities
- Geographic place names
- People's names and nicknames
- Words of family relationship used as names (e.g., Uncle Bill)
- Nationalities, tribes, races, and languages
- Names of historical events, periods, movements, documents, and treaties
- Political parties, organizations, and government departments
- Educational institutions, departments, degrees, and specific courses
- Names of celestial bodies
- Names of ships, planes, and aircraft
- Parts of letters (e.g., Dear John)
- Names of specific software

CAPITALIZING NOUNS	
PROPER NOUNS	**COMMON NOUNS**
Zeus	a god
Book of Mormon	a book
Steen Building	a building

Sixth Street	a street
Ramsey Park	a park
Bell Labs	a company
Nobel Prize	a prize
Young's modulus	theory or formula with a formal name
Albuquerque	a city
Mark	a man
Aunt Agnes	my aunt
Portuguese	a language
Romanticism	a movement
Democratic Party	a political party
Mars	a planet
Queen Elizabeth II	a ship
Microsoft *Word*	a program

Months, days of the week, and holidays are considered proper nouns. The seasons and numbers of days of the month are not considered proper nouns:

Every spring, <u>Victoria Day</u> falls on the first <u>Monday</u> in <u>May</u>.

The meeting is held on the second <u>Tuesday</u> of <u>January</u>, <u>June</u>, and <u>December</u>.

Capitalize the names of school subjects only if they are languages, but capitalize the names of specific courses:

In his final year, he will need to take microbiology, chemistry, biology, <u>English</u>, and <u>Spanish</u>.

Professor Woodman teaches <u>Romanticism</u> to all students majoring in English.

9.3.2 CAPITALIZING TITLES WITH PROPER NAMES

Capitalize the title of a person when it is part of a proper name:

Dr. Norman Bethune   Rev. David Rooke

Maury Hughes, P.E.   Douglas Fairbanks, Sr.

<u>Judge</u> McMurrey gave his decision on the appeal.

Do not capitalize the title when it is used alone:

A <u>judge</u> presided over the inquiry.

Note: In some cases, if the title of an important public figure is used alone, the first letter can appear as either a capital letter or a lower-case letter. Conventions vary:

The <u>prime minister</u> [Prime Minister] dodged the protester's pie.

The <u>president</u> [President] evaded the question.

9.3.3 CAPITALIZING TITLES OF WORKS

Capitalize the first, last, and all other important words in the titles of works such as books, articles, films, and songs.

IMPORTANT WORDS
These important words should be capitalized in titles and subtitles:

- Nouns
- Verbs
- Adjectives
- Adverbs

LESS IMPORTANT WORDS
These less important words should not be capitalized *unless* they are the first or last word of the title or subtitle:

- Articles
- Prepositions
- Coordinating conjunctions

> *Book title:* The Existential Pleasures of Engineering
> *Article title:* "Face Recognition: A Literature Review"
> *Film or series title:* Star Trek: The Next Generation
> *Song title:* "Do You Know the Way to San Jose?"

Use the foregoing guidelines to capitalize chapter titles and other major divisions in a work:

"Of Dullards and Demigods" is Chapter 7 in Samuel C. Florman's *The Existential Pleasures of Engineering*.

9.3.4 CAPITALIZING THE FIRST WORD OF A SENTENCE

Capitalize the first word of a sentence:

Face-recognition systems are now being demonstrated in real-world settings such as bank-card identification, access control, mug-shot searching, security monitoring, and surveillance systems.

If a sentence appears within parentheses, capitalize the first word of the sentence. However, do not capitalize the first word if the parentheses are within another sentence:

The effects of plaque on the heart valves are significant. (See Figure 6.)

The effects of plaque on the heart valves are significant (see Figure 6).

9.3.5 CAPITALIZING THE FIRST WORD OF A QUOTED SENTENCE

Capitalize the first word of a direct quotation, but do not capitalize it if the quotation is blended into the sentence in which the quotation is introduced:

Texas Instruments Chairman Tom Engihous said, "In my opinion, there are only a handful of people whose works have truly transformed the world and the way we live in it—Henry Ford, Thomas Edison, the Wright Brothers, and Jack Kilby."

In his article "Eco-tourism Boom: How Much Can Wildlife Take?" Bruce Obee says that "tour boats . . . are a fraction of the traffic."

If you need to interrupt a quoted sentence to include explanatory words, do not capitalize the first word following the interruption:

"In my opinion," Texas Instruments chairman Tom Engihous said, "there is only a handful of people whose works have truly transformed the world and the way we live in it—Henry Ford, Thomas Edison, the Wright Brothers, and Jack Kilby."

9.3.6 CAPITALIZING THE FIRST WORD AFTER A COLON

When an independent clause appears after a colon, capitalizing the first word is optional; if the content after the colon is not an independent clause, do not capitalize:

> The potential of wind power to help meet America's growing demand for electricity is staggering: if the areas of strong winds, which cover about 6 percent of the mainland states, could be exploited, more than the current U.S. electricity consumption could be supplied.

> We were told to bring the following items for the hike: a compass, a sleeping bag, a tent, and enough food provisions to last seven days.

9.3.7 CAPITALIZING ABBREVIATIONS

Capitalize the abbreviations for government departments and agencies, names of organizations and corporations, trade names, and call letters of television and radio stations.

> CSIS   CIA   NATO   CTV   Magna International   CHCO-TV   CKNW

9.4 ABBREVIATIONS

In most cases, abbreviations should not be used in formal writing, such as academic essays, unless the abbreviations are very well known; for instance, *CBC, CBS,* or *UN.* Abbreviations are more widely used in science and technical writing than in writing for the humanities.

Always consider your reader when deciding whether or not to use any abbreviation. Will he or she understand the abbreviation? Otherwise, you run the risk of confusing the reader. If the type of writing that you are doing requires abbreviations, be consistent in your use of them.

9.4.1 ABBREVIATIONS FOR TITLES WITH PROPER NAMES

Abbreviate titles and degrees immediately before and after proper names. Do not abbreviate a title or degree if it does not accompany a proper name:

> *Incorrect:* The rev. gave a very inspiring sermon to launch the congregation's food drive.

> *Correct:* The reverend gave a very inspiring sermon to launch the congregation's food drive.

Do not use titles and degrees redundantly:

Incorrect: <u>Dr.</u> Steven Edwards, <u>M.D.</u>

Correct: Dr. Steven Edwards

OR

Correct: Steven Edwards, M.D.

ABBREVIATED TITLES

BEFORE PROPER NAMES	AFTER PROPER NAMES
Rev. R. W. McLean	Edward Zenker, D.V.D.
Dr. Wendy Wong	Paul Martin, Jr.
Asst. Prof. Tom Simpson	Margaret Barcza, M.B.A.
Ms. Germaine Greer	John Bruner, LL.D.
Mrs. Sodha Singh	Eleanor Semple, D.D.
Mr. Willy Loman	Roy Shoicket, M.D.
St. John	Barbara Zapert, Ph.D.

9.4.2 ABBREVIATIONS FOR ORGANIZATIONS, CORPORATIONS, AND COUNTRIES

Use standard abbreviations for names of countries, organizations, and corporations:

UK (or U.K.)　FBI　NORAD　RCMP　CIDA　TSN　RCA　IBM

To save money, she got a room at the <u>YWCA</u>.

If you need to use a less familiar abbreviation in your paper, such as COMECON for the Council of Mutual Economic Assistance, do the following:

- Write the full name of the organization followed by the abbreviation in parentheses.
- For each subsequent reference to the organization, use the abbreviation on its own.

9.4.3 B.C., A.D., a.m., p.m., no., $

Use the standard abbreviations *B.C., A.D., a.m., p.m., no.,* and $ only with particular years, times, numbers, or amounts.

The abbreviation *B.C.* ("before Christ") or the acceptable alternative *B.C.E.* ("before the Common Era") always appears after a specific date:

156 B.C. (or B.C.E.)

The abbreviation *A.D.* (*Anno Domini*) or the acceptable alternative *C.E.* ("Common Era") always appears before a specific date:

A.D. (or C.E.) 65

Use *a.m., p.m., no.,* or $ only with a particular figure:

5:15 a.m. (or A.M.) 8:30 p.m. (or P.M.) $175 no. 16 (or No.)

In formal writing, do not use the following abbreviations without particular figures:

Incorrect: We arrived for the dance in the early p.m.

Correct: We arrived for the dance in the early afternoon.

Incorrect: It is impossible to estimate the no. of fish in the stream during spawning season.

Correct: It is impossible to estimate the <u>number</u> of fish in the stream during spawning season.

9.4.4 LATIN ABBREVIATIONS

Since some readers may be unfamiliar with Latin abbreviations, keep use of these abbreviations to a minimum or use the English equivalent.

LATIN ABBREVIATIONS

ABBREVIATION	LATIN	ENGLISH MEANING
c.	*circa*	approximately
cf.	*confer*	compare
e.g.	*exempli gratia*	for example

et al.	*et alii*	and others
etc.	*et cetera*	and the rest
i.e.	*id est*	that is
N.B.	*nota bene*	note well
P.S.	*postscriptum*	postscript
vs.	versus	*versus*

In informal writing, such as personal e-mails, it is acceptable to use Latin abbreviations:

This Tuesday, it's the Raptors vs. the Sonics.

Structural engineering is the field of civil engineering particularly concerned with the design of complex structural systems such as buildings, bridges, walls, dams, tunnels, etc.

Structural engineers ensure that their designs are safe (i.e., do not collapse without due warning) and are serviceable (i.e., vibration and sway are not uncomfortable).

Entry-level structural engineers design simple beams, columns, and floors of a new building, including calculating the loads on each member and the load capacity of various building materials (e.g., steel, timber, masonry, concrete).

ANALYSIS Remember that *i.e.* means "that is" or "in other words" and that *e.g.* means "for example."

In formal writing, use the full English words or phrases:

Incorrect: The Sumerians came down to the bank of the Euphrates and Tigris rivers c. 3500 B.C.E. Many artifacts provide evidence of their cultural advancement e.g. the bronze mask portrait of King Sargon and the headdress of Queen Sub-ad.

Correct: The Sumerians came down to the bank of the Euphrates and Tigris rivers approximately 3500 B.C.E. Many artifacts provide evidence of their cultural advancement; for example, the bronze mask portrait of King Sargon and the headdress of Queen Sub-ad.

9.4.5 MISUSES OF ABBREVIATIONS

Abbreviations are generally not appropriate in formal writing:

Incorrect: Henry Petroski and Samuel C. Florman are popular authors in eng. lit. because they have written outstanding books about the profession.

Revised: Henry Petroski and Samuel C. Florman are popular authors in <u>engineering</u> <u>literature</u> because they have written outstanding books about the profession.

TYPES OF ABBREVIATIONS TO AVOID IN FORMAL WRITING

CATEGORY	FORMAL	INFORMAL
Names of Persons	Jennifer	Jen
Holidays	Christmas	Xmas
Days of the Week	Tuesday to Thursday	Tues. to Thurs.
Months	from January to August	Jan. to Aug.
Provinces	Saskatchewan	Sask. or SK
States	Texas	TX
Academic Subjects	Biology and English	Bio. and Engl.
Units of Measurement*	6 ounces	6 oz.
Addresses	Madison Avenue	Madison Ave.
Subdivision of Books	chapter, page	ch., p.**

* except metric measurements

** except as part of documentation

Metric abbreviations are often permitted in formal writing, as in 25 *kg* or 15 *mm*. However, do not use a number written in words with an abbreviation, as in *twenty cm*.

Abbreviations are acceptable in company or institution names only if the abbreviation is part the company's or institution's official name, as in *Jack's Windows & Roofing Co.*, or *Writer's Inc. Consulting*. Never arbitrarily abbreviate a company's name. For example, if a company's name is *Randolph Architectural Group*, do not shorten it to *Randolph Arch. Gr.* When corresponding with any company, use the full company name that appears on company stationery, in the firm's advertising, or on its web site.

9.5 NUMBERS

9.5.1 SPELLED-OUT NUMBERS

Spell out numbers of one or two words, or if a number starts the sentence; use figures for all other numbers and amounts:

Incorrect: It has been 8 years since we last heard from him.

Revised: It has been <u>eight</u> years since we last heard from him.

Incorrect: In a single section of Biology 101, there are three hundred and fifty-six first-year students.

Revised: In a single section of Biology 101, there are <u>356</u> first-year students.

Incorrect: 721 collapsible chairs are required for the wedding reception.

Revised: <u>Seven hundred and twenty-one</u> collapsible chairs are required for the wedding reception.

You might also consider recasting the sentence if it begins with a figure:

For the wedding reception, we require <u>721</u> collapsible chairs.

In some instances, if numbers follow one another, you may wish to write one as a figure:

During the Olympic trials, she swam <u>four 100</u>-meter heats.

Note: In business and technical writing, figures are sometimes preferred for all numbers except one to nine because they provide clarity and brevity. However, usage varies, so it is best to check with your instructor.

9.5.2 FIGURES FOR NUMBERS

Figures are acceptable in the following writing situations:

Dates
January 16, 1952   21 B.C.   A.D. 400

Time
3:51 a.m.   7 p.m.   If *a.m.* and *p.m.* are not used, write the time in words.

one o'clock in the morning

12 midnight

eight-thirty in the evening

Addresses
31 Bloor Street West

75 West Broadway

Exact amounts of money
$15.99 $30 $72,300.68

Unnecessary: I was stunned to learn that the price of a movie ticket had gone up to twelve dollars and seventy-six cents with tax.

Revised: I was stunned to learn that the price of a movie ticket had gone up to $12.76.

Percentages, Fractions, Decimals
92 percent   1/5   3.75

Unnecessary: The poll indicates that the president has a ninety-three percent approval rating.

Revised: The poll indicates that the president has a 93 percent approval rating.

If a paper is heavily statistical, however, use the % sign. It is appropriate with a list of figures, but not in a paper where words predominate.

Statistics, Scores, Surveys
In Canada, 14 babies are born each year for every 1,000 people.

Argentina won the game against Germany by a score of 1–0.

According to the study, 1 out of every 10 residents was out of work.

Measurements and counts
4.5 meters clearance

19,800 people at the game

Divisions of books
Chapter 7, page 381

Divisions of plays
Act V, Scene ii, lines 10–15

Identification numbers
Highway 427, Room 311, Channel 2, #73321

9.6 ITALICS (UNDERLINING)

Italics is the typeface in which letters slant to the right and appear like handwritten script. Italics is a typeface option on most word-processing programs. When writing by hand, use <u>underlining</u> to indicate italics.

Note: The *MLA Handbook* recommends that students use underlining, not italics, in their papers. Check with your instructors to determine their preferences.

9.6.1 ITALICS FOR TITLES OF WORKS

Convention requires that you use italics (or underlining) when making reference to certain types of work or material, listed in the following table. Use quotation marks to identify the titles of

- Short stories, poems (except long poems published independently), and essays
- Journal, magazine, or newspaper articles, including titles of reviews, interviews, and editorials
- Unpublished material such as theses, dissertations, or papers read at meetings or published in conference proceedings; lectures, speeches, or readings
- Manuscripts in collections; published letters
- Chapters in a book
- Laws and treaties
- Songs and television episodes

TITLES OF WORKS IN ITALICS (OR UNDERLINED)

PRINT

Books, Plays, Long Poems	*To Engineer Is Human: The Role of Failure in Successful Design*
Journals, Magazines	*Journal of Automobile Engineering*
Newspapers	the *Winnipeg Free Press*
Conference Proceedings	*Proceedings of the Institution of Mechanical Engineers*
Published Dissertations	*Port-Based Modeling and Control for Efficient Bipedal Walking Robots*[1]
Maps, Charts	*Michelin Germany*
Comic Strips	*Dilbert*

ART

Visual Works of Art	Picasso's *Guernica*
Musical Compositions, Scores	*Symphony in C Major*
Ballets, Operas	*Don Giovanni*
Performances	*Nothing Sacred*

ELECTRONIC

Films, Videotapes	*The Sweet Hereafter*
Sound Recordings	*John Wesley Harding*
Radio and Television Programs	*Cross Country Checkup, Venture*
Software Programs, Games	Adobe *Acrobat*

ON-LINE

Web Sites	*Civil Engineering Virtual Library*
Books, References	*Structural Engineering Reference Manual*
Projects, Services, Databases	*Aerospace and High Technology Database*
Discussion Lists, Newsgroups	*Alliance for Computers and Writing Listserv*

TITLES OF WORKS IN ITALICS (OR UNDERLINED)

GOVERNMENT PUBLICATIONS

Acts, Statutes	*Canadian Charter of Rights and Freedoms*
Court Cases	*Roe v. Wade*
Debates	*Chemical and Biological Engineering Debates*
Papers, Hearings, Reports	*Digital Emission Spectrum Model*

[1] **Duindam, Vincent**, PhD, Universiteit Twente (The Netherlands), 2006.

Do not italicize, underline, or place in quotation marks the following:

- Names of sacred religious works; for example, the Bible or names of books within it
- Laws
- Unpublished letters
- The title of your own essay or report

9.6.2 ITALICS FOR NAMES OF SHIPS, TRAINS, ETC.

Italicize or underline the names of ships, trains, planes, and spacecraft:

Bluenose II   *The Orient Express*   *Spirit of St. Louis*   *Columbia*

During the *Apollo 11* lunar-landing mission, Mission Commander Neil Armstrong became the first person to walk on the moon.

9.6.3 ITALICS FOR FOREIGN WORDS

Italicize or underline foreign words that have not become part of the English language:

Given her advocacy of oil production in national parks, she is *persona non grata* among environmentalists.

You do not need to italicize or underline words that have become part of the English language, such as the following:

café au lait, bon voyage, habeas corpus, per se

Remember that English is an evolving language and new words borrowed from other languages are regularly being accepted into common English usage. If you are unsure about whether or not to italicize or underline, check in a recent edition of a comprehensive dictionary.

9.6.4 ITALICS FOR WORDS, LETTERS, NUMBERS AS THEMSELVES

Italicize or underline letters, words, or numbers mentioned as themselves:

> The second *E* in IEEE stands for "Electronics."

> Be careful how you use the words *can* and *may*.

It is also acceptable to use quotation marks to set off words mentioned as words.

9.6.5 MISUSE OF ITALICS (UNDERLINING)

Writers occasionally use italics or underlining to emphasize important words in their work. Such an emphatic technique is effective only when it is not overused:

> The residential development in the Golden Horseshoe of southern Ontario was *rampant*.

10

FINDING INFORMATION

10

Finding Information

10   Finding Information

Much has changed about how you can find information. Although the computer and the Internet have made the process easier, you still need the physical library. The following sections discuss strategies for using libraries, the limitations of Internet-based information, strategies for information searches, methods for finding different types of information resources, and guidelines for evaluating the information resources you find.

10.1 RESEARCHING: INSIDE AND OUTSIDE THE LIBRARY

Learning how to retrieve information efficiently is an important skill for any engineering career. Become familiar with the local college's or university's library services even if you are not a student. Find out whether there is more than one library that you can use. Many universities have a general library, an undergraduate library, and libraries for various disciplines or fields of study. Is there an engineering library on your campus?

Find out what is available in these libraries, where that information is located, how materials are arranged, and whom to ask if you need help. College and university libraries often provide free handouts, pathfinders, and guides, either in paper or online. These resources can be as specific as "How to Research and Write a Technical Report." It is not uncommon to find "How Do I . . .?" or FAQ (frequently asked questions) pages on library web sites to provide information even when the physical library is closed. Many libraries have online tutorials that cover the entire research process.

Many government agencies and corporations also have library services, which may have some of the same characteristics described in the preceding.

10.1.1 ONLINE RESEARCHING

The arrival of the graphical user interface (GUI) to access the Internet around 1994 drastically and forever changed library services and collections. College and university libraries have been viewed as storehouses of information in such forms as books, journals, magazines, newspapers, microfiche and microfilm, videocassettes, films, and filmstrips. These storehouses, in most libraries, are no longer just physical but "virtual," because the information has been digitized. Nearly one-third or more of a library's resources may be online in some way. The contemporary online library enables you to do such things as the following:

- Read content online
- Send content to yourself via e-mail
- Download content to save or to print
- Use computerized search tools, such as online library catalogs, to find books
- Use electronic databases for online access to reference works such as dictionaries, encyclopedias, handbooks, guides, and almanacs
- Use online periodical indexes or databases to find citations to articles in academic journals, popular magazines, and newspapers, as well as the full text of the article

10.1.2 ON-SITE RESEARCHING

Even though you use a computer to access the information you need, the physical library should remain the focus of most of your research efforts. You may have heard the expression "information wants to be free," attributed to Stewart Brand, at the first Hackers' Conference in 1984 (Clarke, Roger), but often it is not. Quality information is usually not available free of charge on the World Wide Web. Instead that quality information is available only from databases for which publishers charge for access to content. Fortunately, you can avoid this expense by using libraries that subscribe to these databases; libraries make this digitized content available to their users. These library collections of e-resources are available 24 hours a day, 7 days a week, through Internet access, even when the physical library is closed. Your library-user identification number and your library's login process will get you to the resources you need when you are not physically in the library.

Since most public and nearly all college and university libraries provide access to information through computer searches, you should familiarize yourself with finding tools and search techniques before you have a specific information-research project. By doing so, you avoid frustration and save precious research time. Reference librarians and other library staff are available to answer your questions and will assist you in finding resources and information you need. Many libraries provide an "Ask a Librarian" service where you can communicate with a reference librarian online in real time or request assistance by email, usually with a 24- to 48-hour response time.

10.2 RESEARCH PROCESS

To ensure that your information search is efficient and rewarding, follow the research process described here.

10.2.1 START WITH A SPECIFIC QUESTION

Research begins with a good question. Often, that question leads to other, related questions. Spend some time developing your research question—broadening or narrowing its focus—before you start looking for materials to answer that question. As you research your topic, you may find that you must revise your research question: there may not be enough, or there may be too much information. Refining the question is one of the most important steps in doing research.

10.2.2 MAKE A LIST OF SEARCH TERMS

The Milwaukee School of Engineering's Library Research Tutorial states that "effective library research begins before an actual search is undertaken." One "presearch" step includes creating a list of concepts associated with your topic. Three methods are suggested for generating this list of terms:

- Generate a list of search terms by writing either a description of your topic using *professional jargon* understood by knowledgeable professionals or by writing a description of your topic that a *non-specialist* could understand.

- Generate a list of questions related to your topic, and circle the words that identify the important concepts. These are your key-words, but try adding to this list.
- Use concept mapping or brainstorming to think about alternative wording, synonyms, or related terms for your research question. Think of broader and narrower terms. Ask yourself, "Are there unique words, specific phrases, names, or acronyms I might use to search?"

Do any of your terms have spelling variations (for example, *labor* vs. *labour*) or other, different meanings and contexts that you will need to consider (for example, Turkey, the country vs. turkey, the bird)? Make sure there are no **stop words** on your list. These are common words—usually prepositions—that are not searchable (for example, *an, and, as, at, by, for, from, in, not, of, on, or, the, to,* and *with*). Some of these have specific uses in searching and cannot be used as search terms. Others are too common and would retrieve too many results.

10.2.3 FIND ADDITIONAL SEARCH TERMS

Once you've generated a good list of search terms, use them to search online and print library resources and with Web search engines. Start simple. Begin with specific words that describe exactly what you're looking for. Broader, general terms will retrieve more results—perhaps too many—when you search online resources, so try to target your search.

Take notes on the search statements you actually use; doing so will save you time. Try variations of the search statements that you have used and record them in your notes as well. One good way to collect variations of your search terms is to examine titles and abstracts of the resources you find. For each new term that you discover, go back and search again using that new term.

10.2.4 FINDING VS. DISCOVERY

Once you have your list of search terms or statements, you are ready to begin looking for answers to your research questions. You will probably use two different approaches—"finding" and "discovery"—to proceed with your research.

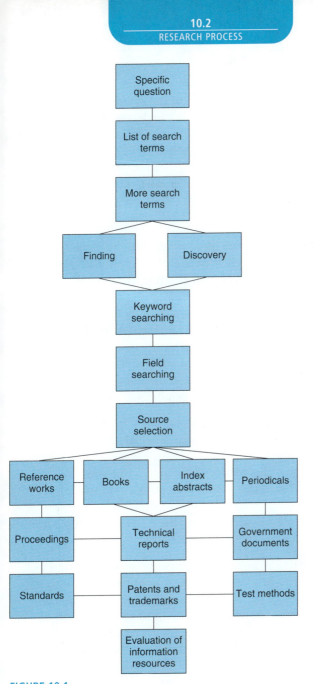

FIGURE 10.1
Overview of the information-search process.

FINDING

Think of **finding** as locating a known item—something specific you are searching for where the outcome of the search is fairly certain. Examples of finding would be searching for a book title, a definition, a formula, an address, or the answer to a question such as "What is the viscosity of motor oil?" Finding techniques also include limiting a search by date or other options or going directly to a page containing the information you need. You know (or are pretty sure) these exist— you just need a pointer or a resource with the answer.

DISCOVERY

Think of **discovery** and "dumb luck" or "serendipity" ("the faculty of making fortunate discoveries by accident"—*The American Heritage Dictionary of the English Language*, 4th ed.). It is a process that results in locating items you did *not* expect to find or that you weren't sure existed. The outcome of the search is uncertain. A lot of research involves searching of the discovery type. Original, scholarly research, for instance, usually begins with a "literature review" to discover what has been already been done in an area and which questions or issues have been raised, discussed, or answered.

Examples of discovery searching include:

- Browsing by subject in an online database without being sure what terms will be in the subject list or which ones might pertain to your topic
- Perusing the table of contents of a book
- Browsing by call number in a library's stacks or in the online catalog
- Answering questions such as "Does the ASME (American Society of Mechanical Engineers) have a web site?"

Discovery searching involves using the resources you find to identify still other search terms. When you are in discovery-searching mode, stop and skim what you've found. You're likely to find clues for other search terms. Table 10.1 provides some examples of resources you might use for each type of search. The library catalog can be used as a discovery tool as well as a finding tool.

Searching usually involves both finding and discovery tools and activities. Finding mode gives back fewer results, but those

TABLE 10.1
Tools for Finding Library Resources

FINDING TOOLS	DISCOVERY TOOLS
Library catalog	Library catalog
Index in back of book	Web search engines
Index volumes of print encyclopedias	Web directories
Dictionary	Metasearch engines
Handbook	Cross-database search applications
"How to" manual or cookbook	Subject links in databases

results are much more precise. In discovery mode, results are undeterminable, usually have a broader range, and produce more items than finding mode does. In the search process, you switch from one mode to the other frequently. For example, in discovery mode you might find a useful citation in the bibliography in a reference work. You would then switch to finding mode, using the library catalog to search by author or by title to find that item on the shelf. The electronic search environment often blurs finding and discovery; you sometimes use the same tools for both purposes.

10.2.5 KEYWORD SEARCHING

While search terms can yield more precise results and enables access to group information resources, you still may not know the formal subject terms used by the database you are searching. For this reason, many librarians suggest beginning with **keyword searching**, sometimes called "quick keyword" or "basic search." Keyword searching is more a discovery than a finding search. It is the most useful technique for new topics (for which authoritative subject terms have not yet been assigned) and unusual terms. A keyword search, as diagrammed in Figure 10.2, looks for your search terms in all fields of an item record and anywhere in the text, which can result in an unusually large return.

Fortunately, you can use several techniques to increase the precision of a keyword search, as shown in Table 10.2.

FIGURE 10.2
Different Methods of Keyword Searching.

TABLE 10.2
Methods of Narrowing Keyword Searches

TECHNIQUE	WHAT IT DOES	EXAMPLE
Truncation—adding a symbol to the root of the word to retrieve related terms and variant endings for the root term. Some databases have left- and right-hand truncation.	Expands your search	*structur** finds *structure, structuring, structures*, etc. **elasticity* will find *elasticity, aeroelasticity, viscoelasticity*
Boolean **AND**—retrieves only those records containing *all* your search terms	Narrows your search	*finite* AND *element* AND *methods*
Boolean **OR**—retrieves records containing *any* of your search terms; especially useful for synonyms, alternate spellings, or related concepts	Broadens your search	*energy* OR *fuel* *pollut** OR *contaminat** *sulfur* OR *sulphur*
Boolean **NOT, AND NOT**—attempts to exclude a term that is not useful or relevant	Narrows your search	*"Advanced Materials"* AND *composite* NOT *wood*
Proximity—retrieves terms within a specified distance of one another; variations of proximity searches are **phrase**	Narrows your search	*"Styrenic Block Copolymers"* (quotation marks ensure that the multiple-word term

TABLE 10.2 (*Continued*)

searches, where the terms must be retrieved exactly as entered; **NEAR, ADJACENT, WITH,** and **WITHIN** searches		is searched as a phrase, but is not required for all databases)
Parentheses ( )—groups terms with Boolean for more complex searches	Combines searches	"*mechanical engineering*" AND (*handbook* OR *dictionary*)

Truncate your search terms whenever it seems useful. Computers are literal: if you enter the singular form of a word—for example, *optic*—the database may search only for that specific form of the term (although some databases automatically search for variants) and will not retrieve any other variants: for example, *optic, optical, opticals*. You can force a search for variants like these by using a truncation symbol to expand your search: your search will then yield items with different word endings such as -*s*, -*es*, -*ed*, and -*ing*. You can also retrieve alternative spellings by truncating *within* a word: wom*n retrieves both *woman* and *women*. Truncation within a word is useful if there are British and American spellings to consider: *hemoglobin* and *haemoglobin*. The truncation symbol is usually an asterisk (*), but other symbols—for example, the question mark (?)—may be used as well. Check the database instructions or search tips to see what options are available. Take care in deciding where to truncate your terms. If you truncate *manufacture* to *man**, you will retrieve far too many results and most will be irrelevant. Consider how many words begin with "man"!

With keyword searching, you can count on having a few "false drops"—search results that are not relevant to your topic questions even though they contain the keywords you searched for. For example, a search for female* AND alcohol* may retrieve "treatment of *female alcoholics* with psychotherapy," but it may also retrieve "treatment of *alcoholism* by *female* doctors," which is quite different. Another example: art AND history AND (China OR Chinese), may retrieve works on the history of Chinese art, but may also retrieve the art of war and the history of China. These latter are examples of false drops.

10.2.6 FIELD SEARCHING

Field searching lets you focus the search strategy on a particular part, or "field," of the record (author, title, journal, date, or descriptor). You can use field searching for more precise searching in various engineering and science databases. Table 10.3 shows some common examples of field abbreviations.

Some databases provide an advanced or guided search form that lets you combine field and keyword searches and that assists you in creating effective search logic and language (see Figure 10.3).

You can choose to search only the subject terms or descriptor field of a database. To do so, you must use *controlled vocabulary*—the language of the database—to locate relevant results. Subject terms are precisely defined and may have references or links to broader or narrower terms that you can use. Databases usually have a thesaurus or list of subject terms that you can consult to find the best terms for your search. Once you know which terms to use, subject searching will lead you directly to relevant results. For example, a database might prefer the subject term "integrated circuits" to "computer chips." If you perform a subject search on "integrated circuits," you'll find everything relevant that database has to offer. As you can see, searching with subject terms as defined by the database can be "efficient, effective, and precise" ("Preparing to Research").

TABLE 10.3
Abbreviations (Field Codes) for Field Searching

FIELD	CODE	FIELD	CODE
Abstract	AB	Language	LA
Article title	TI	Patent application data	PA
Author name	AU	Publisher	PB
Descriptor	DE	Publication type	PT
ISBN	IB	Publication year	PY
ISSN	IS	Subject	SU
Journal name	JN	Title	TI

FIGURE 10.3
Combined field and keyword searching.

The advanced or guided search form allows you to have the flexibility of a keyword search with the precision of database-defined subject terms. Limiting your search to a particular part of the record—for example, the author field—increases the precision of your search. Consider the difference, for instance, between these searches:

- Field search: $(AU = Einstein, Albert)$ AND $(SU = relativity)$. This field search might retrieve a very small set of results with only Einstein's writings on relativity.
- Keyword search: einstein AND relativity. This keyword search might include the preceding results but within a very large set of results with all items that include both the terms *einstein* and *relativity*.

10.3 INFORMATION CYCLE

Understanding how information is created and communicated is important in helping you decide what resources you should use for the various research questions you may have formulated at various times during the research process. Ask yourself these questions before you select resources to search:

- What is the time frame of these topic questions?
- Do I need a historical perspective or only information published in the past few years?
- Is this topic a fairly new one or one that has been discussed over a period of time?

Information goes through several phases of communication, development, and dissemination. Knowing at which phase you need to retrieve it will help you decide which tool to use. Look at the overview of coal gasification technology presented in Figure 10.3.

In the excerpt in Figure 10.4, you can see that publication on this issue has a long history and that the impetus for the research and application is different from one decade to the next. If your focus is historical, use books to develop a timeline and detailed descriptions of the phases of research in this area. If your focus is how coal gasification fits into current research for alternative fuels, use recently published articles that discuss current efforts in this area.

Knowing how information is created, communicated, distributed, and incorporated into the body of knowledge of a field or discipline will make it easier to decide how to find a specific type of information. The following discussion shows how information is communicated at various phases and how you can access it at each one of them. The process is diagrammed in Figure 10.5.

Scientists began to look at the possibility of coal gasification, a process for converting coal partially or completely into combustible gases, over 200 years ago. The process was developed by Scottish engineer William Murdock in 1792 (Clark). "Many U.S. cities and towns had their own coal gasification plant in the early to mid-1900s. Coal was heated in the presence of steam and a carefully controlled amount of air to produce a moderate-BTU gas that could be burned for heat and light. This so-called water gas was used in residences, businesses, and street lamps. The process produced a lot of waste and pollution, however. More efficient gasification processes were developed in the 1940s; Germany, for example, used gasification to produce gasoline and other liquid fuels during World War II. But the technology gradually gave way in the 1950s and 1960s to the use of natural gas, which was then cheaper than coal" (A Second Act for Illinois Coal?). Coal reserves are much greater than oil and gas reserves combined, making coal more viable as an energy resource. Gasification technologies are making coal more environmentally friendly than other resources. "The first coal gasification electric power plants are now operating commercially in the United States and in other nations, and many experts predict that coal gasification will be at the heart of the future generations of clean coal technology plants for several decades into the future" (U.S. Department of Energy).

FIGURE 10.4

Phases in the communication, development, and dissemination of information about a research topic.

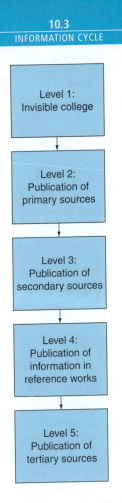

FIGURE 10.5
Process of the development, communication, and publication of research information.

10.3.1 LEVEL 1. THE INVISIBLE COLLEGE

The invisible college is "ground zero," the phase of information creation where an idea is developed and research is in progress. Researchers working on new ideas are communicating only with colleagues doing or interested in similar research or with experts on the subject. At this phase, information is usually recorded in lab notes, on computer files, in letters, or in e-mail messages to colleagues, in discussion lists, or in newsgroups. This network of informal information sharing among colleagues is known as the **invisible college**. A more formal phase is reached when ideas are presented at conferences that do not have published proceedings. A still more formal phase is reached when the ideas are published online as preprints or as technical reports through the authors' department. You may have heard the term *gray literature*, which refers to this sort of shadowy publication that it is difficult to find in libraries.

You can find this information using the following methods:

- Periodical database searches for letters published in journals
- Web search-engine searches using your topic words and the terms *research* or *research projects*
- Searches of databases that index e-print and preprint documents and technical reports.

E-prints are scholarly and professional works electronically produced and shared by researchers with the intent of communicating research findings to colleagues. They may include preprints, reprints, technical reports, conference publications, or other means of electronic communication.

10.3.2 LEVEL 2. PUBLICATION OF PRIMARY SOURCES

In the "invisible college," information is never formally published. Instead, the first formal publication of original research findings, referred to as **primary sources**, usually occurs at least two years after the research begins. These sources provide very detailed information. They can include conference papers, peer-reviewed journal articles, student theses, and technical or government reports.

You can find this information using the following methods:

- Subject-specific periodical database searches for articles, conference papers, and reports
- Web search-engine searches for freely available, peer-reviewed electronic journals on the Web
- Library catalog searches to find technical and government reports

10.3.3 LEVEL 3. PUBLICATION OF SECONDARY SOURCES

As research continues in a field, new journal articles are published that summarize research that has gone before and that present new developments. Similarly, books that summarize and provide perspective on the research will also be published. These publications are called **secondary sources**. They summarize and evaluate the main findings in primary sources. They provide more depth but perhaps less detail than the primary sources, and they provide historical perspective of the research. These may include reviews (e.g., the *Annual Review of Biomedical Engineering*), scholarly books, textbooks, journal articles with references, and popular magazine articles (which often summarize scholarly literature).

You can find this information using the following methods:

- Database searches for articles and the library catalog for new books
- Chronological sequencing of search results with the most recently published title first—a facility offered by many library catalogs
- Searching for terms such as *review, annual review,* or *advances* as a way of finding reviews on the topic

10.3.4 LEVEL 4. PUBLICATION OF INFORMATION IN REFERENCE WORKS

Resources such as dictionaries, encyclopedias, handbooks, manuals, and data books are called **reference works**. They consolidate and summarize information from primary and secondary sources. Reference works used to be published within two to seven years after the original publication on a subject, but online reference works and e-books often provide more frequent updates.

You can find this information using the following methods:

- Library-catalog searches for broad subject terms in conjunction with the term for the type of reference work you want to find—for example, `mathematics—tables`, `engineering—handbook` or `engineering AND dictionary`.
- Library–web site searches for such terms as *electronic reference resources, reference resources,* or *subject guides*. Librarians create and make available guides of web resources and print reference works for various subject areas or disciplines.

10.3.5 LEVEL 5. PUBLICATION OF TERTIARY SOURCES

Guides to the literature of a discipline or subject area enable you to find major primary and secondary sources, including reference works, books, journals, databases, and web sites. These guides, often called **tertiary sources**, enable you to get an overview of the resources available on your topic.

You can find this information with library-catalog searches using broad subject area in conjunction with terms such as *bibliography, research*, or *information sources*. For example, `engineering–bibliography` or `engineering AND information sources` would find Charles R. Lord's *Guide to Information Sources in Engineering*.

10.4 TYPES OF INFORMATION RESOURCES

When you begin your research, you may find that you use the resources discussed in the preceding section in the opposite order. Begin with a tertiary source, such as a guide to the literature related to your topic, which will point you to secondary and primary sources. However, if you know that you need recently published information, begin with a primary source, such as an article in a journal—one that reports original research—or a technical report that describes a project. The following sections describe each of these resources in turn.

10.4.1 REFERENCE WORKS

Reference works have the oldest, established information. Use them only for preliminary research to obtain an overall understanding of your topic. You cannot rely exclusively on general reference works; they can only get you started. You'll need to explore specialized reference works offering more specific information.

Reference works are usually labeled `REF` or `Reference Coll.` and shelved in a location in the library separate from materials that can be borrowed and taken out of the library. Because reference works cannot be taken from the library, you must use them there, taking notes or making photocopies of needed information. Don't forget to jot down information about the reference work itself so that you can cite the work properly.

An increasing number of reference works are available online in electronic format as e-books. Your library may provide access to online reference collections or to individual works. Many of the reference works you find in print may be available online where you can print information from them or e-mail it to yourself.

As already noted, reference works can be located through the library's computer catalog. You can search for reference works with the type of work, Boolean AND, and the subject or discipline: for example, `encyclopedia AND chemistry` or `handbook and engineering`. Table 10.4 lists major types of reference works and examples of each type. These can be in book or electronic, online formats.

TABLE 10.4
Major Categories of Reference Works

TYPE	EXAMPLE
General encyclopedias	*The Canadian Encyclopedia; Encyclopaedia Britannica Online*
Subject-specific encyclopedias	*AccessScience (online version of McGraw-Hill Encyclopedia of Science and Technology) Encyclopedia of Materials Science and Engineering*
General and subject dictionaries	*Comprehensive Dictionary of Engineering and Technology*
Biographical dictionaries	*The Biographical Dictionary of Scientists*
Almanacs	*The World Almanac and Book of Facts*
Atlases	*The New Atlas of Planet Management*
Yearbooks	*Kempe's Engineers Yearbook*
Gazetteers	*Columbia Lippincott Gazetteer of the World*
Books of quotations	*Practically Speaking: A Dictionary of Quotations on Engineering, Technology, and Architecture*
Indexes to articles in magazines, journals and newspapers	*Applied Science and Technology Abstracts Academic Search Premier (electronic)*
Databases	*Compendex (Ei Village 2) (electronic)*

TABLE 10.4 (Continued)

Handbooks (concise overviews of key topics; formulae and definitions, etc.)	*CRC Handbook of Chemistry and Physics Standard Handbook of Engineering Calculations Mark's Standard Handbook for Mechanical Engineers*
Guides (lists types of information sources)	*Information Sources in Engineering*
Data and tables	*CRC Handbook of Tables for Applied Engineering and Technology*
Test methods	*ASME Boiler and Pressure Vessel Code National Electric Code*
Standards	ASTM, ISO, etc.

10.4.2 PERIODICALS AND NEWSPAPERS

Periodicals are publications issued on a regular basis, but less frequently than daily publications such as newspapers. Popular magazines and academic, scholarly, and professional journals are examples of periodicals. Periodical publications can differ widely in content, readership, and frequency of publication.

- *National Geographic*, Scientific American, and *Discover* are examples of magazines. Magazines can offer good sources of opinion and current information.
- The *Journal of Applied Mechanics* is an example of a professional journal. Scholarly journals report findings of original research, and professional journals discuss current developments in a particular field.

Within libraries, current issues of periodicals may be displayed in separate areas, and back issues may be bound and placed on general stacks. Some libraries catalog these bound issues and shelve them by call number.

For many libraries, however, the bulk of the periodical collection may be online. Libraries subscribe to online periodical indexes, which contain citations (used to find articles), abstracts (brief summaries of articles), and some full text (the entire content of the article). Online periodical indexes not only enable you to find articles directly related to your research tropic, but also provide the full text of those articles as well. Thus, online periodical indexes enable you to use these resources when the physical library is closed.

Figure 10.6 illustrates the retrieval of an article's contents from a periodical index or database. The full text of the article content is stored digitally in the database, which is accessed via the Internet, then retrieved through a keyword, title, author, or subject term search.

10.4.3 INDEXES AND ABSTRACTS

An **indexing tool** enables you to find articles in the primary sources via an author index, a subject index, or both. What you get from an index is called a **bibliographic reference** or **citation**. It enables you to

FIGURE 10.6
Retrieval of an engineering article through a keyword search of a database.

find the actual article. Most indexing tools cover journal articles; some also cover conference papers. An **abstracting tool** (also known as an "abstract") is an indexing tool that also supplies an **abstract** or summary of the content of the articles listed. Use abstracts so that you can know what to expect to find in the article content and thus save time.

Table 10.5 lists periodical indexes and reference databases that are commonly found on university and engineering library web sites and that are useful for engineering research. Some of the general (for example, *Academic Search Premier*) and business indexes may also be available on public and college library web sites. Most of these are available only by subscription. **TRIS** and **TRB** are free to the public.

TABLE 10.5
Engineering Periodical Indexes and Reference Databases

DATABASE	USEFUL FOR
ABI/INFORM A business-oriented article database for subjects such as advertising, marketing, company information, industry trends, human resources, economic policy, health care, consumer products and services.	general engineering; engineering business
Academic Search Premier (1965–Present) A multidisciplinary database, offering information in nearly every area of academic study, provides full text for more than 4,600 scholarly publications, including more than 3,500 peer-reviewed publications. In addition to the full text, this database offers indexing and abstracts for all 8,040 journals in the collection.	general engineering
ACM (Association for Computing Machinery) Digital Library A vast collection of citations and full text from ACM journal and newsletter articles and conference proceedings.	computer science
Agricola (1970–Present) A database of literature citations for journal articles, monographs, proceedings, theses, patents, translations, audiovisual materials, computer software, and technical reports pertaining to all aspects of agriculture and land use.	bioresearch and environmental engineering

TABLE 10.5 (*Continued*)

Applied Science and Technology Abstracts covers over 350 key international English-language periodicals in the applied sciences and technology, and abstracts interviews, meetings, conferences, exhibitions, discussions, new product reviews, conference proceedings, and book reviews. It is a good source of information on management, careers and employment, and financial trends in scientific and technological fields.	general engineering; civil engineering
Biological Abstracts (1980 to present) An index with abstracts to articles in biology and the life sciences. Provides access to biological and medical research findings, clinical studies, and discoveries of new organisms. Coverage is international, and includes agriculture, biochemistry, biomedicine, biotechnology, genetics, botany, ecology, microbiology, pharmacology, and zoology.	bioresearch and environmental engineering
BIOSIS Previews® (1969–Present) The online version of *Biological Abstracts*® and *Biological Abstracts/RRM*® (Reports, Reviews, and Meetings), is the largest collection of biological sciences records in the world. BIOSIS Previews contains references to primary journal literature on vital biological research, medical research findings, and discoveries of new organisms. It contains references to traditional biology (botany, ecology, and zoology), interdisciplinary areas (biochemistry, biomedicine, and biotechnology), and related areas (instrumentation and methods).	environmental engineering
Business Source Premier (1965–Present) A major international database for business and management providing the full text of about 3,300 electronic journals with abstracts for over 300 additional titles. Many of the top scholarly journals are available from 1965, with some going back even further. BSP covers a wide range of journals, including over 1,000 peer-reviewed titles as well as trade and weekly journals and full text information such as country economic reports and detailed company profiles for the world's 5,000 largest companies.	engineering business

CAB Abstracts Covers the applied life sciences: agriculture, forestry, human nutrition, veterinary medicine and the environment, molecular biology, genetics, biotechnology, breeding, taxonomy, physiology and other aspects of pure science relating to organisms of agricultural, veterinary, or environmental importance.	bioresearch and environmental engineering
Chemical Abstracts (1907–Present) The CAplus database covers worldwide literature from all areas of chemistry and chemical engineering and information on commercial sources of chemical substances and regulations. The search engine allows for chemical structure searching.	chemical engineering
Compendex (1970–Present) This primary scholarly index for engineering is a comprehensive multidisciplinary engineering database with over 3 million summaries of journal articles, conference proceedings, and selected technical reports in all areas of engineering, including metal and materials science, mechanical engineering, electrical and computer engineering, and civil engineering.	general engineering; computer science; civil, electrical, industrial, mechanical, and nuclear engineering
Engineering Village 2 The computerized equivalent of the printed *Engineering Index*, covering all aspects of worldwide technology literature from 1970.	general engineering; computer science; civil, electrical, industrial, mechanical, and nuclear engineering
ENGnetBASE CRC Press Engineering Handbook and Reference Collection.	general engineering; nearly all engineering disciplines
Environmental Sciences A collection of databases including *Aquatic Pollution* and *Environmental Quality, Environmental Engineering Abstracts, Health and Safety Abstracts, Pollution Abstracts, Risk Abstracts, Toxicology Abstracts*, and *Water Resources Abstracts*.	environmental engineering

(Continued)

TABLE 10.5 (*Continued*)

IEEE (1988–Present) Contains the full text of all IEEE and IEE journal articles, conference papers, and standards from the Institute of Electrical and Electronic Engineers (IEEE) and the Institutions of Electrical Engineers (IEE).	computer science; electrical engineering
INSPEC via Engineering Village 2 (1969–Present) A database accessed through the Ei Village/ Engineering Village 2 interface. Includes scientific and technical journals and conference proceedings in physics, electrical engineering and electronics, computing and control, and information science.	general engineering; computer science; engineering physics; electrical and nuclear engineering
Knovel Library Engineering and Science Handbook and Reference Collection.	general engineering; nearly all engineering disciplines
Lexis/Nexis full-text trade and scholarly journals and newspapers	engineering business
Materials Sciences A collection of databases including *Metadex* (Metals Abstracts), *Engineered Materials Abstracts, Mechanical Engineering Abstracts, Weldasearch*, and *World Ceramics Abstracts*.	mechanical engineering
NTIS (1964–Present) Indexes and abstracts U.S. government-sponsored research, development, and engineering reports plus analyses prepared by 300 federal agencies (including NASA, DOD, DOE, EPA, DOT, and the Department of Commerce), their contractors, or grantees. Truly multidisciplinary, this database covers a wide spectrum of subjects including business, environment, health, and military science.	general engineering
ProceedingsFirst (1993–Present) Includes 19,000 citations to publications and proceedings from worldwide meetings, conferences, expositions, workshops, congresses, and symposia received by the British Library.	general engineering
Proquest Direct Full-text trade and scholarly journals and newspapers.	engineering business

SciFinder Scholar (1907–Present) Comprehensive database of chemistry information. Includes information such as patents, structures, reactions, and registry numbers. The Sub-Structure Module enables chemical sub-structure searches.	chemical engineering
TRB Research in Progress (http://rip.trb.org/) Contains over 8800 current or recently completed transportation research projects. Most of the RiP records are projects funded by Federal and State Departments of Transportation.	transportation engineering
TRIS Transportation Research Information Service (1968–Present) (http://tris.trb.org/about/) The world's largest and most comprehensive bibliographic resource on transportation information. TRIS is produced and maintained by the Transportation Research Board at the National Academy of Sciences with over 600,000 records of published and ongoing research. TRIS covers all modes and disciplines of transportation.	transportation engineering
Web of Knowledge Science Citation Index (1997–Present) SCI covers over 3,300 of the world's most significant scientific and technical journals in more than 160 engineering and scientific disciplines (the top 25–30% of journals in each discipline determined by citation frequency), with over 700,000 new items indexed each year. Find articles by keyword/author searching or by following "citation-webs" to other important articles that don't necessarily share similar keywords.	general engineering; environmental engineering
Web of Science (1990–Present) The ISI Web of Science Citation Databases (*Science Citation Index Expanded, Arts & Humanities Citation Index, Social Sciences Citation Index*) are multidisciplinary databases of bibliographic information gathered from thousands of scholarly journals. One can search for specific articles by subject, author, journal, and/or author address. Each article includes the article's cited reference list (often called its bibliography).	bioresearch and general engineering; engineering business; computer science; chemical, civil, electrical, environmental, industrial, mechanical, and nuclear engineering

10.4.4 PROCEEDINGS

Papers delivered at conferences are often published in documents called **conference proceedings** or as articles in professional journals. Some may appear on the author's web site or in a web-based conference summary. To find papers from conference proceedings, use the name (and abbreviation) of the sponsoring organization, the name of the conference series, the title of the conference in a particular year (which may be different each year and based on the conference theme), and the location and date of the conference. Because proceedings are typically published in the years following a conference, the publication year may not match the date of the conference.

To search for proceedings in library online catalogs, indexes, or databases

- Use the term search term `proceedings`.
- Choose the important words in the conference title or the name of conference series.
- Use the organization name or abbreviation (e.g., American Society of Mechanical Engineers or ASME) as the author name.

For example, this search, `proceedings AND robotics AND IEEE`, may yield 19 results, including *Proceedings/IEEE International Conference on Robotics and Automation*.

You can also use *ProceedingsFirst*, a database of the conference proceedings held by libraries worldwide, if it is available in your library. If you find proceedings unavailable in your library, you can request those proceedings through your library's interlibrary loan service.

10.4.5 TECHNICAL REPORTS

Technical reports typically describe the progress of research and development projects in universities, laboratories, institutes, and government agencies. Technical reports are published on a variety of topics, covering such subject areas as computer science, engineering, chemistry, energy, and all the science disciplines. Because these reports document ongoing projects, technical reports are not formally published. Many producers of technical reports publish the full texts of their reports online, and many more provide abstracts and citations. With a citation (which identifies the author, title, subject, and date of the report), you may be able to find the full text of a report. Table 10.6 lists some online resources that can provide either citations or the full text of technical reports.

TABLE 10.6
Online Resources for Finding Technical Reports

TITLE	DESCRIPTION
arXiv.org e-Print Archive (http://arxiv.org/	Provides keyword search, subject search and browse, open access to 400,420 e-prints in physics, mathematics, computer science, and quantitative biology.
Contrails Aerospace History—Illinois Institute of Technology (http://www.gl.iit.edu/wadc/)	This collection, formerly known as the Wright Air Development Center Digital Collection, consists of rare, declassified technical reports that chronicle and reference resources that discuss the history of flight.
DOE Information Bridge (http://www.osti.gov/bridge/)	Provides free public access to full-text documents and bibliographic citations of Department of Energy (DOE) research report literature in physics, chemistry, materials, biology, environmental sciences, energy technologies, engineering, computer and information science, renewable energy, and other topics from 1995 forward.
E-print Network: Research Communication for Scientists and Engineers (includes PrePRINT Network) (http://www.osti.gov/eprints/)	A searchable gateway to preprint servers that deal with scientific and technical disciplines of concern to DOE. Such disciplines include the great bulk of physics, materials, and chemistry, as well as portions of biology, environmental sciences, and nuclear medicine.
Energy Citations Database (http://www.osti.gov/energycitations/)	Energy Citations contains bibliographic records for energy and energy-related scientific and technical information from the Department of Energy (DOE) and its predecessor agencies; topics include chemistry, physics, materials, environmental science, geology, engineering, mathematics, climatology, oceanography, computer science, and related disciplines

TABLE 10.6 (*Continued*)

EnergyFiles (http://www.osti.gov/EnergyFiles/)	Over 500 databases and web sites containing information and resources pertaining to science and technology of interest to the Department of Energy, with an emphasis on the physical sciences. A subject listing is available.
Fire Research Information Services (FRIS) (http://www.bfrl.nist.gov/fris/)	55,000-item document collection related to fire research; includes published reports, journal articles, conference proceedings, books, and audiovisual items; some full text access.
GrayLIT Network (http://graylit.osti.gov/)	Collections include DTIC Report Collection, NASA Jet Propulsion Lab Reports, DOE Information Bridge, NASA Langley Technical Reports, and EPA Reports—NEPIS.
Hewlett Packard Technical Reports (http://www.hpl.hp.com/techreports/index.html)	Research is focused in five areas: service-centric infrastructure, access devices (from handhelds to cameras and printers), intelligent enterprise, digital consumer and foundation technologies. Reports range from 1990 to the present and are searchable by keyword.
Jet Propulsion Laboratory Technical Report Server (NASA) (http://trs-new.jpl.nasa.gov/dspace/)	A database of abstracts, citations and full text technical reports written by and for the scientific and technical community; searchable by keyword or browsable by publication year.
NASA Technical Reports Server (NTRS) (http://ntrs.nasa.gov/search.jsp)	Allows users to search the many different abstract and technical report servers maintained by various NASA centers and programs. NTRS includes access to the full-text of recent reports. Includes NASA RECONSelect.
National Technical Information Service (NTIS) (http://www.ntis.gov/)	U.S. government sponsored research and worldwide scientific, technical, and engineering information; unclassified and publicly available from research reports, journal articles, data files, computer programs and audio visual products from federal sources. Fee-based service.

TABLE 10.6 *(Continued)*

Networked Computer Science Technical Reference Library (NCSTRL) (http://www.ncstrl.org/)	Say "ancestral." A collection of computer science technical reports from a number of universities with CS and related programs. Abstracts and citations only, searchable by keyword.
On-line CS Techreports (http://www.cs.cmu.edu/~jblythe/cs-reports.html)	International collection, broken down by country and searchable by simple keyword.
Online Technical Reports (MIT) (http://libraries.mit.edu/guides/types/techreports/#online)	Topics include: aeronautics and astrophysics; civil and environmental engineering; computer science and electrical engineering; earth, atmospheric and planetary sciences; energy; mechanical engineering; nuclear engineering; ocean engineering.
Scientific and Technical Information Network (STINET) (http://stinet.dtic.mil/)	Public STINET provides access to all unclassified, unlimited citations to documents added into DTIC from late December 1974 to the present. There are also full-text versions of all unclassified, unlimited documents recently added into DTIC technical reports collection from September 1998 to the present.
Virtual Technical Reports Center—University of Maryland (http://www.lib.umd.edu/ENGIN/TechReports/Virtual-TechReports.html)	Institutions listed here provide either full-text reports, or searchable extended abstracts of their technical reports on the World Wide Web. This site contains links to technical reports, preprints, reprints, dissertations, theses, and research reports of all kinds.

Subscriptions to most databases are so expensive that individuals cannot afford them, although you can purchase individual articles from most of these databases. The most economical strategy is to access technical reports through a library that subscribes to them. However, a number of high-quality free databases are available on the web. Table 10.7 lists those most useful for engineering topics.

TABLE 10.7
Engineering-Related Free Online Databases

DATABASE	USEFUL FOR
Agricola (http://agricola.nal.usda.gov/) Has citations for journal articles, monographs, proceedings, theses, patents, translations, audiovisual materials, computer software, and technical reports pertaining to all aspects of agriculture and land use.	bioresearch and environmental engineering
Edinburgh Engineering Virtual Library (EEVL) (http://www.eevl.ac.uk/eese/) The "Engineering E-journal Virtual Library" covers the full text of over one hundred engineering e-journals, which are listed in the EEVL catalogue of engineering resources. In order to be selected, e-journals must be free, full text (or offer most of their content as full text) and available without registration.	general engineering; civil, electrical, environmental, industrial, and mechanical engineering
MatWeb (http://www.matweb.com/) MatWeb's database of material properties includes thermoplastic and thermoset polymers such as ABS, nylon, polycarbonate, polyester, and polyolefins; metals such as aluminum, cobalt, copper, lead, magnesium, nickel, steel, super alloys, titanium and zinc alloys; ceramics; plus a growing list of semiconductors, fibers, and other engineering materials.	civil engineering; industrial engineering; mechanical engineering
NIST Web Search Engine (http://www.boulder.nist.gov/) (National Institute of Standards and Technology) Lots of property data is published by the U.S. government and is available on the web. Use this site to search all NIST websites. Use the phrase "property data" and then the type of data you're looking for. Results can be very specific, but once you've identified a site useful to you . . .	general engineering; industrial engineering; mechanical engineering

NIST Manufacturing Engineering Laboratory (http://www.nist.gov/search.htm) National Institute of Standards and Technology. Developing many of the underpinning components of automated intelligent-processing systems that soon will be the core of all world-class manufacturing operations. These include intelligent machines; advanced sensors for real-time in-process measurements; software for precision control of machine tools; and information technology for integrating all elements of a product's life cycle.	industrial engineering; mechanical engineering;
NSSN Search Engine for Standards (http://www.nssn.org/) is a free online information service providing access to information about more than 225,000 approved standards. Search by title, abstract, keyword, or document number to find out if the standard you need exists, and if so, where to find it. If your library does not own the standard, it can be ordered through this website.	civil engineering; industrial engineering; mechanical engineering
Scirus (http://scirus.com/) The earliest coverage for some disciplines starts in 1973 but coverage for most disciplines begins with 1995. Scirus searches both free and membership sources on the web that contain scientific content, such as university web sites and author home pages. For information that is not freely accessible online, Scirus will present the title of the article, the author of that article, the source and some lines of text indicating the content. Scirus filters out nonscientific sites and finds peer-reviewed articles such as PDF and PostScript files, which are often invisible to other search engines.	general engineering; chemical engineering; environmental engineering

(Continued)

TABLE 10.7 (*Continued*)

Thomas Register (http://www2. thomasregister.com) Provides information on industrial products and services, including company and product information for more than 170,000 manufacturers, online catalogs, links to company web sites. Individual registration is required; each user must complete a registration form and receive a personal login and password.	engineering business; manufacturing; mechanical engineering;
TRIS Online (http://ntlsearch.bts.gov/tris/ index.do) Bibliographic database funded by sponsors of the Transportation Research Board (TRB), primarily the state departments of transportation and selected federal transportation agencies. TRIS Online is hosted by the National Transportation Library under a cooperative agreement between the Bureau of Transportation Statistics and TRB.	transportation engineering;

10.4.6 BOOKS

Books in the library's collection will probably be one of the main sources of information for your research projects, but you should remember that the collection is not limited to books on shelves. For example, many libraries include Safari Tech Books Online as part of their collections, as well as collections of full-text e-books such as NetLibrary (covering a variety of subjects) and Books 24 × 7 (covering computing and information technology). A search for books on drinking water quality in Canada might look like Figure 10.7.

Usually, you can search for books by subject, by title, by author, or by a combination of keywords. If available, advanced search options may allow you to limit by fields, search specific libraries within a university system, or search for particular formats such as DVD or videocassette tapes. Results of library-catalog searches look similar to what is shown in Figure 10.8.

If books on the subject are more than 10 years old, consider finding more current information in magazine and journal articles.

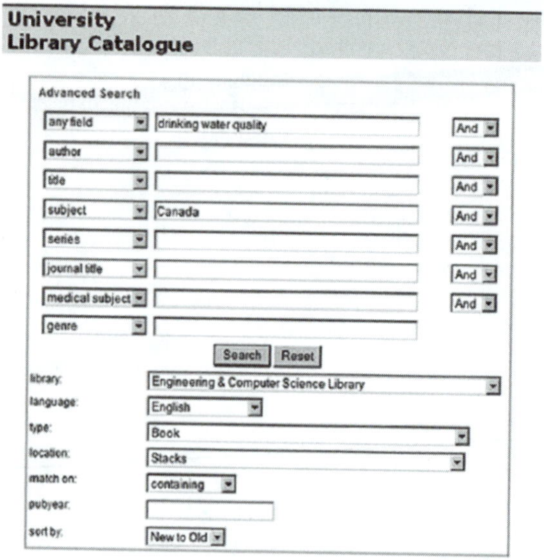
University
Library Catalogue

Advanced Search

any field	drinking water quality	And
author		And
title		And
subject	Canada	And
series		And
journal title		And
medical subject		And
genre		

Search Reset

library: Engineering & Computer Science Library
language: English
type: Book
location: Stacks
match on: containing
pubyear:
sort by: New to Old

FIGURE 10.7
Search form from an online library catalog.

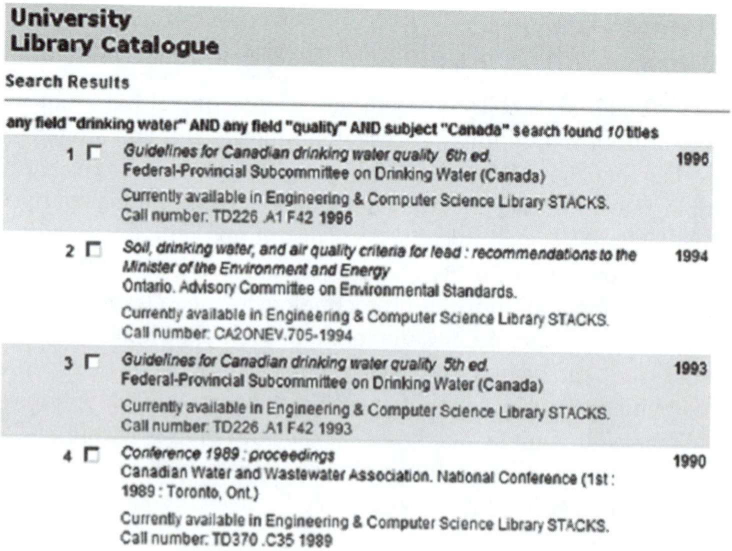
University
Library Catalogue

Search Results

any field "drinking water" AND any field "quality" AND subject "Canada" search found *10* titles

1 ☐ *Guidelines for Canadian drinking water quality 6th ed.* 1996
Federal-Provincial Subcommittee on Drinking Water (Canada)
Currently available in Engineering & Computer Science Library STACKS.
Call number: TD226 .A1 F42 1996

2 ☐ *Soil, drinking water, and air quality criteria for lead : recommendations to the Minister of the Environment and Energy* 1994
Ontario. Advisory Committee on Environmental Standards.
Currently available in Engineering & Computer Science Library STACKS.
Call number: CA2ONEV.705-1994

3 ☐ *Guidelines for Canadian drinking water quality 5th ed.* 1993
Federal-Provincial Subcommittee on Drinking Water (Canada)
Currently available in Engineering & Computer Science Library STACKS.
Call number: TD226 .A1 F42 1993

4 ☐ *Conference 1989 : proceedings* 1990
Canadian Water and Wastewater Association. National Conference (1st : 1989 : Toronto, Ont.)
Currently available in Engineering & Computer Science Library STACKS.
Call number: TD370 .C35 1989

FIGURE 10.8
Results of a keyword search of an online library catalog.

Consider also visiting government web sites that may provide additional information on the topic.

When you are looking for books in the library, print out the results displayed by your catalog search so that you can locate those book in the stacks. The catalog-search systems of some libraries actually enables you to create and print a nicely formatted bibliography. Check the search tips associated with the catalog you are using to see what your options are.

Table 10.8 lists the U.S. Library of Congress (LC) Classification System call numbers used for engineering topics. Once you have a call number, you can browse for other books and materials on your topic in the stacks, because books on the same or similar topics will be shelved together.

TABLE 10.8
Engineering-Related Call Numbers Used in the Library of Congress Classification System

Q	Science	TC	Hydraulic engineering
Q180	Operations research	TD	Environmental engineering
Q300–390	Artificial intelligence	TE	Highway engineering
QA	Mathematics	TF	Railroad engineering and operations
QA75–76	Computer science	TG	Bridge engineering
QA801–939	Analytical mechanics. Fluid mechanics	TH	Building construction
QC	Physics	TJ	Mechanical engineering and machinery
QD	Chemistry	TJ163	Power resources
QE	Geology	TJ210.2–225	Robotics
R	Medicine	TK	Electrical engineering. Electronics. Nuclear engineering

TABLE 10.8 (Continued)

R856–857	Biomechanical engineering	TK5101–6720	Telecommunications
T	Technology	TK7800–8360	Electronics
T55.4–60.8	Industrial engineering	TK9001–9401	Nuclear Engineering
T385	Computer graphics	TL	Motor vehicles. Aeronautics. Astronautics
TA	Engineering (general). Civil engineering	TN	Mining engineering. Metallurgy
TA349–359	Mechanics of engineering	TP	Chemical technology
TA401–492	Materials of engineering and construction	TS	Manufacturing
TA630–820	Structural engineering. Geotechnical engineering	VM	Naval architecture. Marine engineering

The Dewey Decimal Classification (DDC) System for cataloging materials is used in some college and university libraries. You will also find the DDC system used in public libraries and in some web resources. BUBL Information Service (http://bubl.ac.uk/), for instance, uses the Dewey Decimal Classification system as the primary organizational structure for its catalog of Internet resources. Table 10.9 shows the call numbers in the DDC that may be useful for engineering.

Some libraries may shelve books and other media together in the stacks, others may have separate media areas, but usually the same classification system is used for all materials. **Other media** include records, tapes, compact discs, DVDs, films, videos, visuals, and multimedia. Because you may not be able to borrow these materials from academic libraries, you will have to use them in the library.

TABLE 10.9
Dewey Decimal Classification Call Numbers Used in Engineering

500	**Science and mathematics**	551	Geology, hydrology, meteorology
510	Mathematics	**600**	**Technology**
530	Physics	608	Invention & patents
531	Classical mechanics Solid mechanics	610	Medicine
532	Fluid mechanics Liquid mechanics	620	Engineering & allied operations
533	Gas mechanics	621	Applied physics
534	Sound and related vibrations	622	Mining and related operations
535	Light and paraphotic phenomena	623	Military and nautical engineering
536	Heat	624	Civil engineering
537	Electricity and electronics	625	Engineering of railroads, roads
538	Magnetism	627	Hydraulic engineering
539	Modern physics	628	Sanitary and municipal engineering
540	Chemistry and allied sciences	629	Other branches of engineering
541	Physical and theoretical chemistry	630	Agriculture
542	Techniques, equipment, materials	631	Techniques, equipment, materials
543	Analytical chemistry	660	Chemical engineering
544	Qualitative analysis	661	Industrial chemicals technology
545	Quantitative analysis	662	Explosives, fuels technology
546	Inorganic chemistry	670	Manufacturing

TABLE 10.9 (*Continued*)

547	Organic chemistry	680	Manufacture for specific uses
550	Earth sciences	681	Precision instruments & other devices

10.4.7 STANDARDS

Standards are often located in a special section of the library. One way to find standards is to search the American Society for Testing and Materials (ASTM) site at http://www.astm.org/ (click on "Standard Search"). Use a phrase or keyword search (e.g., *piston pump*), browse by interest area, or view alphanumeric listings of standards. However, ASTM standards are not available for free online. Some university and engineering libraries purchase the 70+ volume *Annual Book of ASTM Standards*, containing ASTM's 12,000+ standards. This work is available in print, CD-ROM, and online formats. You can order individual documents from the ASTM web site.

Another way to search for standards is to use the NSSN Search Engine for Standards (http://www.nssn.org/). The National Resource for Global Standards (NSSN) is a search engine that provides standards-related information from a wide range of developers, including organizations accredited by the American National Standards Institute (ANSI), other U.S. private-sector standards bodies, government agencies, and international organizations. Full text of these standards is not available free either. Instead, check a large library, which should have most of the current standards in print in its collections.

10.4.8 TEST METHODS

You can find books on test methods in the reference area in university and engineering libraries. You can also use WorldCat to find local libraries that hold titles on test methods. A sample Google search for "`find in a library`" "`test methods`" "`engineering`" (including the quotation marks) resulted in two titles with twelve library locations. You can search for actual titles—for example, the *NIOSH Manual of Analytical Methods*.

10.4.9 INTERLIBRARY LOAN

If a book that you need is not available at your local college or university, see if it is available through interlibrary loan. Ask a reference librarian about procedures for this service, and get an estimate on how long it will take. If your research project has a tight deadline, interlibrary loan may not be an option.

Another option may be OCLC WordCat (http://www.worldcat.org/), self-described as "the world's largest network of library content and services." Using WorldCat, you can search many libraries at once for an item, possibly locating it in a nearby library. You can get WorldCat results using one of the following methods:

- Include `find in a library` in a Google search.
- Click a library link in Google Scholar (scholar.google.com) and Google Books (books.google.com).
- Add `site:worldcatlibraries.org` to a Yahoo! search.

10.4.10 GOVERNMENT DOCUMENTS

Many federal, provincial, and local government departments provide a wealth of useful research information that may be relevant to your research. To find out what these departments have available, call them directly, using the numbers listed in the blue pages of the phone book. An all-encompassing government web site for federal and state agencies in the United States is USA.gov at http://www.usa.gov/ (formerly FirstGov.gov). You can search the entire site, choose from "Government Information by Topic," select from the "A-Z agencies index," or use the Reference Center for "Data & Statistics" and other resources.

A useful example for the federal government of Canada is 1-800-OCANADA. Often, operators will connect you with the specific department and possibly the person you need to speak with to obtain certain information. If you require statistical information, a good first step is to visit the Statistics Canada website, at http://www.statcan.ca.

10.4.11 PATENT AND TRADEMARK DOCUMENTS

Patents are good sources of information about new technology and inventions. Usually, you can search by keyword or browse subjects in patent searching databases. Table 10.10 lists Canadian, U.S., and other patent web sites.

TABLE 10.10
Selected Internet Sites on Patents, Trademarks, and Other Intellectual Property

---

The Canadian Intellectual Property Office (CIPO)
(http://strategis.ic.gc.ca/sc_mrksv/cipo/youlcome/youlcom-e.html) is the patent, trademark, and copyright administration body of Canada.

The Canadian Patent Database (http://Patents1.ic.gc.ca/intro-e.html) provides access to over 75 years of Canadian patent descriptions and images from more than 1,400,000 patent documents. The web site is maintained by the Canadian Intellectual Property Office (CIPO).

United Kingdom Patent Office Home Page (http://www.patent.gov.uk/).

United States Patent and Trademark Office Home Page
(http://www.uspto.gov/) is the starting point for the U.S. Patent and Trademark Office's free patent databases, including both the U.S. Patent Bibliographic Database (1/1/76-present) and the AIDS Patent database. The following tutorials can help you search the databases efficiently:

- **The Patent Searching Tutorial** (http://www.lib.utexas.edu/engin/patent-tutorial/index.htm) University of Texas—Austin.
- **Research U.S. Patents on the web** (http://library.ucf.edu/GovDocs/PatentsTrademarks/Research.asp) University of Central Florida Libraries.
- **Schreyer Business Library's Patent Search Tutorial** (http://www.libraries.psu.edu/instruction/business/Patents/index.html) Pennsylvania State University Libraries.
- **Web Patent Searching Tutorial** (http://scilib.ucsd.edu/howto/guides/patsearch/) University of California—San Diego.

European Patent Office (esp@cenet)

(http://ep.espacenet.com/?locale=EN_ep) provides a searchable database. The site indexes over 30 million patent documents from European and other patent offices worldwide, including Japanese, German, and U.S. patents. Records provide basic patent information and often an English abstract and image of the first page. Some records provide full-text descriptions, claims, and drawings. German and U.S. patents are indexed from 1920 (providing very basic patent information and drawings) but are available with full text and images from 1972 to present.

Free Patents Online (http://www.freepatentsonline.com/) provides free access to U.S. and European patent data, free PDF downloading, free account features that let you organize and store documents and searches, and more.

Patent Offices around the World (http://www.pcug.org.au/~arhen/).

World Intellectual Property Organization
(http://www.wipo.int/portal/index.html.en) This website includes the full text of the *Patent Cooperation Treaty* and provides links to national patent offices.

WIPO Intellectual Property Digital Library (http://www.wipo.int/ipdl/en/).

10.4.12 WEB SEARCH ENGINES AND DIRECTORIES

The World Wide Web, the graphical user interface to Internet resources, can provide you with a feast of information or meager pickings, depending on what you are seeking. The web is a diffuse and relatively unorganized collection of documents and information stored in many computers all over the world. It is not carefully organized and searchable the way databases are. Documents that you find on the web are linked to one another by a tangled web of hypertext links. Finding a specific document or information about a specific topic of interest can often be a challenge, despite advanced search algorithms. The web may be a good source for current news and material on popular culture. However, no matter what you are looking for, your best strategy is to start with library sources. That way, you have a better understanding of the value of what you find.

There are hundreds of World Wide Web search engines that try to index as many web documents as possible, either by mechanical means—"robots" and "spiders"—or through human submissions. The search engines search their own databases, and *none* of them, no matter how comprehensive, indexes more than about one-third of the public ally available web pages that make up the entire World Wide Web.

To achieve as comprehensive a search as possible, use several search engines, or use one of the multiple search-engine services. *Dogpile* (http://www.dogpile.com/), for instance, is a multiple search engine that provides a Search Comparison Tool (http://comparesearchengines.dogpile.com/) showing the differing results from the three major search engines—*Google*, *Yahoo!*, and *MSN*.

To make the most efficient use of any search engine, read its help documentation to learn its particular functions. Experience and experiment will show you which engines give you the best results. Table 10.11 provides URLs and brief descriptions of some web search engines you might consider.

Use a type of field searching (discussed previously) to increase your precision with search engines. Google and Yahoo!, for example, provide an advanced search that lets you limit your search by site, URL (web address), title, domain, or a combination of these modifiers. You can learn about all of these modifiers at Google using the Query Modifiers link at http://www.google.com/help/operators.html. You can select some of them automatically on the Google Advanced Search

TABLE 10.11
Web Search Engines

Alltheweb (http://www.alltheweb.com/) Uses Yahoo!'s index to search billions of web pages, images, video, audio and news, as well as PDF and MS Word® files.
Altavista (http://www.altavista.com/) Searches the web, images, MP3/audio, video and news.
AltaVista Canada (http://ca.altavista.com/) Searches the web, images, MP3/audio, video, and news and provides access to an index of millions of Canadian-specific web pages.
Google (http://www.google.ca/) Searches web pages, PDF, PowerPoint, and Word documents among other formats. Search options include image search, group search, a directory, news search, Google Scholar, and Google Books.
Google Scholar (http://scholar.google.com/) Like all Google searches, the results will be by relevance and page rank system, meaning that articles or books cited by other works will be at the top of the list. You can search for books, articles, and web sites at one time. This is a good "discovery" search technique but requires extra steps. You usually cannot get full-text articles, but will have to perform another search to see if a library subscribes to the journal or magazine the article is in, or if it is available in full text through a subscription database. You can click on "Find in a Library" to locate a book. You cannot sort Google results. This means you cannot arrange results by publication date or in other ways that might be more meaningful.
MSN (http://www.msn.com) Searches the web, images, news, and maps.
Yahoo! (http://search.yahoo.com) is one of the three major Web search engines and probably has the oldest Web subject directory. Yahoo! Search searches web pages, documents in various formats, images, news, video, audio, shopping, and Creative Commons licensed content.
Yahoo! Canada (http://ca.yahoo.com/) Canadian version of Yahoo.

page in the Occurrences drop-down menu (http://www.google.com/advanced_search).

One particularly useful modifier is likely to involve web domains. Restricting your search to education and government web sites using *edu* or *gov* modifiers may produce better results for research projects. Table 10.12 shows the most common web domains.

Sometimes, rather than using a search engine, you may want to make use of web directories, subject trees, portals, or online libraries whose compilers have organized information on the web into categories—for example, mechanical engineering. The advantage of these collections is that their compilers have already filtered web sites for you, supposedly separating the wheat from the chaff. These sites are usually searchable by keyword, which is

TABLE 10.12
Web Domains

.edu or .ca	Educational institutions in the United States and Canada, respectively
.gov	Government body
.org	Organization with a specific mission or philosophy
.com	Commercial site, used to sell a product or service

the best way to find information on them. Drilling down through hierarchies of web links is less efficient, although this method may provide an overview of "what's out there." Table 10.13 lists collections of links for science and technology web information that are especially comprehensive.

TABLE 10.13
Subject Directories, Portals, and Online Libraries

Bartleby Library (http://www.bartleby.com) An open-access, free, online library of reference sources, including thesauri, dictionaries, quotation sources, and more.
Bubl Link (http://bubl.ac.uk/) Hosted by the Andersonian Library at the University of Strathclyde-Glasgow, this is a highly maintained subject directory of selected Internet resources covering all academic subject areas. Bubl Link uses the Dewey Decimal Classification system as the primary organization structure for its catalog of Internet resources. Each site is evaluated, cataloged, and briefly annotated.
Canada Institute for Scientific and Technical Information (http://www.nrc.ca/cisti) CISTI, the Canada Institute for Scientific and Technical Information, is one of the world's major sources for information in all areas of science, technology, engineering and medicine. CISTI began over 75 years ago as the library of the National Research Council of Canada, the leading agency for R&D in Canada, and became the National Science Library in 1957. The change to CISTI came in 1974 to reflect the wide scope of services provided and its increasing role in the development of electronic information products and services for the sci/tech community. Through this site a library catalog may be searched, and document delivery is available.
CiteSeer (http://citeseer.ist.psu.edu/) A scientific literature digital library, focusing primarily on the literature of computer and information science

TABLE 10.13 (*Continued*)

that indexes the full text of entire articles and provides citations of Postscript and PDF research articles on the Web. CiteSeer uses search engines and crawling plus document submissions to harvest papers on the Web; allows full Boolean, phrase, and proximity searching; and provides full source code at no cost for non-commercial use.

Edinburgh Engineering Virtual Library (EEVL) (http://www.eevl.ac.uk/eese/) The "Engineering E-journal Virtual Library" covers the full text of over 100 engineering e-journals, which are listed in the EEVL catalog of engineering resources. In order to be selected, e-journals must be free, full text (or offer most of their content as full text) and available without registration.

Engineering: Selected Internet Resources (Science Reference Services, Library of Congress) (http://www.loc.gov/rr/scitech/selected-internet/engineering.html) Links to many sites related to engineering, including specialized government sites, national laboratories, societies, and the like.

Infomine (http://infomine.ucr.edu/) Developed by the Library of the University of California-Riverside, this is a librarian-built, virtual library of useful Internet resources such as databases, electronic journals, electronic books, bulletin boards, mailing lists, online library catalogs, articles, directories of researchers, and many other types of information. The directory can be searched by keyword or browsed by keyword, author, title, Library of Congress Subject Heading, or LC call number. Some content is fee-based.

Resource Discovery Network (http://www.rdn.ac.uk/) The UK's free national Internet gateway: a collaboration of educational and research organizations. Unlike web search engines, RDN, using the expertise of specialists from participating institutions, selects catalog and delivers more than 100,000 high-quality Internet resources, as well as access to the "Invisible Web," through a series of subject-based information gateways (or hubs). EEVL: the Internet Guide to Engineering, Mathematics and Computing, is one such linked hub. Though primarily aimed at Internet users in UK higher education, RDN is freely available to all.

Science.gov (http://www.science.gov/) A gateway to authoritative selected science information provided by U.S. Government agencies, including research and development results.

SciCentral: Gateway to the Best Science and Engineering Online Resources (http://www.scicentral.com/) Over 50,000 resources containing over 120 specialties in science, medical research, and engineering, created and maintained by professional scientists.

WWW Virtual Library—Engineering (http://vlib.org/Engineering.html) Embraces all of the engineering disciplines listed in this excellent online library.

10.5 EVALUATING WHAT YOU FIND

When doing library research, you can't just automatically use the information you've found directly in a document. You must evaluate that information. Use the table of contents, the abstract, and the opening and concluding paragraphs to evaluate whether the material you have located is pertinent to your topic. Even if a resource you have found meets this first criterion, you still must consider these other criteria:

TABLE 10.14
Criteria for Evaluation of Information Sources

Point of view	Does this article or book seem objective, or does the author have a bias or make assumptions? What was the author's method of obtaining data or conducting research? Does the web site aim to sell you something or just provide information? What is the author's purpose for researching and writing this article or book?
Authority	Who wrote the material? Is the author a recognized authority on the subject? What qualifications does this author have to write on this topic? Is it clear who the intended audience is? What is the reputation of the publisher or producer of the book or journal? Is it an alternative press, a private or political organization, a commercial press, or university press? What institution or Internet provider supports this information? (Look for a link to the homepage.) What is the author's affiliation to this institution?
Reliability	What body created this information? Consider the domain letters at the end of a web address (URL) to judge the site's quality or usefulness. What kind of support is included for the information? Are there facts, interviews, and statistics that can be verified? Is the evidence convincing to you? Is there any evidence provided to support the author's conclusions, such as charts, maps, bibliographies, or documents? Compare the information provided with other factual sources.
Timeliness	Has the site been recently updated? Look for this information at the bottom of a web page. How does the copyright of a book or publication date of an article affect the information contained in it? Do

TABLE 10.14 (*Continued*)

	you need historical or recent information? Does the resource provide the currency you need?
Scope	Consider the breadth and depth of an article, book, web site, or other material. Does it cover what you expected? Who is the intended audience? Is the content aimed at a general or a scholarly audience? Based on your information need, is the material too basic, too technical, or too clinical?

You should discard information resources:

- If the author has an ax to grind, seeks to sell something, is unaffiliated with a reputable organization, or is an unknown.
- If the information has been produced by an alternative press or a private or political organization.
- If the assertions in the information cannot be verified.
- If the information lacks a bibliography.
- If the information is old and likely to be outdated.

10.6 SAVING INFORMATION YOU FIND

It is often a good idea to download material from a web site onto a diskette or flash drive. Web sites often change from day to day, and the material you find today may not be accessible tomorrow. Keep a log so you know when you referenced the web sites you use. You can print, download, or save articles from library subscription databases or e-mail them to yourself. To avoid plagiarism, you can use a different font or a different font color when copying and pasting content from online or electronic resources. That way, it is easy to tell where your words stop and the author's words begin. Include enough of the citation in brackets after the borrowed material that you will be able to cite it properly if you use it in your research project.

Bibliography

The American Heritage Dictionary of the English Language. 4th ed. Boston: Houghton Mifflin, 2006.

Assessing Website Quality. 2006. TRIO Virtual Center, University of Washington. September 12, 2006 <http://depts.washington.edu/trio/center/howto/design/site/assess/index.html>.

Clark, E. L. "Coal Gasification." N.d. <u>Troubled Times</u>. September 12, 2006 <http://www.zetatalk.com/energy/tengy11a.htm>.

Clarke, Roger. <u>Roger Clarke's Information Wants to Be Free</u>. February 24, 2000. November 11, 2006 <http://www.anu.edu.au/people/Roger.Clarke/II/IWtbF.html>.

<u>Databases/Article Indexes for Engineering</u>. April 10, 2006. Drexel University Libraries. September 12, 2006 <http://www.library.drexel.edu/resources/dbsubjects/engineering.html>.

<u>Engineering, Scientific & Technical Databases</u>. Updated October 24, 2006. Engineering Library, University of Washington. September 12, 2006 <http://www.lib.washington.edu/Engineering/guides/englibdb.html>.

<u>Evaluate Web Pages</u>. N.d. Widener University, Wolfgram Memorial Library. September 12, 2006 <http://www3.widener.edu/Academics/Libraries/Wolfgram_Memorial_Library/Evaluate_Web_Pages/659/>.

Lord, Charles R. *Guide to Information Sources in Engineering*. Englewood, CO: Libraries Unlimited, 2000.

"Preparing to Research." <u>EV 800 Library Research Tutorial</u>. 2005. Milwaukee School of Engineering. July 25, 2006 <http://www.msoe.edu/library/EV800/>.

<u>A Second Act for Illinois Coal</u>? 2005. Southern Illinois University. September 12, 2006 <http://www.siu.edu/~perspect/05_sp/coal1.html>.

U.S. Department of Energy. <u>Fossil Energy: DOE's Coal Gasification R&D Program</u>. Updated June 27, 2006. U.S. Department of Energy. September 12, 2006 <http://fossil.energy.gov/programs/powersystems/gasification/index.html>.

<u>WorldCat [OCLC]</u>. 2001–2006. OCLC Online Computer Library Center, Inc. September 12, 2006 <http://www.worldcat.org/>.

11

ENGINEERING DOCUMENTS

11

Engineering Documents

11   Engineering Documents

The first thing to know about engineering documents is that they go by many different and conflicting names. What is a proposal to one professional engineer is a recommendation report to another. It is important to know the purposes, structure, contents, and strategies related to the engineering documents presented in this chapter, but don't insist on the names used here. Listen carefully to the details of what your client, your supervisor, your manager, or your instructor wants and write accordingly.

11.1 BASIC ENGINEERING REPORTS

Engineering reports that might be called "basic" are the most variously named. Among the many names are inspection reports, accident reports, field reports, investigation reports, trip reports, site reports, and reconnaissance reports, to name a few. Perhaps the only common feature of these reports is that they are short, two to four pages usually. And they are "reports" in the strict sense of the term—the engineer–writer goes somewhere, investigates something, and reports back.

As an engineer, you might need to visit a site and inspect something—for example, the use of solar power at Austin Bergstrom International Airport or the use of maglev trains in

Shanghai. You might have to inspect a project being handled by a contractor working for your engineering firm. You might have to inspect damage caused by a flood or tornado or investigate a problem involving malfunctioning equipment. Perhaps you might go to an engineering conference; you might be expected to write a brief narrative of what you did, what you saw, and what you learned.

All of the "basic" engineering reports can be reduced to one or a combination of essential types of writing: description, narration, process analysis, and cause–effect analysis.

11.1.1 DESCRIPTION

Site, inspection, and accident reports all contain some description— of the site inspected or of the accident. Examples of description are shown in Figures 11.1 and 11.2.

11.1.2 NARRATION

Trip and accident reports, and certainly others, are likely to contain **narration**. You are narrating when you describe the events of something that actually happened. You may have gone to a conference on new photovoltaic technologies and were required to write a report on what you did and what you saw. As an engineer specializing in

---

Hospitals

There was no structural damage reported at any hospital as of July 7, except for some displacement of a precast concrete balcony facade at the Eisenhower Medical Center in Rancho Mirage. Nonstructural damage included a broken water line in the kitchen at the JFK Hospital in Indio. An elevator at Desert Hospital in Palm Springs shut down due to damage in the shaft. Counterweights at 5 of 8 elevators came out of their tracks at Loma Linda University Medical Center, and seismic expansion joints were damaged. Two elevators were disabled because of damage to counterweights at St. Bernardine Medical Center. Superficial cracking of interior drywall was widely reported.

FIGURE 11.1
Example from an accident report describing damage done by an earthquake.

The Lewis Center is a two-story, 13,600-ft² (1,260-m²) building with classrooms, offices, an auditorium, an atrium, and an on-site wastewater treatment system used by the students in the Environmental Studies Program. The design team incorporated numerous energy-saving design characteristics into the building. A prominent feature of the building is a roof-integrated, 60-kW photovoltaic (PV) system that produces electricity on-site. The system, which covers the entire roof, is connected to the local utility grid and does not have a battery backup system. The PV system exports power to the utility grid when the PV system produces more power than the building is currently using. Likewise, the building imports electricity from the utility when the PV system cannot meet the load. Electricity meets all energy needs, including mechanical systems and domestic hot water. This all-electric system was a requirement in order to meet the future net energy-producing vision for the building.

FIGURE 11.2
Example from a site inspection report, describing a photovoltaic installation at a school.

earthquakes, you may have written a narrative on the events of and the destruction caused by an earthquake. An example of narration is shown in Figure 11.3.

11.1.3 PROCESS ANALYSIS

Inspection reports, as well as some of the others, may explain one or more **processes** at a site—that is, step-by-step operational discussion. An example of process analysis is shown in Figure 11.4.

Earthquake Sequence

The Landers earthquake had an unusually shallow preliminary focal depth of 1–3 km. The earthquake began on the north side of the Pinto Mountain fault, resuming the northward rupture of the Johnson Valley fault but with a 2 km westerly offset relative to the Joshua Tree rupture. The Landers rupture propagated northward on the Johnson Valley fault, but then began a series of easterly steps across to the Homestead Valley, Emerson, and Camp Rock faults, with the strike of each successive fault bending further to the west.

The Big Bear earthquake was not preceded by any precursor events similar to those preceding the Landers event. The Big Bear earthquake occurred at a focal depth of about 10 km and ruptured northeastward from near Yucaipa toward the Camp Rock and Emerson faults. The rupture zones of the two earthquakes form a triangle about 70 km on a side, with the Landers rupture on the right side, the Big Bear rupture forming part of the left side, and two strands of the San Andreas fault (the Mission Creek and Banning faults) on the base.

FIGURE 11.3
Example from an accident report narrating the events of the July 22, 1998 Landers and Big Bear earthquakes.

In the Lewis Center classrooms and the conference room, ventilation air is controlled through an outdoor air damper, which is opened to 50% when occupancy sensors detect motion. Occupancy sensors also control the heating and cooling set points for the classrooms. When a classroom is unoccupied, no outdoor air is supplied to the space, and the temperature is controlled to a setback position. When the energy-management system senses that the space is occupied, the temperature set point is switched to an occupied comfort position. The ventilation, heating, and cooling set points in the offices are manually controlled.

Ventilation supply air for the classrooms, offices, and corridors is handled by HP-5, a single large, standard range water source heat pump. According to the manufacturer's specifications, HP-5 has a nominal cooling capacity of 120,000 Btu/hr (35.2 kW) with an EER of 11.4 for cooling and a nominal heating capacity of 135,000 Btu/hr (39.6 kW) with a COP of 3.8 for heating at ARI-320–93 standards. The HP-5 supply fan provides 3500 cfm (4.7 m³/s) of ventilation air to occupied classrooms and offices. Ventilation supply air to the classrooms and offices is 100% outdoor air (there is no mixing of supply and return air streams).

FIGURE 11.4
Example containing process discussion.

The damaged bowling alley in Yucca Valley is a steel frame structure with tapered roof girders enclosed by CMU walls on the west and north sides, a lobby along the front, and a stud wall along the east side. During the earthquake, the stud wall pulled away from the building and fell on the adjacent vacant lot, *apparently due to inadequate roof-to-wall connections.*

In the Big Bear area, there were several notable failures related to cripple wall foundations in relatively new houses. In several instances failure was due to *improper nailing of the plywood.* In a large two-story house with a cripple wall failure, calculations showed the failed plywood shear walls to be *severely overloaded,* even though they were sufficient to meet the requirements of the single-family dwelling code. One building, lacking adequate connections to concrete piers, slid about four and a half feet down a slope impacting a structure below.

Roof damage from collapsed masonry or stone chimneys was common. Interior chimneys confined by roof and ceiling framing appear to have performed better than exposed chimneys. Some of the exposed chimneys crumbled to the ground, although the majority were observed to fail near the roof line. Of 25 damaged masonry chimneys looked at closely by the San Bernardino County Department of Building and Safety, 24 were *improperly constructed.* Among the flaws noted were *absence of reinforcing steel, discontinuous reinforcing, poor grouting, and inadequate ties to the wood framing.*

FIGURE 11.5
Excerpt from an investigation report in which not only damage is described but causes are explained. (Italics included to indicate causal explanations.)
EERI Special Earthquake Report, August 1992

11.1.4 CAUSE–EFFECT ANALYSIS

Investigation and accident reports, as well as some of the others, may seek to explain why the problem occurs or why the accident occurred.

But what does a complete basic engineering report look like, once fully assembled? To answer that question, first consider overall format. Such a report can be housed in a memo, an e-mail, a business letter, or even a formal though brief report framework. Figure 11.6 shows an internal memo reporting describing the status of a construction project.

Date: Fri, 6 June 2009
To: Patrick Hughes
From: Jane A. McMurrey
Subject: Downtown construction update

Patrick, the following is an update on downtown construction work:

SD42 Downtown Sioux Falls
The work downtown is complete except for sealing the joints. This work would be quick, but traffic should be alert to workers and equipment when they are present.

18th Street bridge over I229
Modifications are nearing completion on the structure over I229. The steps on the south half of the structure for pedestrian movements are in place. Bridge rail work along the sidewalk on the north half of the structure has begun and should be completed next week. Sidewalk on the NE corner has been removed and replaced to provide a flatter slope onto the bridge sidewalk. Two-way traffic will be maintained; however, the shoulder area of 18th Street through the bridge will be closed. Traffic should be alert for construction equipment.

Interstate 29 and Madison Street in Sioux Falls—Minnehaha County
Industrial Builders, Inc., continues building footings for the new interstate bridges over Madison Street. They have also started construction on the south-bound bridge over the Ellis and Eastern Railroad.

Runge Enterprises is continuing the grading on the new embankment for the southbound interstate lanes. Grading has started on Madison Street west of I-29.

Metro Construction has started installing a storm sewer on Madison Street east of I-29 and will continue installation through next week.

FIGURE 11.6
Engineering report using a memo format.

11.2 INSTRUCTIONS AND USER GUIDES

If you write for nonengineers—for example, customers, clients, or employees who have no engineering background—you're likely to write consumer-oriented instructions, which are covered in the next section. If you write instructions to be used with a product or service, you are likely to write a user guide, which is a formal published document. And finally, if you write instructions for fellow engineers, you're likely to use the format for operating specifications shown at the end of this section.

11.2.1 CONSUMER-ORIENTED INSTRUCTIONS

Think of "consumers" as customer, clients, or employees who have no engineering or technical background. As a consumer yourself, you have probably read plenty of instructions but may not have paid much attention to their structure, organization, format, and style. The following provides a quick review of the essentials of writing instructions along with the extras that turn a set of instructions into a user guide.

APPROACHES TO INSTRUCTION WRITING

To begin, instructions provide steps and supplemental explanation that enable readers to operate, construct, maintain, or repair something. Instructions tell readers exactly how to do something— exactly which actions to perform. However, there exist at least two fundamental approaches to writing instructions, one of which can easily cause problems and the other of which almost always works the best:

- **Feature- and function-oriented approach:** One method is to explain what each button, switch, knob, menu, and option—associated with either hardware or software—does. This method works if the product is simple, the labels are adequately self-descriptive, and combinations of features are not needed to accomplish tasks. For example, the Exit button or the On–Off switch clearly indicate their function. However, the crop function in image-editing software requires that you first select an area then press Crop. The function approach breaks down at this point: you have to combine multiple functions in order to achieve the task; the button by itself does not convey that idea.

- **Task-oriented approach:** The better method for writing consumer-oriented instructions is to explain how to accomplish tasks related to the product. Instead of writing a section on the crop function or button, you write a section entitled "Cropping an Image" or "How to Crop an Image." The steps then direct the user to outline an area using the selection tool and then click **Crop**.

WRITING STYLE FOR INSTRUCTIONS

The core of any set of instructions is the **imperative mood**: commands such as "Load paper print side down into the printer." Additional explanatory material can use **second-person indicative mood**: "You may want to fan the paper to loosen it before loading it into the printer." These actions are typically formatted as numbered lists. They are numbered because they are likely to be in a required sequence. Sometimes, the actions are in no required order and in this case bulleted lists are used—for example, when the copy machine is jammed, you look for the problem in different areas of the machine but in no required order. For more on numbered and bulleted lists, see Chapter 2.

ELEMENTS OF AN INSTRUCTION

An individual instruction—a specific action that readers must take—actually has a number of elements that potentially must be included:

- **Specific action or actions:** At the very core of instructions is the imperative statement: for example, "press Enter." Sometimes, an instructional step can have several closely related actions: for example, "type your name in the blank, and press Enter." Combining brief, closely related steps helps maintain continuity and readability of the instructions.
- **Prerequisites:** You may need to tell readers about things they should have done prior to the current step or things that they should know and understand.
- **Locations:** You may need to help readers to find the interface elements that they must work with—for example, the "quick-access tray at the bottom right corner of the screen."
- **Before and after:** It may be helpful to readers to explain how things should look before the step, after the step, or both.
- **Reasons and importance:** Sometimes, readers need a little motivation as to why the instructional step must be performed and why it is important to do it carefully.

- **Anticipatory troubleshooting:** Some instructional steps involve common mistakes ("gotchas") that you may need to warn readers about.
- **Notices:** Some instructional steps carry with them the potential for damage to equipment, failure of the procedure, or injury to people. In these cases, use the appropriate notice format discussed in Chapter 2, "Professional Document Design." Although industry practices vary, common notice types are notes, attention, caution, warning, and danger. Each type has a special format that increases in noticeability with the severity of the notice.

COMPONENTS OF INSTRUCTIONS

Now that you've read about the essential inner workings of a set of instructions, consider the individual components of an instructions document:

Title. The title of a set of instructions should indicate the task as well as the object. "Microwave Oven" does not work as a title for a set of instructions; "Operating the Microwave Oven in the Break Room" does. Typically, titles for instructions use **gerund phrasing** (-ing) or "how to" phrasing.

Introduction. In the introduction to a set of instructions

- Tell readers exactly what overall procedure will be presented (even though it is stated in the title).
- Tell readers what skills, knowledge, or experience they need to have in order to perform the instructions. If no skills, knowledge, or experience are needed, tell them that.
- Consider including a scope statement, indicating what your instructions will not cover.
- Consider including some motivation: why the procedure is important, why it must be done exactly as described, and what can happen if it is not.
- Include an overview of the main instructional phases that your instructions will cover—for example, unpacking, setting up, installing, configuring, testing, and troubleshooting.

The introduction to your instructions may not need to include each of these elements, and these elements can be expressed in a neat condensed way, as you can see in the example in Figure 11.7.

> **Operating a Pressurized Barrier Fluid System on a Double Mechanical Seal**
>
> In a pressurized barrier fluid system (PBFS), double mechanical seals are used on pump shafts to provide an additional layer of protection against spills or releases of hazardous process material.
>
> This document describes (1) pressurized barrier fluid systems and the equipment used in them and then (2) provides operating techniques for proper use of PBFSs, which include (a) preparing the PBFS for operation and (b) monitoring the PBFS while in operation. Each of these sections is discussed in detail below.
>
> Entry-level technicians should use these instructions during their training. Both entry-level technicians and experienced technicians should use these instructions to standardize the best demonstrated practices. Proper operating practices pertaining to a PBFS are vital in preventing release of hazardous chemicals and to ensuring reliability of pump seals. Any changes or deviations from this procedure should be approved by manufacturing staff prior to execution.

FIGURE 11.7
Introduction to a set of instructions. Notice that procedure, background, overview, audience, and motivation are all included here.

Background. Your instructions may or may not need background discussion. For example, the instructions for operating the microwave oven in the break room may need only one statement—that the oven needs to be operated and cleaned properly—and that statement can go in the introduction. However, more technically oriented instructions may need some background, which will, in turn, enable readers to understand what they are doing when they perform the actual steps. Image-editing software, such as Adobe Photoshop, is a good example. Concepts such as layers, channels, bevel, and skew may require some background explanation.

Equipment and supplies. Instructions often need lists of equipment (the tools needed to perform the instructions) and supplies (the "consumables") that readers must gather before they start the procedure. As you can see in Figure 11.8, these are presented in simple or bulleted vertical lists.

Instructions. The core of any set of instructions is the step-by-step instructional actions that readers must perform. You've already read about these. Interwoven with this text are headings to help readers find the information they need, notices to alert readers to

Tools and Supplies

To create an electrode, you will need the following:

6" length of tungsten wire
Jar of epoxy
Jar of zapping solution
Pair of gloves
Protective plastic electrode box

The following equipment will be needed:

Zapper
Profiler
Dipping ring
High-powered microscope

Equipment and Materials

A PBFS consists of the following components:

- *Seal pot.* 1- to 2-gallon stainless steal vessel equipped with a site glass, pressure gauge, vent line, drain line, and fill line.
- *Stainless steel tubing.* $\frac{1}{2}$" to $\frac{3}{4}$" tubing used to transfer the liquid from the seal pot to the seal and from the seal back to the seal pot.
- *Nitrogen regulator.* Pressure control device on a nitrogen supply line used to adjust and maintain pressure on the seal pot.

FIGURE 11.8
Equipment and supplies lists—examples.

potential problems, and illustrations that enable readers to visualize important actions and objects, including orientations.

To review some essential guidelines for instructions, most of which are detailed in Chapter 2, "Professional Document Design":

- Make sure you understand the needs and limitations of the readers of your instructions. Adapt your discussion accordingly.
- Use the task-oriented approach; identify the key tasks that your readers will perform.
- Use imperative phrasing for individual instructional steps; use second-person indicative for additional explanation.
- Use numbered or bulleted lists for all steps.
- Introduce all lists with lead-in phrases or clauses; don't use headings as list lead-ins.
- Use headings to enable readers to find the information they need quickly. Design and subordinate those headings properly.
- Use notices to alert readers to potential problems.

- Use illustrations to help readers visualize essential actions, objects, and orientations.
- Make sure the introduction indicates the topic, the purpose, required audience characteristics, and overview of what the instructions cover. Keep background information to a minimum.

Headings. Chapter 2, "Professional Document Design," provides details on creating and formatting headings. With brief instructions you can bend the rules about outlining and parallelism. For example, instructions typically have a section with the heading "Equipment and Supplies" and a heading for procedure sections such as "Setting Up the Client" or "Editing a Protocol." Strictly speaking, these headings are not parallel in phrasing. Parallelism could be forced—for example, "Gathering Equipment and Supplies," but doing so may seem pedantic. Also, a sequence of headings such as "Equipment and Supplies," "Setting Up the Client," and "Editing a Protocol"—all at the same level—would violate traditional outlining rules. Traditional outlining would require "Equipment and Supplies" and another same-level heading such as "Procedures" or "Instructions," followed by "Setting Up the Client" and "Editing a Protocol" as lower-level, subordinated headings. Once again, forcing such precise subordination on such a small scale may seem rather pointless and pedantic.

Notices. As mentioned in the preceding, you must alert readers to the potential for major and minor injury, damage to equipment or data, problems with success of the procedure, exceptions, and points of emphasis. You do this by using the notice format presented in Chapter 2—not by applying garish combinations of bold, larger font, color, and all caps to the existing text! Remember that in a commercial environment, notices protect your organization from lawsuits. In Figure 11.9, notice the use of the caution; in the computer industry, caution notices are used for situations in which equipment or data could be damaged or in situations in which the outcome of the procedure could be ruined.

Highlighting. **Highlighting** refers to the use of bold, italics, color, alternative fonts, different font sizes, or icons to emphasize points in the text or to cue readers—all of which enable readers to anticipate actions they must take. Because highlighting is easily overused, you must have a highlighting "scheme," as explained in Chapter 2. Too much highlighting creates a busy-, annoying-looking text that readers prefer not to read. In Figure 11.9, notice that bold is used for interface elements that cause something to happen when clicked. Notice too that the example is shown in a monospace, typewriter-style font; this font cues readers that they should not use the text in this font in this procedure. In Figure 11.10, notice that no highlighting other than initial capital letters is used for the name of an on-screen window, IP Address window. It can't be clicked; it doesn't cause anything to happen;

Setting Up the Novell Client

Before you can configure an NDPS print environment, you must install and configure the Netware client on a workstation. The client allows the workstation to connect and communicate with a Netware server.

Install the client. Novell creates and supports clients for most Windows desktop operating systems such as Windows 9x, NT, 2000, and XP. To load the client:

1. Map a drive to (for example, `\\Spy-2K\H\Novell\EnglishClients\`) using a Windows operating system.
2. Open the folder that corresponds to your operating system.
   Caution:
   Loading the 9x client on a Windows ME operating system may cause your PC to lock up or exhibit unexpected behavior because Novell does not develop or support a client for Windows ME.
3. Double-click **setupnw.exe** (**setup.exe** for 9x clients), and click **Yes** on the Software License Agreement screen.
4. On the Novell Client Installation screen, choose the Custom Installation option, and then click **Install**.
5. Click the checkbox next to the Novell Distributed Print Services option, and then click **Next**.
6. In the following screen, do not select any additional options to install; just click **Next**.
7. Choose the IP and IPX protocol option, and click **Next**.
8. On the Login Authenticator screen, keep the default setting and click **Next** and then click **Finish**.
9. When prompted to remove the Microsoft NetWare Client Service, click **Yes**.

Configure client login settings. Once the client has installed, you must supply login information in order to connect to the server . . .

FIGURE 11.9

Instructions example. Notice the use of headings and subheadings, numbered steps, notices, and highlighting.

Editing a Protocol

To edit the protocol:

1. Select the protocol you wish to edit.
2. Select the **Properties** button.
3. In the IP Address window, select the **Specify an IP Address** radio button.
4. Enter the IP address and subnet mask. Requirements are as follows:

   - Use a regular IP address; static IP addresses are recommended but not required.
   - Set the subnet mask as `255.255.255.0` (same for all computers).
   - Create a unique IP address for each computer.

     The following is an example of an IP address setup for three workstations and a server:

     – Server: `10.0.0.1`
     – Workstation 1: `10.0.0.2`
     – Workstation 2: `10.0.0.3`
     – Workstation 3: `10.0.0.4`

FIGURE 11.10

Instructions example. Notice the use of nested bulleted items for requirements that are in no necessary order.

and bolding it would create visual clutter. In this same figure, notice the use of the monospace font for text that readers are expected to enter verbatim.

Illustrations. Diagrams, photographs, and schematics are essential in instructions. They enable readers to orient themselves to the objects they are working with, to visualize the actions they must take, and to understand what things should look like before and after important steps in the procedure. See Chapter 2 for details on illustrations.

11.2.2 USER GUIDES

A **user guide** is a set of instructions with supporting materials such as covers, an edition notice, a table of contents, and back matter surrounding it. Find a user guide for practically any commercially produced hardware or software product, and look for the following elements:

- **Front and back cover:** These should be nice glossy covers that show the title of the user guide; these elements indicate the product name and model or version number if necessary and include a graphic or logo. The back cover usually has very little material: book number, company address, the recycling logo, an indication where the book was printed, a bar code.
- **Title page:** The first page inside the user guide, this element repeats some of the material on the front cover.
- **Edition notice:** Usually on the back side of the title page, this element specifies date of publication, copyright owner, and other legal matters.
- **Table of contents:** This is the list of chapter, section, and subsection titles.
- **List of figures and tables:** Few user guides include a list of figures. If you believe your readers will need to find specific figures and tables, include this list.
- **Preface:** This section provides some brief description of the product, contents of the document, audience characteristics, highlighting conventions, and other such introductory material.
- **Chapters and appendixes:** This is the main text of the user guide, which uses the structure and format of instructions, explained in the preceding section.
- **Glossary:** Some user guides include a glossary of important, potentially unfamiliar terms along with their definitions. Typically, glossaries use a two-column format.
- **Index:** User guides of more than 20 pages include an index, which is a list of page pointers to the occurrence of key topics within the user guide.

User guides also contain other, non-instructional text such as parts lists, specifications, error codes, and other reference material. Obviously, an "in-house" user guide, one written for fellow employees, may not need some of these elements.

11.2.3 OPERATING SPECIFICATIONS

Another form of instructions is commonly called **operating specifications**, and sometimes engineering instructions. They resemble the

> 5.4.3 Scheduling of examinations is the responsibility of WFM. Equipment users are responsible for ensuring that equipment is made available for examination and is not used beyond the examination date.
>
> **5.5 REPAIRS, MAINTENANCE AND MODIFICATIONS**
>
> 5.5.1 Pressure vessels shall be repaired or modified only in accordance with the appropriate standards. Any proposed repair or modification shall be subject to the same design process as set down in 5.2.1.
>
> 5.5.2 Work on relief streams, including the fitting and removal of relief devices, shall be carried out only by persons appointed in writing by the Gowan Center Engineering Manager to carry out such work.

FIGURE 11.11
Operating instructions (also known as operating specifications and engineering instructions)—example.

kind of instructions discussed in the preceding section. However, there are some essential differences.

Operating instructions have the following distinguishing characteristics:

- Use third-person point of view and *shall*:

    Relief streams shall be subject to initial inspection to a written scheme of examination before being put into use.

PRESSURE SYSTEMS

AUTHOR:	Gowan Center Engineering Manager
APPROVED BY:	Gowan Center Engineering Manager
ISSUING AUTHORITY	Gowan Center Site Manager
REVIEW DATE:	1 MAY 2008

REVISION	DATE	BY	APPROVED BY	SIGNATURE
1	01/06/01	ROS	D Williams	D Williams
2	01/06/03	ROS	D Williams	D Williams
3	01/05/04	ROS	D Williams	

FIGURE 11.12
Common sections at the beginning of operating instructions (also known as operating specifications and engineering instructions).

1 PURPOSE

To define the engineering and operational requirements for pressure systems (including pressure vessels and relief streams) in Gowan Center.

2 SCOPE

The whole of Gowan Center.

3 REFERENCES

3.1 Pressure Systems Safety Regulations 2000—HSE

3.2 ASG/WCE/2007 Control of Modifications to Plant

3.3 Labprac 03 Gas Cylinders

4 DEFINITIONS

4.1 Pressure System. A system comprising one or more pressure vessels, any associated pipework and relief devices, or a similar system of pipework and relief devices to which a gas cylinder is connected (see further definitions in Ref 3.1)

4.2 Relief Device. A relief valve, bursting disc or similar device designed to protect a pressure system against overpressure.

FIGURE 11.13
Common body sections at the beginning of operating specifications.

- Use a decimal-numbering system for all headings and sections for ease of reference:

- Start with a formal system of document-owner name, approver name, and review and revision dates.

- Begin with standard purpose, scope, references, and definitions sections:

- Use a standard outline format with major sections single-enumerated and subsections double- or even triple-enumerated (as shown in Figure 11.13). In the example shown in Figure 11.14, the entire outline looks like this:

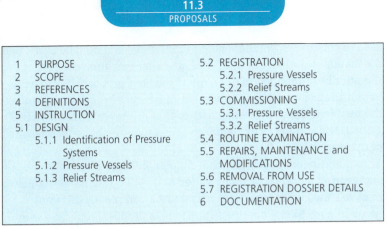

1	PURPOSE	5.2	REGISTRATION
2	SCOPE		5.2.1 Pressure Vessels
3	REFERENCES		5.2.2 Relief Streams
4	DEFINITIONS	5.3	COMMISSIONING
5	INSTRUCTION		5.3.1 Pressure Vessels
5.1	DESIGN		5.3.2 Relief Streams
	5.1.1 Identification of Pressure Systems	5.4	ROUTINE EXAMINATION
	5.1.2 Pressure Vessels	5.5	REPAIRS, MAINTENANCE and MODIFICATIONS
	5.1.3 Relief Streams	5.6	REMOVAL FROM USE
		5.7	REGISTRATION DOSSIER DETAILS
		6	DOCUMENTATION

FIGURE 11.14
Full outline of an example of operating specifications.

See "Specifications" in this chapter for other format and style issues relating to operating specifications.

11.3 PROPOSALS

Professional engineers obtain a good deal of their work through proposals. A request for proposals (RFP) will be issued by some corporate or governmental organization. The engineer or engineering firm will receive the RFP and send a proposal back to that organization, offering to do the project described in the RFP. The corporate or governmental organization will review this proposal as well as others and select the proposal and its engineering firm that has the best plan, talent, financials, and other elements, as described in the proposal.

As you can see, a proposal is a competitive thing. Multiple engineering firms are competing to do the project. These firms hear about the project in a variety of ways: direct contact from the requesting organization; announcement in some widely available publication such as the *Business Commerce Daily*; and of course through networks, grapevines, and rumor.

11.3.1 TYPES OF PROPOSALS

Not all proposals are external, where the requesting organization and the engineering firm are separate entities. Some proposals are internal; in many corporate or governmental organizations, projects must also go through the proposal process. Nor are all proposals written in response to a formal RFP. In an unsolicited proposal, you seek to convince a potential client that a project should be done, that the results would be advantageous, and that you and your firm are the ones best qualified to do it. An unsolicited proposal has to work harder than the solicited proposal; it must work harder to convince the recipient that a problem exists or that improvements are to be gained, and it must work harder to accentuate the beneficial results of the project.

As mentioned throughout this chapter, document types in the engineering world go by many varied and conflicting names. What may be called a proposal here may go by a completely different name in a given engineering firm. There, a proposal may be a completely different type of document. The core definition of the proposal, as it is discussed here, is the effort to win approval to do a project, to convince a client—whether external or internal—that you or your firm is the best qualified to do the project.

The "great idea" document is often mistakenly considered a proposal. For example, imagine that you think it would be a great idea to set up everybody in your globe-spanning corporation with audio and video equipment that would be tied into sophisticated meeting software. You do the research on functions and costs, you set up working models, you get employees interested—you work it all out. Then you write a document describing the system, its advantages, its costs, and the other such details that management will want to know. That's not actually a proposal in the traditional sense of the term. The missing element is the persuasive effort to get you or your firm hired to do the project.

Proposals vary greatly in their format and contents; they can range from a brief two-page memo or business letter to a 100-page bound document. However, most proposals do follow a basic core structure, as described in the following subsections.

11.3.2 FORMAT AND CONTENTS

A proposal can be a bound formal-looking document or a simple business letter or memo. A bound proposal is rather long—over 20 pages—and may have a cover letter or memo attached to the front. A really long proposal—over 50 pages, for example—will be bound and look like a book. Brief proposals can be presented as business letters or memos, in which case the content of the proposal is integrated right into the body of the letter or memo.

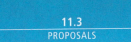

COVER LETTER OR MEMO

A proposal should include a cover letter or memo addressed to the client. Whether the cover letter or memo is attached to the front, bound into the document as the first page, or not used is a matter of organizational preference. Figure 11.15 shows an example of a cover memo that is separate from the proposal proper.

INTRODUCTION

For any introduction in a professional document, you should consider including indicators of topic, purpose, audience, situation,

San Marcos Photovoltaic Systems, Inc.
1124 State Blvd., Suite 300
San Marcos, TX 78666

March 09, 2009

Mr. Fernando dos Marias
EDEL—Luanda
1214 Saldanha
Luanda, Angola 2442 335437

Subject: Proposal to add photovoltaic cells to the EDEL distribution system

Dear Mr. Dos Marias:

The following is a proposal that will guide your staff in the addition of photovoltaic cells to the Empresaa de Distribuiçao de Electricidade (EDEL) system. As an electrical engineer working in Lubango, I observed the need for photovoltaic systems. Like many other areas in Africa, Lubango experiences frequent power shortages. The system outlined in this proposal will go a long way toward resolving this problem.

The following proposal describes the benefits of using photovoltaic cells for the production of electricity, the installation process, output projections, and costs. Included is a schedule of completion and my company's qualifications.

Please feel free to contact me at 517-000 0000. I appreciate your time and consideration.

Sincerely,

Helena Genao
Attached: proposal

FIGURE 11.15
Cover letter for a proposal. An alternative is to integrate the letter with the proposal and send it as a formal business letter.

and contents and provide just enough background to get readers oriented and interested (but not necessarily in this order). Don't be hesitant to start off with "The following is a proposal to . . . . ." Include some indication of the situation—for example, a reference to the RFP or a meeting in which the project was discussed. Be sure and overview the contents; list the main sections of the proposal similar to what you see in Figure 11.16. Avoid including too much background (if any is needed at all); a sentence or two will suffice, but anything more should be moved to a body section.

PROBLEM OR OPPORTUNITY

By definition, proposals address a problem or an opportunity to design, build, or otherwise improve things. Levees may be deteriorating and in need of overhaul. New design for lighter, longer-lasting, more powerful batteries may be needed for the next generation of hybrid electrical vehicles. Free wireless communications in the downtown area would be an excellent enhancement. Whatever the case, state or describe the problem or opportunity that your proposal addresses—in a section just after the introduction, under its own heading (as you see in Figure 11.17). Remember too that if your proposal is unsolicited (the client has not requested proposals), you must spend extra time motivating the client to get concerned about the problem or interested in the opportunity. (This section might also be called "Statement of the Problem.")

Proposal: Installation of a Photovoltaics System at the EDEL Distribution Facility

The following is a proposal to design and implement a photovoltaic-based electricity-production for Empresaa de Distribuiçao de Electricidade (EDEL) in Lubanga, Angola. Included in the following is a discussion of the benefits of photovoltaic systems, the production of electricity, the system-installation process, output projections, costs, a schedule of completion, and my company's qualifications. My consulting team and I have successfully installed the photovoltaic cells system in two other provinces (Lubango and Benguela).

FIGURE 11.16
Introduction to a proposal. Notice that it intentionally repeats some of the content of the cover letter in case the letter gets separated from the proposal.

Background: Opportunities and Challenges

Deep-sea basalt aquifers have the scientific and technical potential for long-term sequestration of CO_2, thus reducing atmospheric build-up of this greenhouse gas. The basalt dissolution and (Ca^{++}, Mg^{++}, Fe^{++}) CO_3 precipitation rates in the oceanic basalt crust potentially have a geological capacity large enough to (a) accommodate a major portion of fossil fuel CO_2 which is produced globally (~22 billion tons as CO_2 or 6 Gt-C/yr); and to (b) produce (Ca^{++}, Mg^{++}, Fe^{++}) CO_3 infilling minerals from the chemical precipitation reaction of CO_2 and basalt, permanently sequestering carbon in a chemically stable and nontoxic form.

The rate of this reaction in realistic field environments cannot be sufficiently reproduced in the laboratory or via computer models (although both are useful to understand the boundary conditions); instead, a field study is required. By injecting and monitoring CO_2 into the well-sedimented crust on the flanks of the Juan de Fuca ridge, key scientific questions can be answered concerning the rates of the basalt-brine-CO_2 reactions and the immobilization of CO_2 as non-toxic and stable Ca-Mg carbonate crystals by these reactions. Any aquifer-rock system such as this one must be tested in the field in order to evaluate the effects on the in situ hydrogeology and potential for CO_2 sequestration.

FIGURE 11.17
Problem–opportunity section of a proposal. This example presents a method for reducing greenhouse gases (opportunity) and seeks funding to do field testing to determine the potential of this method.

ACTUAL PROPOSED PROJECT

Proposals have a way of getting out of control and sounding as if they will solve all the world's problems. To protect yourself and your firm, include a section in which you state specifically what you will and will not do in the project. (This section might also be called "Objectives.")

Proposed Objectives

We propose to core sediments deposited on the Wilkes Land margin with the following objectives:

1. to obtain the onset of glaciation (Eocene or older) by drilling strata across the glacial onset reflector (regional unconformity WL2) in two depositional environments, shelf progradational wedge foreset (Sites WLSHE-07A or alternate WLSHE-09A) and lower continental rise/abyssal plain hemipelagic strata (Sites WLRIS-02A);

FIGURE 11.18
Specifics on the proposed work. Proposals need to be specific about what the proposed work will and will not accomplish.

FIGURE 11.18 (*Continued*)

2. to obtain a high-resolution Neogene-Quaternary record of glacial/interglacial cycles from continental rise mounded deposits (Sites WLRIS-01A);
3. to date major changes in shelf prograded wedge geometry (below and above the regional WL1 unconformity) that document large fluctuations in the glacial regime, possibly through much of the Miocene (Site WLSHE-08A);
4. to help assess the main controls on sediment transport and deposition on ice-dominated continental shelves and rises in order to test present architectural models of glacial processes and facies for high-latitude margins; and
5. to constrain the timing and the nature of changes in glacial regime and paleoceanography that result in the development of large mounded deposits (i.e., up to 700 m relief), and large upper-fan channel-levee complexes (i.e., 900 m relief) on the continental rise.

DESCRIPTION OF THE COMPLETED PROJECT

In some cases, you may need to include a description of the finished project in your proposal. Limit the description in the proposal proper to an overview. Shift extensive descriptions, elaborate drawings, diagrams, maps, and other such detail to the appendix. Figure 11.19 shows the plan for a study of telecommuting issues. The proposal seeks to win a contract to advise the recipient on the feasibility and implementation of a telecommuting plan.

Report Description

The following is a tentative outline for the proposed report:

I. Introduction
II. Telecommuting Data and Statistics
    Yearly Telecommuting Statistics
    Increasing Numbers of Telecommuters
III. Telecommuting Issues for Employers
    Advantages of a Telecommuting Program
        Increased Productivity
        Business Continuity
        Facility Cost Savings
        Reduced Sick Leave
    Disadvantages of a Telecommuting Program
        Increased Set-up Cost
        Added Administration Considerations
        Lack of Employee/Manager Interaction
IV. Best Practices
    Case Studies
    Company Success Stories

V. Conclusion
VI. References

Graphics and Tables

The report will contain various graphics and tables to help illustrate the data and information provided. The following is a tentative listing:

Cisco Telecommuting ROI	Chart
AT&T Telecommuting Savings	Graph
Nortel Employee Relocation Costs	Graph
Dow Chemical Administrative Cost Savings	Pie chart
Comparison: Commuting-Hours vs Productive-Work Hours	Graph

FIGURE 11.19
Project description section of a proposal.

RESULTS, BENEFITS, FEASIBILITY

Particularly in unsolicited proposals, you should get potential clients excited about the benefits and advantages of the project. The city will be safe from floods; your company will take the lead in hybrid-vehicle battery technology; downtown will be more attractive to businesses. But remember not to sound as if the project will solve all the world's problems; discuss the feasibility of the success of the project. You can't guarantee the improved levees against a category 5 hurricane; you can't guarantee your engineers will be able to develop lighter, longer-lasting, more powerful batteries for the next generation of hybrid electrical vehicles; you can't guarantee people will use the downtown wireless system or that it will attract business to the downtown area.

Benefits of the Proposed Project

By installing a high-speed IEEE 802.11b wireless local area network (WLAN):

- A pervasive, always-on access to the GUNET2000 campus network will be available to faculty and students.
- An infrastructure will be available for fast and inexpensive upgrade expansion.
- Students and faculty will be able to explore new research opportunities; new ways will be available for ideas to flow between faculty and students.
- With wireless network becoming standard at many universities, this university will stay competitive and not be left behind.

FIGURE 11.20
Results, benefits, feasibility section of a proposal.

TECHNICAL BACKGROUND

In some proposals, you must discuss technical aspects of the problem, the opportunity, or both. Just what's the problem with those levees? How are levees designed; how do they work? What are the essential design features of current hybrid-vehicle batteries? What advances in this technology are needed or possible? At a time when wireless communications were not as commonplace as they are today, you would need to provide some background on that technology. When you provide well-written, understandable technical background, you increase your client's perception of your professionalism.

PROCEDURES AND METHODS

In some proposals you may need to explain how you will perform the project—which methods you will use and why. The client may like your approach to the project and select you or your engineering firm on the basis of that alone! In a procedures and methods section, explain step by step or milestone by milestone how you will carry out the project. (This section may be called "Plan of Action.")

IV. Plan of Action

This section addresses our plan for achieving the previously described objectives. The proposed project will follow the design process as outlined in *Product Design and Development* [Ulrich and Eppinger, 2000]. The steps in that process include identification of customer needs, establishment of target specifications, concept generation and testing, development of product architecture and industrial design, prototyping, and a presentation of final results.

Identifying Customer Needs

Based on the project objectives stated in the previous section, the first step is to communicate with our clients to establish *customer needs*. Some of these customers include past, present, and future solar decathletes as well as the engineering, architecture, and education departments of Virginia Tech. In order to elicit information from these customers, a survey has been created (see Appendix A). When possible, interviews will be conducted as well.

FIGURE 11.21
Method, procedure section of a proposal. Notice that the subsection is the first of five subsections, detailing how the team will go about the project.

(Adapted with permission from *A Proposal to Implement a Monitoring and Control System into Virginia Tech's 2005 Solar House*, Michael Christopher, Ken Henderson, Dan Mennitt, Josh McConnell of the Solar Decathlon Team, Mechanical Engineering Department, Virginia Tech. http://www.writing.eng.vt.edu/design/sample_proposal.pdf (2003). Accessed February 23, 2007.)

In some cases, proposals must include a management plan, which as Figure 11.22 shows, is a combination of qualifications, schedule, and costs.

SCHEDULE, MILESTONES

Clients will also expect to see your schedule for the project—that is, the dates at which you expect to complete the major phases of the project. Instead of exact dates, you can project the total days or months to complete each phase. A Gantt chart, such as the one shown in Figure 11.22, might be in order if the phases of the project overlap each other. (In some proposals, this section may be called "Management Plan.")

QUALIFICATIONS

Include a summary of your qualifications in the proposal—your experience, training, and education along with references from former clients. If you plan to work as a team, include summaries for each team member. More elaborate proposals include full résumés of team members in the appendix, with only brief summaries in the body of the proposal proper.

VII. Management Plan

Personnel

The control project will be broken into three main categories: LabView programming, web development, and installation of sensors and other components into the 2002 house or a test environment. Team members' specialized skills and background have been matched with these categories:

- Dan Mennitt and Michael Christopher will work on the Lab View program development.
- Ken Henderson will design the web site.
- Josh McConnell will install sensors in the solar house and work with the 2003–2004 Heat-Pump Group.

FIGURE 11.22

Management-plan section of a proposal. Required in some proposal contexts, this section combines information on personnel qualifications, schedules and costs.

(Adapted with permission from *A Proposal to Implement a Monitoring and Control System into Virginia Tech's 2005 Solar House*, Michael Christopher, Ken Henderson, Dan Mennitt, Josh McConnell of the Solar Decathlon Team, Mechanical Engineering Department, Virginia Tech. http://www.writing.eng.vt.edu/design/sample_proposal.pdf (2003). Accessed February 23, 2007.)

FIGURE 11.22 (*Continued*)

See Appendix C for the résumés of team members.

Schedule

The Gantt chart shown below displays the project steps and their respective dates of completion.

	Aug	Sept	Oct	Nov	Dec	Jan	Feb	Mar	Apr	May
Identify customer needs	■	■								
Establish specifications		■								
Generate and test concepts		■	■							
Plan product architecture				■						
Plan industrial design				■						
Design and test prototype					■	■	■	■		
Present final results										■

Notice that planning, design, and prototyping will be completed by the end of January and that the project will be completed by the end of May 2004.

Budget

Although a budget of $1,000 has been allotted for this project, donations may be needed for some of the resources, which include shunt resistors, thermocouples, flow meters, and a data acquisition system, a display device, and a computer. The team will seek donations of these items.

COSTS, FEES, EXPENSES

Only after you have convinced your potential clients that you understand their problem or that they have a problem; only after you have impressed them with the benefits of solving their problem and solving it your way; only after showing them your excellent qualifications—only then is it to time to reveal the costs of doing the

Qualifications

Relevant experience includes:

- Clients include GQM Energy, Inc.; Two-Phase Engineering and Research, Inc.; Raluska Geothermal, Walls Construction; Encee Geothermal Venture 1; Hanks Energy International; University of New Mexico; and CBD Corrosion Control, Ltd.

- Provided technical evaluation of combustion turbines and FQD contractors for Hanks Energy, International. Services provided to Two-Phase Engineering include a technical review of available backpressure and total flow geothermal steam turbine technology.

- Raluska Geothermal tasks include conversion of Organic Rankine Cycle turbines from fluorocarbon to iso-pentane working fluid, supervision of submersible production pump installation and plant optimization.

- Provided engineering to Walls Construction, Inc., at the UCSF Medical Center Central Utilities Plant combined-cycle cogeneration installation to resolve NOx system control problems. Also provided engineering to Encee Geothermal Venture 1 to resolve turbine seal and control system problems.

- Managed a plant-betterment program for Bermuda Private Power Co. Ltd., at its 60 MW slow speed diesel/steam turbine combined-cycle plant at Kingston, Jamaica, which included improvements to O&M procedures, modifications to the boiler-level control system, pipe-support improvements, relief valve system and oily water separation system.

(See attached résumé for details.)

FIGURE 11.23
Qualifications section in a proposal.

project. This strategy is known as the **indirect approach**. Present positive information first; save the bad news for later. And usually, project costs feel like bad news!

Specifying costs and fees may not be appropriate in the context of the proposal you write. But if it is, break out the costs in their various categories such as research, travel, per diem, and so on. Your sense of professional practice will tell you whether to reveal projected hours and your hourly rates.

CONCLUSION, ENCOURAGEMENT

Whether it's a simple memo or a formal bound document, end with a summary of the best points about your bid to do the project. Express hopes that the client will select you or your firm to do the project. In the letter or memo, mention how the client can contact you with questions, and indicate that you will follow up in 10 days or so.

CATEGORY	RATE	UNITS	QUANTITY		COST
			Task 1	Task 2	
Supervising engineer	75	$/hr	30	96	13,520
Technician	40	$/hr	50	288	9,450
Other (vehicle)	0.31	$/mi	500	5400	1,829
Other (supplies)	51			12	612
Other (equipment)	105			12	1,260
Subtotal, direct					26,671
Indirect	20%				5,334
TOTAL					**$32,000**

FIGURE 11.24

Combined costs and schedule section from a proposal to perform comprehensive analyses of multijurisdictional use and management of water resources on a watershed or regional scale.

Conclusion

The 2002 Solar House had its strong points, but it was plagued by a number of problems that resulted in poor scores. This proposal has illustrated how our plan for a well-functioning monitoring and control system can prevent these problems in the upcoming 2005 Solar Decathlon. Our proposed design will build on the spirit and innovation of the previous house while seeking to overcome its weaknesses. Specifically, the monitoring and control system will monitor conditions, diagnose problems, and evaluate performance of the house. It will provide an interface between the user and this critical information. Along with an in-house graphical display, the team will create an interactive web site to expand public awareness and educate future generations on solar energy. Our team has invested much in the preliminary design and has much confidence that we have the experience and know-how to design, construct, and evaluate such a system within the given time frame of nine months.

FIGURE 11.25

Conclusion section of a proposal. Notice how this conclusion seeks to build readers' confidence that the team has a good plan and can do the job.

(Adapted with permission from *A Proposal to Implement a Monitoring and Control System into Virginia Tech's 2005 Solar House*, Michael Christopher, Ken Henderson, Dan Mennitt, Josh McConnell of the Solar Decathlon Team, Mechanical Engineering Department, Virginia Tech. http://www.writing.eng.vt.edu/design/sample_proposal.pdf (2003). Accessed February 23, 2007.)

11.4 PROGRESS REPORTS

After you or your firm has landed a project through the proposal, you use progress reports to keep the client apprised of the status of the project.

11.4.1 PURPOSE AND FREQUENCY

A **progress report** describes the status of a project: what has been accomplished, what's going on right now, and what work is still ahead. It also describes any problems, any changes, or anything unexpected relating to the project. And it sums up the overall status of the project.

Progress reports are an expected part of all large, lengthy, expensive projects. They may be required by the contract for the project, and in any case they are a standard part of good business practice and professionalism. After all, if your client or your management is spending millions of dollars on a multiyear project, they want to know how it's going. Progress reports can be weekly, monthly,

Work Completed

The following is a summary of the efforts during this reporting period:

1. We completed the two short-term projects for manufacturing and delivering carbon fiber roving to meet the immediate needs of the Automotive Composites Consortium development programs.
2. We developed and provided partially oxidized polyacrylnitrile (PAN) fiber to Oak Ridge National Laboratory (ORNL) to support the microwave research efforts at ORNL.
3. We requested and we were granted a no-cost extension by ORNL.
4. We completed and submitted the mid-year and quarterly reports as requested by ORNL.
5. We completed and submitted the current project final report to ORNL in September 2004.

Future Work

Research-scale trials and preliminary cost-model estimates have led to three viable routes for the development of continuous-tow low-cost carbon fibers with mechanical properties meeting low-cost carbon fiber targets:

- *Chemical modification of textile acrylic fibers.* Uncollapsed textile acrylic fiber gels will be exposed to an aqueous NaOH solution during the spinning process to induce functional group hydrolysis and hence speed up the stabilization process.
- *Radiation pretreatment of textile acrylic fibers.* This task will involve exposing textile acrylic fibers to an E-beam radiation dose of 30 Mrad prior to oxidation to induce stabilizing cyclization reactions.
- *Sulfonation of polyethylene fibers.* Commercial polyethylene fibers will be passed through a bath of hot concentrated sulfuric acid in order to induce cross-linking reactions that render the fibers infusible and carbonizable.

These three development routes will be the focus of our work in the next reporting period.

FIGURE 11.26
Progress-report example using the time-periods approach.

quarterly, yearly—whatever the situation, the contract, or the standard practice requires.

11.4.2 ORGANIZATIONAL APPROACHES

Progress reports typically use a time or task approach or some combination of the two. In the **time approach**, you describe what you have accomplished to date, what work is going on currently, and what work is still ahead. In the **task approach**, you describe each

Scale-up and Verification of Chemical Modification Technology
Objectives in this task include demonstrating and verifying production technologies for producing the textile fiber with chemical modification and conversion into carbon fibers.

Modification of Textile Acrylic Line. This subtask consists of modifying the textile line to implement and incorporate the application of chemical modification in-line. Within the scope of this task, the engineering design requirement for installation will be developed and modification equipment will be integrated in the line before verification and scale-up trial run are performed.

Verification of Scaled-up Chemical Modification Process. Once the modification of the textile line is completed and equipment is debugged to ensure operational performances, the next step will be to verify scale-up of the chemical modification processing on the large textile fiber. Several trials will be performed to verify processing conditions, to develop manufacturing procedures, and to manufacture materials for conversion into carbon fiber at our facilities.

FIGURE 11.27
Progress-report example using the tasks approach.

task in terms of what you have completed, what is the current focus, and what lies ahead. Obviously, these two approaches are mirror images of each other

11.4.3 STANDARD CONTENTS

Apart from the essential work of summarizing work accomplished, current work, and planned work, progress reports should describe the project. After all, the progress report might be circulated to individuals in the client organization who are unfamiliar with the project.

Requirements for progress reports may include financial data, such as costs for the reporting period. Most progress reports conclude with an overall evaluation of the project.

An interesting example of a large-scale progress report entitled *Progress Report on the Federal Building and Fire Safety Investigation of the World Trade Center* is available at http://wtc.nist.gov/progress_report_june04/progress_report_june04.htm. In particular, look at Chapter 2, "Progress on the World Trade Center Investigation."

Project Overview

In October 1999, the U.S. Department of Energy's Office of Transportation Technologies (DOE-OTT), through the Oak Ridge National Lab (ORNL), awarded this corporation a multiyear contract to define and develop technologies needed for the commercialization of low-cost carbon fibers (LCCFs) to be used in automotive applications. Lighter-weight automotive composites made with carbon fibers can improve the fuel efficiency of vehicles and reduce pollution. However, for carbon fibers to compete more effectively with other materials in future vehicles, their cost must be reduced. Therefore, this project targets the production of carbon fibers with adequate mechanical properties, in sufficiently large quantities, at a sustainable and competitive cost of $3 to $5/lb.

Project Deliverables

At the end of this multiyear project, technologies for LCCF production will be defined. This definition will specify required materials and facilities and will be supported by detailed manufacturing cost analyses and processing cost models. To achieve this definition, laboratory trials and pilot-scale demonstrations will be performed.

Planned Approach

Initially, this project was divided into two phases:

Phase I: Critical review of existing and emerging technologies, divided into two tasks:

- Literature review and market analysis, which led to further refinement and down-selection of the most promising technologies.
- Laboratory-scale trials and preliminary LCCF manufacturing cost assessments of the proposed technologies.

Phase II: Evaluation of selected technologies using pilot-scale equipment and cost models. Phase II was divided into three tasks:

- Pilot-scale design for the evaluation of selected LCCF technologies. This included modifications of a polyacry-lonitrile (PAN) spinning pilot line and two different carbon fiber conversion lines (a single-tow research line and a multi-tow pilot line) and the construction of continuous sulfonation processing equipment.
- Experimental evaluation of down-selected LCCF technologies, including commodity textile-tow PAN (with chemical modification and radiation and/or nitrogen pretreatment) and poly-olefins linear low-density polyethylene [(LLDPE) and polypropylene (PP)].
- Large-scale feasibility studies of selected LCCF technologies.

FIGURE 11.28
Progress-report excerpt providing an overview of the project.

(Excerpts adapted by permission from Hexcel Corporation's progress report to Oak Ridge National Laboratory on a project to define technologies needed to produce a low-cost carbon fiber (LCCF) for automotive applications.)

Overall Project Evaluation

To date, this project has demonstrated the technical viability of the textile acrylic fiber-based technologies through pilot-scale experiments. Through engineering feasibility studies, we have successfully defined the manufacturing facility and processing costs needed to make 1,820 MT/year (4×10^6 lb/year) of LCCF in two carbon fiber production lines.

Our detailed economic analyses have indicated that the carbon fiber manufacturing cost can be reduced from ~$13.20/kg (~$6.00/lb) for large tow textile PAN to ~$9.9/kg (~$4.50/lb) with either chemical or radiation pretreatments. These mill cost estimates are based on standard carbon fiber manufacturing model parameters and exclude any return on investment (ROI). If you require the inclusion of ROI, we will need to research additional processing improvements in order to reach the LCCF cost targets in an economically sustainable manner.

FIGURE 11.29
Progress-report example providing an overall evaluation of the project.

11.5 FEASIBILITY, RECOMMENDATION, AND EVALUATION REPORTS

An important group of reports that professional engineers write includes feasibility, recommendation, and evaluation reports. Of course, as with most documents such as these, the names vary and conflict with each other. However, these reports are unified by one common thread: they provide the engineer's professional opinion along with comparisons and conclusions to back that opinion up.

11.5.1 OVERVIEW

Feasibility, recommendation, and evaluation reports are all similar in that comparison is—or should be—their foundation. Notice how comparison is the fundamental component in the excerpt from an outline of a feasibility report shown in Figure 11.30. Each option is compared according to three criteria.

FEASIBILITY REPORTS

Feasibility reports compare a proposed project against requirements involving some combination of technical, financial, legal, and administrative practicality and then state whether the project is feasible—that is, whether it will be practical, useful, or successful. For example, imagine that city officials have approached your engineering firm for consultation on whether implementing wireless

FIGURE 11.30
Excerpt from a table of contents of a feasibility report.

(Adapted with permission from U.S. Environmental Protection Agency and TAMS Consultants, Inc., *Hudson River PCBs Reassessment RI/FS Phase 3 Report: Feasibility Study* (December 2000).)

capability for the entire downtown area is feasible. Your feasibility report would address questions such as the following:

- Is it technically feasible—is it technically possible?
- Is it financially feasible—can the city afford it?
- Is it socially feasible—will people use it?

Your feasibility report would provide advice and recommendations on these and other areas along with supporting data.

RECOMMENDATION REPORTS
Recommendation reports compare options against each other and against requirements and then recommend one of the options, a

subset of the options, or none of the options. For example, imagine that the city had decided to implement wireless capability in the downtown area. The next issues involve the design of the system and the equipment to use. A recommendation in the area might be a comparison of different design options and a recommendation of one. In the second area, a recommendation report might compare different vendors of wireless technology and recommend one.

EVALUATION REPORTS

Evaluation reports compare something (such as a product or program) against a set of criteria or requirements and then state a conclusion as to whether that product or program has met its expectations. A recommendation might follow, stating whether to continue with the program or to purchase the product. For example, imagine that at long last the wireless system for the downtown area is up and running. At some point, the system must be evaluated: Is it working? Do people like it and use it? Can its cost be justified?

In the real world, of course, few people use these terms with any precision at all. In fact, some people might use the term "proposal" for what is defined above as a feasibility or recommendation report. Before you rush off to write the kind of document you think your clients want, ask them a few questions.

11.5.2 FEASIBILITY REPORTS

As a consultant, you structure your feasibility report according to your clients' needs and according to the nature of the project. For example, the technical feasibility of the project may already be assured; you needn't discuss that. The social feasibility of the project may be irrelevant. There may be other areas of feasibility that your clients are interested in. Also, your clients may demand that you also compare some alternative ways of doing the project, in which case you import parts of the recommendation report into your feasibility report.

The structure of the feasibility report for the downtown area wireless system might look like this, depending of course on your clients' requirements:

1. *Executive summary*. Documents such as these must place key details, conclusions, and recommendations up front, just after

the transmittal letter and cover page. An executive summary for this project would include the following:

- Review of the project's origins: who requested it, when, who did the work
- Brief description of the project, the questions that this report addresses
- Brief narrative of what was done
- Summaries of key detail and conclusions from the technical, financial, and social feasibility sections in the body of the report
- Recommendations as to the feasibility of the project

2. *Introduction.* The introduction does what an introduction ought to do, even repeating some of the contents of the executive summary. This sort of repetition is typical and expected in reports like these; you design these reports so that readers can access the information in a variety of ways:

- Background on the project. State who requested it, who performed it, what was done, and when the project began and ended.
- Purpose of the project and the report. In an introduction, you needn't state your final conclusions as to the feasibility of

ABSTRACT

This feasibility study provides a general technical description of several types of floating platforms, mooring systems, and anchor types for offshore wind turbines. A rough cost comparison is performed for two technically viable platform architectures using a generic 5-MW wind turbine. One platform is a Dutch study of a tri-floater platform using a catenary mooring system; the other is a mono-column tension-leg platform developed at the National Renewable Energy Laboratory. Cost estimates showed that single-unit production cost is $7.1 M for the Dutch tri-floater, and $6.5 M for the NREL TLP concept. However, value engineering, multiple unit series production, and platform/turbine system optimization can lower the unit platform costs to $4.26 M and $2.88 M, respectively, with significant potential to reduce costs further with system optimization. These foundation costs are within the range necessary to bring the cost of energy down to the DOE target range of $0.05/kWh for large-scale deployment of offshore floating wind turbines.

FIGURE 11.31
Abstract of a feasibility report. Notice that this abstract states that the two designs compared are technically feasible and that design modifications make them economically feasible.

the project. You also must indicate the purpose of this report—to report findings and conclusions and to make recommendations.

- Target audience. Indicate who are the expected readers and what background those readers need to read the report intelligently.
- Overview. Include a list of what your report contains, its major sections.
- Scope. If applicable, state what your feasibility report does *not* contain.

3. *Technical feasibility*. Use your engineering expertise to assess the technical possibility and practicality of the project. Report these details from an engineering perspective, which, of course, the client's engineers will want to consider. But also provide a translation of this information for nonspecialists, the nonengineers who will be reading your report. If you want to use the indirect

---

29th Steet Trolley Feasibility Study

Introduction

The following study of the feasibility of a 29th Street trolley is the result of a nine-month analysis of the engineering, economic, and social feasibility of reintroducing an electric heritage trolley to the 29th Street corridor in Brykerwoods, Texas. Included in this report is information on the:

- identification of the potential usage of a trolley system in the study area;
- evaluation of a number of potential trolley-route and trolley-station alternatives to serve the potential usage identified in the first step;
- identification of potential physical constraints and physical impacts on existing residential and commercial areas;
- estimates of the potential ridership, construction, and operating costs, including the need for subsidies; and
- evaluation of development potential of real estate adjacent to the trolley system; and presentation of design models of trolley and station alternatives.

The following report summarizes these analyses, recommends a system design, and includes construction and operating cost estimates and various system design elements. These recommendations are presented to the city council, the transit system's board of directors, and the mayor's office.

FIGURE 11.32
Introduction to a feasibility report. In most report formats, the introduction comes right after the executive summary (abstract).

approach, save a conclusion as to the technical feasibility of the project for the end of this section. Otherwise, you can begin the section with this conclusion.

4. *Financial feasibility.* You may need to work with financial experts to develop this section. In fact, feasibility reports are typically team written. In this section, you present cost projections for the project, perhaps with return-on-investment details. Once again, make sure that some part of this discussion can be read by people who are not financial experts. Present essential financial data, probably in the form of tables. Put large unwieldy tables in

---

Engineering Feasibility

The feasibility of reinstating trolley operations in the 29th Street corridor was studied through:

- Analysis of the grades along 29th Street relative to trolley capabilities
- Evaluation of engineering and safety implications of different locations of the trolley
- Analysis of trolley turning radius requirements
- Impacts on traffic at key intersections in the corridor
- Impacts on the storm sewer, waste water sewer, and water service

Our engineering analyses showed that it is feasible to design a trolley without compromising current rail engineering practice and standards. It should be noted that the minimum radius of a trolley turn does not allow the vehicle to operate inside curb to inside curb. Thus, whenever the trolley is turning left from the outside lane or right from the inside, the operations shift to the opposite side of the street when exiting the curve.

As for the issue of curb lane alignment versus a roadway center line alignment, curb lane operations is the preferred alternative for the following reasons:

- Conflicts between trolley riders and vehicle traffic would be reduced, because riders enter and exit the trolley car at the curb instead of at the center of the street platform.
- Placement of the platform in the middle of the vehicle travelway raises safety concerns.
- Providing the overhead power source for the roadway center line alignment is more difficult because the wire must be suspended over the roadway on cables.

In the peak period, the minimum trolley headway is estimated at 15 minutes. Thus, the maximum number of trolley passes through any intersection would be 8 in an hour. This low number of trolley pass-bys would not likely have a noticeable effect on the traffic along the route.

FIGURE 11.33
Technical-feasibility section from a feasibility report.

the appendix. As mentioned previously, you can state the conclusion as to the financial feasibility of the project either at the end of this section or at the beginning.

5. *Social feasibility*. In the case of the downtown wireless system, you might be expected to survey public opinion. You might even contract this survey out to a firm specializing in such surveys and use its results in your feasibility report. Present the essential survey data in this section, probably in the form of tables. Once again, put the big tables in the appendix. And also once again, state the conclusions as to the social feasibility of the project at the beginning or end of the section.

6. *Recommendations*. This final section is a detailed exploration of whether the project is feasible. Here you work your way through the good and bad points about the project and explain how you arrive at your final conclusion as to the feasibility of the project. Remember: you can't just declare that the project is feasible and walk away; you must present facts, logic, and requirements to support that declaration.

7. *Appendixes*. Put large tables and diagrams, lists, forms and other less essential materials in the appendix. Don't force readers to wade through material that only individual specialists are interested in. However, do provide cross-references to these appendixes so that readers know they exist.

11.5.3 RECOMMENDATION REPORTS

At its simplest, a recommendation report resembles a *Consumer Reports*–style document. It analyzes several products belonging to a category, compares them to a set of requirements, draws conclusions, and then recommends one (or none, or several based on different preferences). On the staff of an engineering firm, you might be asked to do a recommendation report on some particularly expensive, strategic product that the firm must purchase. You study the market, draw your conclusions, and then write your findings in a recommendation report. Your report reviews the need, the requirements, and then the leading contenders in the market; compares those contenders against the requirements; summarizes the conclusions; and then states your recommendations. The recommendation report enables someone else in your firm to review your findings and draw different conclusions.

The structure of the recommendation report resembles that of the feasibility report in a number of ways. Your engineering firm might be hired to recommend equipment and design for that downtown area wireless system:

1. *Executive summary*. Documents such as these must place key details, conclusions, and recommendation up front, just after the transmittal letter and cover page. An executive summary for this project would include the following:

   - Review of the project's origins: who requested it, when, who did the work
   - Brief description of the project, the questions that this report addresses
   - Brief narrative of what was done
   - Summaries of key detail and conclusions from the comparative sections in the body of the report
   - Recommendations as to equipment and design

2. *Introduction*. This introduction also does what an introduction ought to do, even repeating some of the contents of the executive summary. This sort of repetition is typical and expected in reports such as these; you design these reports so that readers can access the information in a variety of ways:

   - Background of the project: who requested it, who performed the project, what was done, when the project began and ended.
   - Purpose of the project and this report: in an introduction you needn't state your final conclusions as to the feasibility of the project. You also must indicate the purpose of this report: to report findings and conclusions and to make recommendations.
   - Target audience: Indicate who the expected readers are and what background these readers need to read this report intelligently.
   - Overview: Include a list of what your report contains, its major sections.
   - Scope: If applicable, state what your feasibility report does *not* contain.

3. *Background of technology and needs*. If the technology involved in the recommendation report is potentially unfamiliar to your

TABLE OF CONTENTS

FIGURE 11.34
Table of contents from a feasibility report. Notice how much comparison is built into a report such as this. This page comes right after the title page.

clients, explain, in a section of its own, how it works and how it is useful. Either before or after the technology section, discuss your clients' need for this technology—this product they are considering. True, they know all about their need, but they may want to know whether *you* understand their need. Another reason for providing this background is that your report may get circulated around the client's organization to those who are less familiar with the project.

LIST OF FIGURES AND TABLES

FIGURE 11.35

List of figures and tables from a feasibility report. Use this same format for the list of figures and tables in any engineering document. This page comes right after the table of contents.

4. *Requirements.* In a separate section, explain how you will evaluate the different options, what your minimum and maximum values are, and which requirements have the highest priority. Requirements can be numerical (for example, a maximum budgetary allowance), yes–no (the presence or absence of an essential feature), or survey based (opinions of those involved in some way), and ranking based (typically, a 1-to-5 scale). Everything you base your recommendation on must be stated in this section. If you discover something else that sways your recommendation later in the report, rewrite this section.

5. *Comparisons.* As with other reports of this type, comparison is the core structural element. You compare each option systematically against one requirement at a time. For example, you might have a cost section in which you discuss and compare the costs of each option. Either at the beginning or end of each comparative section, you state a conclusion: which option is best in relation to that requirement.

6. *Conclusions table and summary*. After the comparative sections, construct a table that summarizes the key conclusions from those sections. This enables people who prefer not to read the whole report to get the key data quickly and reference it easily. The textual discussion in the comparative sections enables those who are more inquisitive to get the extra detail a table cannot provide.

   Also, include a textual summary of the key conclusions— preferably a numbered list. In this list, include primary conclusions (the ones in the comparative sections), secondary conclusions (those that balance conflicting primary conclusions and reject some of the options), and final conclusions (those that state which options are the best based on the requirements).

7. *Recommendations*. It might seem that the final conclusion is the recommendation, but it is not. One option may be the best of those considered, but it may not meet the requirements. You may be forced to recommend none of the options!

8. *Appendixes*. Put large tables and diagrams, lists, forms, and other less essential materials in the appendix. Don't force readers to wade through material that only individual specialists are interested in. However, do provide cross-references to these appendixes so that readers know they exist.

11.5.4 EVALUATION REPORTS

The evaluation report could actually be the final item in a sequence of engineering documents. Your engineering firm may have produced a proposal to get the consulting job to assess the feasibility of a project. Once the feasibility study had demonstrated that the project was indeed feasible, your next research and writing project may have involved a report recommending design and equipment options. If your firm was involved in the construction of the project, you may have had to write progress reports at certain intervals. But then, after the project was complete and operating, you may have been directed to write an evaluation report, studying the success of the project.

The structure of the evaluation report resembles that of the feasibility report except for some important differences. Whereas the feasibility report is essentially a prediction as to the success of a project, the evaluation report looks back over the performance of

Proceedings of the 2000 DOE Hydrogen Program Review
NREL/CP-570-28890

Analysis of the

Sodium Hydride-based Hydrogen Storage System

being developed by

PowerBall Technologies, LLC

Prepared for

The US Department of Energy
Office of Power Technologies
Hydrogen Program

Prepared by

J. Philip DiPietro
and
Edward G. Skolnik,

Energetics, Incorporated

October 29, 1999

FIGURE 11.36
Cover page of an evaluation report.

(Adapted with permission from Enegetics, Inc. and the U.S. Department of Energy. The table of contents usually comes right after the cover page.)

the completed project, seeking to determine whether it has met its requirements:

1. *Executive summary.* As with the feasibility report and the recommendation report, place key details, conclusions, and evaluations up front, just after the transmittal letter and cover

page. An executive summary for this project would include the following:

- Review of the project: description, history, key players
- Summaries of key detail and conclusions from the technical, financial, and social performance sections in the body of the report
- Evaluative conclusions as to the success of the project

2. *Introduction.* Like the introductions for the preceding types of reports, this introduction also does what an introduction ought to do, even repeating some of the contents of the executive summary. You design reports so that readers can access the information in a variety of ways:

- Background of the project: Who requested it; who performed the projects; what was done; when the project began and ended.
- Purpose of the project and this report: In an introduction you needn't state your final conclusions as to the value of the project. You also must indicate the purpose of this report: to report findings and conclusions and to evaluate the project.
- Target audience: Indicate who are the expected readers and what background readers need to read this report intelligently.
- Overview: Include a list of what your report contains, its major sections.

Introduction

Part of the role that Energetics plays in the DOE Hydrogen Program is to provide independent technical assessments of ongoing hydrogen R&D projects. In addition, Energetics performs analyses on hydrogen-related processes and systems. During May 1999–April 2000, Energetics visited five laboratories in order to perform assessments on hydrogen production and storage R&D projects. This evaluation report discusses these assessments. In addition, Energetics has analyzed an alternative regeneration scheme for a hydrolysis-based metal hydride storage system. This subject is discussed in the report entitled *Analysis of the Sodium Hydride-Based Hydrogen Storage System* being developed by PowerBall Technologies, LLC, which follows the present report.

FIGURE 11.37
Introduction to an evaluative report.

(Adapted with permission from Enegetics, Inc., and the US Department of Energy.)

• Scope: If applicable, state what the evaluation report does *not* contain.

3. *Technical performance.* Use your engineering expertise to assess the technical performance of the project. In the case of the downtown-area wireless system, were people able to stay connected? Was the performance reliable and speedy? How does actual performance compare with the performance predicted in the feasibility report? Report these details from an engineering perspective, which, of course, the client's engineers will want to review. But also provide a translation of this information for non-specialists, the nonengineers who will be reading your report. State the conclusion as to the technical performance of the project at the beginning or end of this section.

4. *Financial performance.* You may also need to report on the financial performance of the project. How do the actual operating costs compare with those predicted in the feasibility report? Once again, make sure that some part of this discussion can be read by people who are not financial experts. Present essential financial data, probably in the form of tables. Put large unwieldy tables in the appendix. End this section with a conclusion as to the financial performance of the project. State the conclusion as

---

Financial Evaluation

The base case cost of the sodium hydride system corresponds to a cost of hydrogen delivered to the vehicle power system of $5.30/kg. A low-cost scenario reduces this cost to $4.10/kg. The low-cost case is twice as expensive as costs predicted for compressed hydrogen gas systems. However, because the sodium hydride system is amenable to geographically dispersed hydrogen vehicles and because it promises a relatively low-cost on-board storage system, it may still prove to be an attractive option. Including the on-board equipment, the low-cost case of the PowerBall system is a little over $1/kg more expensive than a comparable liquid hydrogen system, and a little over $2/kg H_2 more expensive than a compressed hydrogen system. The safety features and ease of transportation of the encapsulated NaH pellets increases the attractiveness of the system, making it of interest at least to niche markets.

FIGURE 11.38

Final evaluation. Notice the evaluative phrases "promises a relatively low-cost on-board system," "may make it attractive," and "of interest at least to niche markets."

(Adapted with permission from Enegetics, Inc., and the US Department of Energy.)

to the financial performance of the project at the beginning or end of this section.

5. *Social performance.* In the case of the downtown wireless system, you might be expected to survey public opinion once gain to see whether people have actually used the system, whether they like it, and other such details. Present the essential survey data in this section, probably in the form of tables. Once again put the big tables in the appendix. And also once again, begin or end this section with conclusions as to the social performance of the project.

6. *Final evaluation.* This final section is a detailed discussion of how the project has performed, whether it has lived up to its expectations, and whether it should be continued. As with all the other reports discussed here, your final evaluation should flow logically from the preceding discussion. Your presentation of supporting data should enable readers to come to their own independent evaluative conclusions.

7. *Appendixes.* Put large tables and diagrams, lists, forms, and other less essential materials in the appendix. Don't force readers to wade through material that only individual specialists are interested in. However, do provide cross-references to these appendixes so that readers know they exist.

11.6 PRIMARY RESEARCH REPORTS

Primary research occurs in the laboratory or in the field and produces new or original data. A **primary research report**, as its name indicates, reports that data along with summaries of other research literature, descriptions of the method, and presentation of conclusions and recommendations related to that data. Secondary research, on the other hand, occurs in libraries, on the Internet, and through other methods of gathering existing information. Primary research reports go by many other names: lab reports, field studies, investigative reports, and so on.

Primary research reports have a standard organization as the following subsections describe.

11.6.1 ABSTRACT

Just below the title and the names of the authors, you will see the summary of the research report. Abstracts can be informative or

ABSTRACT: Many field research projects have been conducted to study the effects of natural foliage on the propagation and attenuation of sound. This research takes natural foliage into a controlled laboratory setting to test its low-frequency acoustic characteristics. Absorption of low-frequency components of unwanted noise is of interest to the Army, but has been an unsolved problem due in part to the cumbersome and expensive testing facilities needed to study long wavelengths.

In this research, low-frequency absorption and reflection coefficients were found reliably and consistently. Due to study of the steady state conditions, the methods presented here could constitute a more consistent method than ever before. The procedures described in this paper can serve as a handbook for future research; recommendations are included.

FIGURE 11.39
Abstract. Primarily informative, this abstract summarizes the question, other research, the conclusion drawn from the research, and recommendations.

descriptive. The **informative abstract** is a condensed view of the report in which the research question, literature review, method, key findings, conclusions, and recommendations are presented in a brief paragraph. A **descriptive abstract** may provide some of the initial contents just listed for the informative abstract, but for the essential contents—findings, conclusions, and recommendations—merely states that this information is included in the report.

11.6.2 INTRODUCTION

The introduction to a primary research report can expand or contract depending on the length and complexity of the report itself. In addition

This research was performed to determine whether cogeneration, or some other cooling option, could economically benefit the Air Force medical facility at Davis-Monthan AFB, AZ, where the cost of purchased electrical power is relatively high compared to that of natural gas. A cooling-load profile was developed for the facility by reviewing plant records and interviewing plant operators; boiler logs (daily and monthly) were consulted to determine heating loads; and initial costs and savings were entered into the Life Cycle Cost in Design (LCCID) computer program to determine simple paybacks and savings-to-investment ratios for all options. Based on the results of the investigation, preferred options were recommended for meeting the facility cooling load.

FIGURE 11.40
Abstract. Primarily descriptive, the abstract summarizes the question and the method but does not reveal the conclusions or recommendations.

I. INTRODUCTION

Limited capital investment for major transportation improvements and growth in metropolitan areas require the most efficient use of the existing transportation system. Provisions of the Clean Air Act Amendments and TEA21 further intensify these concerns. One means to improve mobility is high-occupancy vehicle (HOV) lanes. The concept of an HOV lane is to increase the person-carrying capacity of freeways by providing dedicated lanes for multi-occupant vehicles. By doing so, one HOV lane can serve the travel needs of more people than a freeway lane, thereby increasing the efficiency of the entire system. While a variety of types of HOV lanes have been designed and implemented, a number of issues must be considered for an efficient and effective HOV facility. This field study describes a HOV implementation in the Dallas area, describes the data collection methodology used, and presents the data collected along with conclusions and recommendations.

FIGURE 11.41
Introduction to a primary research report.

to revealing the topic and purpose and providing an overview of what will be covered—which any introduction should do—introductions to brief primary research reports may also describe the research question or problem and provide a literature review. In longer, more complex reports, the research problem and the literature review are likely to occupy separate sections of their own.

11.6.3 RESEARCH PROBLEM, QUESTION

Often introduced as a separate section with the heading "Background," this part of a primary research report describes the problem or question that is the focus of the research.

The Air Force Civil Engineering Support Agency (AFCESA) has been actively involved with a cogeneration project at Tyndall AFB hospital. The project uses an absorption chiller to satisfy the hospital's base cooling load. The steam for activating the absorption chiller is obtained from waste heat, which is derived from an engine driving a generator to produce electrical power—which is also used by the hospital. Based on this experience, AFCESA funded USACERL to perform an analysis to see if such a concept, or some other cooling option, could be of economic benefit at the Air Force medical facility at Davis-Monthan AFB, AZ, where the cost of purchased electrical power is relatively high compared to that of natural gas.

FIGURE 11.42
Background on the research problem or question. Information like this is integrated into the introduction of brief primary research reports or presented as a separate section in longer primary research reports.

11.6.4 LITERATURE REVIEW

Sometimes, even the literature review may occur under the heading "Background." It too may be integrated into the introductions of particularly brief primary research reports. In any case, its function is always the same: to summarize what is known about the research question and what is not known. You can read a detailed discussion of literature reviews elsewhere in this chapter. Figure 11.43 shows an excerpt of a literature-review section occurring in a primary research report.

11.6.5 METHOD, MATERIALS, EQUIPMENT

Primary research reports are expected to describe how the research was performed and what equipment was used. These descriptions should be detailed enough that others can replicate the research to

---

II. LITERATURE REVIEW

Prior to this study, aeolian conditions and particle accumulations on Mars were not well understood. Dune fields, time-variable bright and dark streaks associated with topographic obstructions, and other wind-related surface features were revealed by Mariner 9 and Viking Orbiter data [2,3,4,5,6]. Barchan, star, transverse, and longitudinal dunes have been observed, and their morphologies and orientations used to interpret regional wind patterns [7,8,9,10], but it is not clear whether any of these dune forms are currently active. Wind tunnel investigations indicate unrealistically strong winds would be required for direct entrainment of dust-sized particles under Martian conditions [1], so the mechanism for raising massive quantities of dust during regional and global dust storms is not well understood. Accumulations of fine particles deposited and shaped by wind were observed at both Viking Lander sites [11,12,13], and the orientations of drifts correlate reasonably well with highest wind speed directions inferred from wind streaks seen from orbit [14]. Changes to the Viking Lander 1 site (Mutch Memorial Station) during the mission were minor, including trenches and small artificial piles of soil affected by winds perhaps as high as 50 m/sec. Previously undisturbed materials were not perceptibly eroded. No wind-related morphological changes were observed at the Lander 2 site [15,16]. Dust deposition and subsequent removal of dust by light winds occurred at both landing sites [17,15].

FIGURE 11.43

Excerpt from a literature-review section of a primary research report. The bracketed numbers refer to the individual article, reports, and books that address this research question. "Initial Results of the Imager for Mars Pathfinder Windsock Experiment."

(R. Sullivan Greeley, et al. Space Sciences, Cornell University; Department of Geology, Arizona State University; Jet Propulsion Laboratory, Lunar and Planetary Laboratory, University of Arizona. http://mars.jpl.nasa.gov/MPF/science/lpsc98/1901.pdf.)

determine whether similar data and conclusions can be derived. In Figure 11.44 you can see that the windsock is described along with its placement and readings taken from it.

11.6.6 FINDINGS, DATA

After the methods, materials, and equipment section comes the findings section. In this section, also known as results or analysis, you summarize the data you found. As Figure 11.45 shows, this discussion is usually accompanied by tables, charts, and graphs.

III. METHODOLOGY AND EQUIPMENT

Three Imagers for Mars Pathfinder (IMP) windsock units are mounted on the ASI/Met mast at heights of 33.1, 62.4, and 91.6 cm above the solar panel. The IMP windsocks function like conventional terrestrial windsocks in that deflection from vertical is related to wind speed, and azimuth of deflection relates to wind direction. Each windsock consists of a hollow aluminum cone rigidly joined to an aluminum-sheathed steel counterweight spike, which pivot together on a small, lowfriction gimbal mount. The windsocks are counterbalanced for sensitivity to wind at typical Martian surface pressures. Field testing and wind tunnel tests showed the units to be aerodynamically stable at all deflection angles. Each windsock unit is constructed of electrically conductive materials and is grounded to prevent accumulation of static charge from affecting windsock deflection.

Three IMP windsocks were successfully deployed with the ASI/Met mast on Sol 1. A regular program of daily windsock imaging was established on sol 13, and occasional monitoring of likely areas for wind-related changes was begun on Sol 19. A dust devil search sequence was activated on Sol 66. Daily windsock observations from Sol 13 typically involved four six-frame observations of the top windsock, and one or more twelve-frame observations of all three windsocks for profile information. This program was cut back later in the mission as overall downlink allocations declined, ending with final profiles returned on Sol 82. All windsock images were tightly sub-framed and compressed 6:1. Six-frame observations of the top windsock generally were carried out at 0850, 0950, 1400, and 1630 local time with some sol-to-sol variations due to scheduling of spacecraft or other camera activities. Wind profile measurements were obtained near noon, a time when ASI/Met and windsock data indicated relatively stronger winds than in morning or afternoon.

FIGURE 11.44

Method and equipment section from a primary research report. "Initial Results of the Imager for Mars Pathfinder Windsock Experiment."

(R. Sullivan Greeley, et al. Space Sciences, Cornell University; Department of Geology, Arizona State University; Jet Propulsion Laboratory, Lunar and Planetary Laboratory, University of Arizona. http://mars.jpl.nasa.gov/MPF/science/lpsc98/1901.pdf.)

IV. ANALYSIS

One of the objectives of HOV lanes is to increase *person*-throughput rather than *vehicle* throughput in the corridor. It is, therefore, not very useful to analyze the number of vehicles using a facility. It is, however, important to investigate the number of carpool (multi-occupant) vehicles utilizing a facility. An increase in the number of multi-occupant vehicles on a facility indicates an increase in the person-throughput of a facility. Figure 10 shows the number of two-or-more person (2+) carpools on each of the facilities before and after the HOV lane opened. "Before" data consists of six averaged quarterly collection periods prior to HOV lane construction, and "after" data consists of averaged collection periods since HOV lane opening. After each HOV lane was opened, a significant increase in the number of 2+ carpools on each of the facilities resulted. The collected data also show that the percent increase in carpools ranged from 88% on eastbound IH-635 to 238% on IH-35E North. An analysis of the carpool volumes indicates that the implementation of HOV lanes has resulted in a substantial increase in the number of carpools in each corridor.

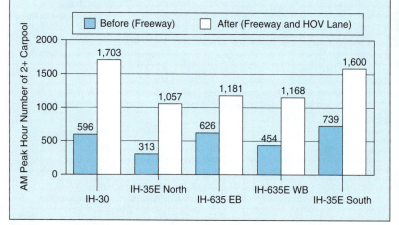

FIGURE 11.45
Analysis (findings, results) section of a primary research report.

(Douglas A. Skowronek, Stephen E. Ranft, and A. Scott Cothron. *An Evaluation of Dallas Area Hov Lanes, Year 2002*. Texas Transportation Institute, Texas A&M University and Texas Department of Transportation, Research and Technology Implementation Office. http://tti.tamu.edu/documents/4961-6.pdf)

11.6.7 CONCLUSIONS AND RECOMMENDATIONS

Primary research reports typically separate the presentation of the findings from the conclusions and recommendations. This way, readers can consider the data and draw their own conclusions without interference from the writer. Recommendations often accompany the conclusions, typically suggesting further research. Once again, primary research reports vary: some put the recommendations in their own separate section with their own heading; other combine conclusions and recommendation as does the example shown in Figure 11.46.

11.6.8 REFERENCES

The references section is typically the last section of a primary research report. It lists all the articles, books, reports, web sites, interviews, and other information sources that the writer/researcher used in the creation of the report. The list uses one of the standard documentation styles, such as IEEE, APA, or CSE, depending on the requirements of the publication or the organization. See Chapter 12, "Documentation," for details.

V. CONCLUSIONS AND RECOMMENDATIONS

The goal of this research was to investigate the operational effectiveness of the new concurrent flow HOV lanes in the Dallas area as well as to assess the effectiveness of concurrent flow (buffer-separated) versus contraflow (barrier-separated) HOV lanes in the Dallas area. As shown in Table 14 and the data summary in Tables 15 through 20, the concurrent flow lanes have generated a substantial number of carpools, have increased the person movement in the corridor, have increased the occupancy rate in the corridor, and have not negatively impacted the operation of the adjacent freeway general-purpose lanes. Experience from Houston, however, indicates that two to four years of operation of a facility are required before a complete and thorough assessment can be made. All five HOV lane projects are cost effective and have attained, or are projected to attain, a benefit–cost ratio greater than 1.0 within the first five years of operation. While this appears to indicate that either type of HOV lane is acceptable, other issues must be considered, such as the safety of a non-barrier-separated lane. Future safety research for pinpointing the most critical factors contributing to crashes in buffer-separated HOV lanes will require a microscopic analysis of particular crashes and their circumstances.

FIGURE 11.46
Conclusions and recommendations section from a primary research report.

(Douglas A. Skowronek, Stephen E. Ranft, and A. Scott Cothron. *An Evaluation of Dallas Area Hov Lanes, Year 2002*. Texas Transportation Institute, Texas A&M University and Texas Department of Transportation. Research and Technology Implementation Office. http://tti.tamu.edu/documents/4961-6.pdf)

11.7 LITERATURE REVIEWS

A **literature review** summarizes what is known about a specific research topic, narrates the milestones of the research history, indicates where current knowledge conflicts, and discusses areas where there are still unknowns. A literature review can be a standalone document or a component of a primary research report (as discussed previously). Research journals often contain articles whose sole purpose is to provide a literature review. As a component of a research report, a literature review can be as long as a whole chapter in a book, only a paragraph in a research article, or as short as a few sentences in an introduction. In all cases, the function of the literature review is the same: to summarize the history and current state of research on a topic. As you know from the preceding section, a primary research report (such as those in engineering research journals) focuses on a question—for example, the effect of weightlessness on growing vegetables. The literature-review section of that report would summarize what is known about this topic, indicate where current knowledge conflicts, and discuss areas where there are still unknowns.

A well-constructed literature review tells a story. It narrates the key events in the research on a particular question or in a particular area. Who were the first modern researchers on this topic? What were their findings, conclusions, and theories? What questions or contradictions could they not resolve? What did researchers following them discover? Did their work confirm, contradict, or overturn the work of their predecessors? Were they able to resolve questions their predecessors could not? You narrate this series of research events in a literature review. You can consider this research process as similar to the thesis-antithesis-synthesis process. You start out with a thesis, then along comes an antithesis to contradict it, and eventually some resolution of this conflict called a synthesis is achieved, which is actually a step forward in the knowledge about that topic. But now the synthesis becomes a thesis, and the process starts all over again.

Hilton Obenzinger of Stanford University in "How to Research, Write, and Survive a Literature Review (http://www.stanford.edu/dept/undergrad/urp/PDFLibrary/writing/LiteratureReviewHandout.pdf) calls this type of literature review a "road map." He identifies

Face recognition, in addition to having numerous practical applications, such as bankcard identification, access control, mug shots searching, security monitoring, and surveillance systems, is a fundamental human behavior that is essential for effective communications and interactions among people.

A formal method of classifying faces was first proposed in [1]. The author proposed collecting facial profiles as curves, finding their norm, and then classifying other profiles by their deviations from the norm. This classification is multimodal, i.e., resulting in a vector of independent measures that could be compared with other vectors in a database.

FIGURE 11.47
Introduction to a literature review.
(Adapted with permission from A. S. Tolba, A. H. El-Baz, and A. A. El-Harby, "Face Recognition: A Literature Review." *International Journal of Signal Processing* vol. 2, no. 2, 2005.)

several other types, most importantly those that review the *methodology* of the research as well as or instead of the research findings. Obenzinger emphasizes that the literature review is not just a passive summary of research on a topic but an evaluation of the strengths and weaknesses of that research—an effort to see where that research is "incomplete, methodologically flawed, one-sided, or biased." In any case, as the following examples show, a literature review is a *discussion* of a body of research literature, not an annotated bibliography. Notice in the following examples that literature reviews use standard bracketed IEEE textual citation style and end with a bibliography (called "References").

  Consider Figure 11.47, which shows the beginning of the review of literature found in A. S. Tolba, A.H. El-Baz, and A.A. El-Harby, "Face Recognition: A Literature Review." *International Journal of Signal Processing* vol. 2, no. 2, 2005.

Progress has advanced to the point that face recognition systems are being demonstrated in real-world settings [2]. The rapid development of face recognition is due to a combination of factors: active development of algorithms, the availability of a large database of facial images, and a method for evaluating the performance of face recognition algorithms.

FIGURE 11.48
Literature review—current status of research on the topic.

In [83], a combined classifier system consisting of an ensemble of neural networks is based on varying the parameters related to the design and training of classifiers. The boosted algorithm is used to make perturbation of the training set employing MLP as base classifier. The final result is combined by using simple majority vote rule. This system achieved 99.5% on Yale face database and 100% on ORL face database. To the best of our knowledge, these results are the best in the literature.

FIGURE 11.49
Literature review—discussion of specific advances and problems.

As you can see, the first paragraph establishes the topic and its importance; the second paragraph goes back to the beginning of modern research that provided a foundation for computer-based face recognition. This literature review moves on to the current status of research in this field.

Notice how the next excerpt describes an important advance in the research on this topic, but then points out its limitations.

The literature review of face-recognition research examines many different methods used in computer-based face recognition. For each, it summarizes the method, the results, and the strengths and weaknesses of that method. This example is not so much the thesis-antithesis-synthesis pattern mentioned previously but rather a collection of efforts all striving after a common goal—increased accuracy of computer-based face recognition. Here's how the summary of that process ends in this literature review:

One of the pioneering works on automated face recognition by using geometrical features was done by [46] in 1973. Their system achieved a peak performance of 75% recognition rate on a database of 20 people using two images per person, one as the model and the other as the test image. . . .[P]recisely measured distances between features may be most useful for finding possible matches in a large database such as a mug shot album. However, it will be dependent on the accuracy of the feature location algorithms. Current automated face feature location algorithms do not provide a high degree of accuracy and require considerable computational time.

FIGURE 11.50
Literature review—conclusion.

11.8 SPECIFICATIONS

If you search the Internet, you will find a wide variety of specifications, and strong opinions about how specifications should be written. Because operating specifications so much resemble instructions, they are discussed in the instructions section of this chapter. In the following text, manufacturing (design) specifications are discussed. The focus here is description of product requirements. Specifications of the type discussed here focus on the design and manufacture of a product and enable people to design and manufacture a product or purchase it accurately. When you write specifications, you must pay close attention to accuracy, precision of detail, clarity, and standard specification-writing practices. Problems in these areas can cause organizations to lose large amounts of money, result in delays in project completion, and lead to expensive lawsuits.

Note: Because of the financial and legal issues associated with specifications, treat the following only as a high-level overview; consult with specification-writing professionals for any specific project. And because the content, organization, and format of specifications vary widely, find out the requirements of the organization for which you are writing.

11.8.1 SPECIFICATIONS STRUCTURE

Specifications use an organizational style that enables readers to find specific details quickly. To make that possible, use multiple-enumerated headings and multiple-enumerated lists as shown in Figure 11.51. Use one of the following organizational methods to facilitate quick retrieval:

- *Introductory general description*. Describe the product, component, or program first in general terms—administrative details about its cost, start and completion dates, overall description of the project, scope of the specifications (what you are not covering), and anything general in nature that does not fit in the part-by-part descriptions.
- *Organization*. Focus the main headings of the specifications on the overall process. Figure 11.51 shows that qualifications and references to other specifications are included, along with details about the actual physical products. Some specifications include main

SAFETY-DISCONNECT SWITCHES

1.0 GENERAL

1.1 The supplier shall manufacture the required (cooling tower) (closed circuit cooling tower)(evaporative condenser).

1.2 The manufacturer of the (cooling tower)(closed circuit cooling tower)(evaporative condenser) shall furnish the low-voltage non-fused switches as specified herein and as shown on the contract drawings.

2.0 REFERENCES

The switches and all components shall be designed, manufactured and tested in accordance with the latest applicable standards:

2.1 NEMA KS-1

2.2 UL 98

3.0 SUBMITTALS—FOR REVIEW/APPROVAL

The following information shall be submitted to the Engineer:

3.1 Dimensioned outline drawing

3.2 Conduit entry/exit locations

3.3 Switch ratings, including:

    3.3.1 Short-circuit rating

    3.3.2 Voltage

    3.3.3 Continuous current

3.4 Cable terminal sizes

3.5 Product data sheets

4.0 QUALIFICATIONS

4.1 The supplier of the assembly shall be the manufacturer of the (cooling tower)(closed circuit cooling tower)(evaporative condenser).

4.2 For the equipment specified herein, the manufacturer shall be ISO 9001 certified.

4.3 The manufacturer of this equipment shall have produced similar electrical equipment for a minimum period of ten (10) years. When requested by the Engineer, an acceptable list of installations with similar equipment shall be provided demonstrating compliance with this requirement.

5.0 PRODUCTS

5.1 Heavy-Duty Safety Disconnect Switches

    5.1.1 Provide switches as shown on drawings, with the following ratings:

        5.1.1.1 30 to 400 amperes

        5.1.1.2 250 volts AC, DC; 600 volts AC (30A to 200A 600 volts DC)

        5.1.1.3 3 pole

        5.1.1.4 Non-fusible

        5.1.1.5 Mechanical lugs suitable for aluminum or copper conductors

FIGURE 11.51

Example from a set of specifications.

FIGURE 11.51 (*Continued*)

5.2 Construction

   5.2.1   Switch blades and jaws shall be visible and plated copper.

   5.2.2   Switches shall have a red handle that is easily padlockable with three 3/8-inch shank locks in the OFF position.

   5.2.3   Switches shall have defeatable door interlocks that prevent the door from opening when the handle is in the ON position. Defeater mechanism shall be front accessible.

   5.2.4   Switches shall have deionizing arc chutes.

   5.2.5   Switch assembly and operating handle shall be an integral part of the enclosure base.

   5.2.6   Switch blades shall be readily visible in the ON and OFF position.

   5.2.7   Switch operating mechanism shall be nonteasable, positive quick-make/quick-break type. Bail type mechanisms are not acceptable.

   5.2.8   Switches shall have clear line terminal shields.

   5.2.9   Embossed or engraved ON-OFF indication shall be provided.

   5.2.10 Double-make, double-break switch-blade feature shall be provided.

   5.2.11 Renewal parts data shall be shown on the inside of the door.

5.3 Enclosures

   5.3.1   All enclosures shall be NEMA 3R rainproof unless otherwise noted.

   5.3.2   Other types, where noted, shall be:
        5.3.2.1 NEMA 4X watertight corrosion resistant
        5.3.2.2 30A to 400A—304 stainless steel

   5.3.3   Paint color shall be ANSI 61 gray.

   5.3.4   30A to 100A NEMA 4X enclosures shall be provided with draw-pull latches.

5.4 Nameplates

   5.4.1   Nameplates shall be front-cover mounted.

   5.4.2   Nameplates shall contain a permanent record of switch type, ampere rating, and maximum voltage rating.

sections on testing, inspection, delivery, maintenance, and whatever else the project requires.

- *Part-by-part description*. In the actual description of the product, present specifications part by part, element by element, trade by trade—whatever is the logical, natural, or conventional way of doing it.
- *General-to-specific order*. Wherever applicable, arrange specifications from general to specific.

11.8.2 SPECIFICATIONS: CONTENTS, ORGANIZATION, STYLE

In general, design specifications as follows:

- Place each individual specification in its own separate sentence and use the decimal numbering system to enable cross-referencing.
- Depending on the requirements of the organization, use either the open (performance) style or the closed (restrictive) style. The *open* style specifies what the product or component should do—that is, its performance capabilities. The *closed* style specifies exactly what it should be or consist of.
- Always cross-reference existing specifications rather than repeating their full detail in your specifications. These existing specifications are published by government agencies as well as trade and professional associations.
- Use specific details to describe the product or component as precisely as possible. Double-check for potential ambiguity. Can any of your words be interpreted in more than one way? Use the same technical jargon that you find in other specifications of the type you are writing—even if you don't like it.
- Use "shall" to indicate requirements in design, manufacturing, construction, or procurement specifications. In specifications writing, "shall" is understood as stating a requirement.
- Provide numerical specifications in both words and symbols: for example, "the distance between the two components shall be three centimeters (3 cm)."
- Use a terse, even "telegraphic" writing style in specifications: incomplete sentences are common as well as the omission of obvious function words such as articles.
- Be careful with pronouns and complex sentences. There should be no doubt about what words such as "it," "they," "which," and "that" refer to. Be careful with sentences containing a list followed by a descriptive or qualifying phrase—does the phrase refer to just one or all of the list items? In cases like these, use additional words to clarify, or break the sentence into two or more separate sentences.
- Use words and phrases that are standard in similar specifications, even if they seem awkward. Specification writing is no place for

creativity or originality. Past usage has most likely proven that standard specification language is reliable.

Make sure your specifications cover everything; imagine yourself as the user of your specifications—in particular, an unreliable one. How might a careless, incompetent contractor misread your specifications? How might a contractor willfully misread your specifications in order to cut cost or time?

11.9 POLICIES AND PROCEDURES

At first glance, policies and procedures resemble instructions as well as specifications, but there are essential differences. The most important fact about this type of document is that it focuses on organizations—a company, a governmental agency, or departments or units within these entities. Policies and procedures define how the organization should operate; they establish the rules and regulations that employees are expected to follow.

11.9.1 KEY TERMS

Policies-and-procedure documents obviously are made up of policies and procedures. A **policy** is a goal or objective that the organization wants to achieve. For example, a sales department might have as one of its objectives "to treat customers with the utmost respect"—that's a policy. **Procedures** are the detailed instructions on how to achieve the goal stated in that policy. The steps in these instructions are typically numbered: for example, the sequence of actions to take when a potential customer calls or walks in the store. However, the steps can be bulleted. If the policy establishes an employee dress code, the procedures would likely be a series of bullets indicating what is and is not acceptable to wear to work.

11.9.2 CONTENT, ORGANIZATION, AND FORMAT

Notice the format of the example policy-and-procedure document in Figure 11.52:

- Multiple enumeration is used—in other words, 1.0, 1.1, 1.1.1.
- The purpose section identifies how and when this document should be used.
- The definition section, a common and important section in policies and procedures, defines key terms in the document.

HANDWASHING FOR HOME HEALTHCARE PERSONNEL

1.0 PURPOSES

The purposes of this policy are to:

1.1 Promote the prevention and control of infection.

1.2 Establish guidelines for handwashing in the home environment.

2.0 DEFINITIONS

2.1 Handwashing is the rubbing together of the surfaces of the hands with soap or antimicrobial agent followed by rinsing the hands in clean water.

3.0 POLICY

3.1 All patient care personnel will wash their hands using the technique outlined in this handwashing procedure.

3.2 Personnel will wash their hands:

3.2.A After entering the patient's home, before rendering care

3.2.B Before performing an invasive or sterile procedure

3.2.C Before contact with an immunocompromised patient

3.2.D After any care or treatment that involves touching organic materials or open wounds

3.2.E After handling contaminated materials, dressings, or equipment

3.2.F After completing patient care

3.3 If gloves are worn, handwashing must follow the removal of gloves.

4.0 PROCEDURE

4.1 Remove liquid soap and paper towels from black bag and place beside a sink that has warm, running water.

4.2 Roll up sleeves and remove jewelry.

4.3 Inspect hands and fingers for cuts or breaks in the skin. Report lesions to your supervisor.

4.4 Stand close to the sink, but refrain from leaning against the sink or touching the sink surface during washing.

4.5 Turn on the water using a paper towel to cover the faucets as you adjust the water flow and temperature. Water should be warm.

4.6 Discard paper towel.

4.7 Wet hands thoroughly, keeping your hands lower than your elbows.

4.8 Apply liquid soap to hands, and rub hands vigorously until a lather is formed.

4.9 Wash hands for at least 15 seconds using circular motions.

4.10 Rinse hands thoroughly, keeping them down so that the water runs from the wrists to the fingertips.

4.11 Dry hands thoroughly using clean paper towels. Start with the fingers and move up to the wrists.

4.12 Use a dry, clean paper towel to turn off the faucet(s).

4.13 Discard paper towels in a proper receptacle.

4.14 After final use, return handwashing supplies to your black bag.

4.15 If you are in a home where there is no running water or where sanitary conditions are such that proper handwashing is impossible, scrub hands thoroughly using disposable antiseptic towelettes. Wash hands with soap and water as soon as possible.

FIGURE 11.52
Policies and procedures example.

- The policy section establishes the objectives. In these policies and procedures, the policies might sound like procedures, but they do not provide specific step-by-step instructions.
- The procedure section of Figure 11.52 explains exactly how to perform what you might have thought was a simple activity, using imperative phrasing, carefully specified sequencing, and precise language.

Although these example policies and procedures do not, some policies and procedures also state actions that management will take if policies and procedures are not adhered to. Good examples are failure to comply with smoking, alcohol, and drug policies or use of business assets for personal reasons.

11.10 HANDBOOKS

As its name implies, a **handbook** acts as a "handy" reference or learning tool for a wide range of information in a specific field. The handbook is a hybrid of other types of technical documents discussed in this chapter.

11.10.1 SCOPE OF HANDBOOKS

The scope of a handbook is defined by how it is used and who uses it:

- Organizations have handbooks. Such a handbook might contain company history, policies and procedures, organizational charts, instructions (how to operate the fax machines, for example), reference tables (for example, contact information and part numbers for reordering supplies), and anything else that employees need to refer to.

- Professions, fields, occupations, and do-it-yourselfers all have handbooks. A handbook for building your own notebook computer would contain not just instructions, but conceptual information explaining how the thing works and defining key terms and reference information on part numbers, specifications, compatibility details, and so on.

11.10.2 TYPES OF HANDBOOK CONTENT

As you can see from these examples, handbooks contain combinations of three basic types of information:

- *Conceptual, background information.* In a civil engineering handbook, you might find information on landfills, incinerators, soil types, stress distribution, and the fundamentals of hydraulics. The purpose of this information is not to teach but to summarize well-known, established knowledge.
- *Instructional information.* In the same civil engineering handbook, you might find instructional information on construction estimating, water and wastewater planning, and designing high-speed ground transportation systems.
- *Reference information.* Handbooks typically contain quick look-up information such as the International System of Units (SI), physical constants, conversion constants and multipliers, and symbols and terminology for physical and chemical quantities. A handbook might include lists of professional societies, academic institutes, journals, conferences, and so on.

Which of these information types to include and in what proportions are established by the needs of your audience and the nature of your subject matter? Imagine a handbook for interns at a manufacturing facility. It would contain information on basic theory and background related to the products and processes of the facility, instructional information on procedures that interns must perform, and reference information on where to find things, whom to contact, and other such vital quick look-up information. Figure 11.53 shows what the table of contents of such a local handbook might look like.

TABLE OF CONTENTS

FIGURE 11.53
Table of contents of a handbook.

12

DOCUMENTATION

Documentation

12   Documentation

Documentation refers to the business of indicating the sources of the information that you borrow. Whether you are in the workplace or in academia, you must cite the sources from which you borrow information. Different fields and disciplines use different documentation styles. Specific engineering disciplines use primarily the IEEE (Institute of Electrical and Electronics Engineers) and the APA (American Psychological Association) documentation style. A few use the CSE (Council of Science Editors) documentation style. A few use the MLA (Modern Language Association) style. All four are covered in the following sections. If you are writing in an academic environment, ask your instructor which documentation style to use. If you are writing an article for a journal or some other professional publication, check with the editor of the journal or other publication.

When you **cite** the sources of your borrowed information, you indicate within the text, in an abbreviated way, from which source and which author you borrowed the information. There are several good reasons to cite the sources of the information you use in your engineering documents:

- Not to do so is **plagiarism**—theft of the intellectual property of others. Getting caught while still in academia can get you a failing grade for the document or for the course—or even expulsion from the institution. Getting caught after you've graduated can seriously curtail your professional opportunities. Getting caught for plagiarism in a commercial environment can get you or your organization sued.

- Citing your sources lends a great degree of professionalism to your engineering documents. Your in-text source citations and your reference lists give your work an air of reliability and authority. Readers can see that you've done your homework.

- When you cite the sources of your borrowed information, you become a part of the conversation regarding the topic of your engineering document. You become part of the process by which knowledge about the topic advances.

What should you cite? Everything that is not common knowledge. Any data, conclusions, recommendations, images, tables, theories, special terminology—you must document anything that you could not have known before you began your research. Difficulties arise, however, with the meaning of "common knowledge." For some, common knowledge is anything found in multiple encyclopedias or dictionaries. Check that assumption with those to whom you will submit your engineering documents. Be careful about something that seems common knowledge to you now: you may have absorbed that knowledge several months ago and have forgotten the source; you're still obligated to go back and find it.

The following sections on the IEEE, APA, CSE, and MLA document styles cover only the highlights. For most of your needs, you can find answers just by scanning the example entries. For issues not covered in these guidelines and not illustrated in the examples, see the published style guides for the documentation style you are using. These are stated at the beginning of each section in the following pages.

12.1 IEEE STYLE OF DOCUMENTATION

The IEEE system uses bracketed numbers (called **citations**) in text which correspond to entries in a references section at the end of the document. To see the source of the bracketed number, you go to the reference section and find the corresponding number:
Documentation guidelines outlined in this section are consistent with those provided in *IEEE Information for Authors* available at

http://www.ieee.org/portal/cms_docs/pubs/transactions/auinfo03.pdf

An excellent resource is available from the University of Toronto's Engineering Communication Centre. It provides guidelines, plenty of examples, and a JavaScript-based "bibliography builder". See http://www.ecf.toronto.edu/~writing/bbieee-f1.html. For questions about plagiarism, see this excellent IEEE resource, http://www.ieee.org/web/publications/rights/Plagiarism_Guidelines_Intro.html.

12.1.1 IN-TEXT CITATIONS: IEEE STYLE

IEEE in-text citations use bracket source numbers to identify sources used in the document. In the text, author names and page numbers are acceptable but rarely provided. The in-text bracketed citations direct the reader to a list of references at the end of the document, where information needed to locate the sources is provided, as shown in Figure 12.1. To create in-text citations:

- Provide the bracketed source number after the phrase or clause that contains the borrowed information. In Figure 12.1, notice that source 6 is the origin of "compensating for time delay include teleoperation with predictive displays" and source 7 is the origin of "real-time generation of low level commands in a graphical predictive display environment." (These two sources are not shown in the references excerpt.)

- However, nothing prevents you from citing author names or page numbers:

According to Cooper, RSVP builds on the lessons learned and the heritage of the Mars Pathfinder mission [1:1373].

In this case, the author wants to single out Cooper and direct readers to page 1373 of source 1.

- To indicate that information comes from two or more authors, use "and" and commas as shown in the following examples:

RSVP builds on the lessons learned and the heritage of the Mars Pathfinder mission [1] and [2].

Several studies, for example [1], [3], and [5], have established that the synchronization of traffic lights has no effect on accident rates.

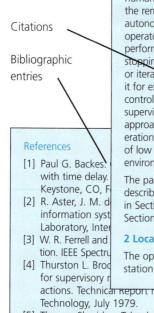

Citations

Bibliographic
entries

In supervised autonomy, commands are generated through human interaction, but sent for autonomous execution at the remote site. These commands are individually autonomous because they are executed independently of operator control, utilizing remote site feedback loops to perform control and monitor sensors for success and failure stopping conditions. A command can be sent immediately or iteratively saved, simulated, and modified before sending it for execution on the real robot. Early works in supervisory control include [3] and [4]. A more complete description of supervisory control can be found in [5]. Alternative approaches to compensating for time delay include teleoperation with predictive displays [6] and real-time generation of low level commands in a graphical predictive display environment [7].

The paper is organized as follows. The local site system is described in Section 2. The remote site system is described in Section 3. Results using the system are shown in Section 4 and conclusions are given in Section 5.

2 Local Site

The operator interface of the local site operator control station

References

[1] Paul G. Backes.
with time delay.
Keystone, CO, F
[2] R. Aster, J. M. d
information syst
Laboratory, Inter
[3] W. R. Ferrell and
tion. IEEE Spectru
[4] Thurston L. Broo
for supervisory r
actions. Technical Report MITSG 79-20, Massachusetts Institute of Technology, July 1979.
[5] Thomas Sheridan. Telerobotics, Automation, and Human Supervisory Control. M.I.T. Press, 1992.

FIGURE 12.1

Excerpts: sample page with bracketed citations and the corresponding reference page.

(Source: P. G. Backes, J. Beahan, M. K. Long, *A Prototype Ground-Remote Telerobot Control System*. http://trs-new.jpl.nasa.gov/dspace/bitstream/2014/35262/1/93-0842.pdf)

- Bracketed in-text citations can be used directly in sentences instead of author or document name:

  As outlined in [3], future flight projects will require efficient use of Deep Space Network (DSN) resources.

  Filter impedance and both open- and closed-loop gain equations are easily calculated and have been provided in Eqs. (1) through (3), respectively. These equations have a key role in PLL phase noise and depend on values chosen for R_1, R_2, and C_f. Full derivations of these formulas have been provided by [2].

FIGURE 12.2
Customizing numbering style in Sun Microsystem's OpenOffice Writer to create
IEEE-style bibliographic entries.

12.1.2 IEEE REFERENCE LIST

In IEEE style, the list of works cited appears at the end of the
document and is called "References." The following presents
only the essential guidelines for setting up a reference list and
example entries for common types of sources cited. For other
variations, see *IEEE Information for Authors*, mentioned
previously.

GUIDELINES FOR THE IEEE REFERENCE LIST

- Place the reference list at the end of your document on a separate
  page. Left-align the title, *References*.
- Number the page or pages of your reference list sequentially with
  the rest of the research paper.
- Double-space all entries in your reference list.
- Use the hanging-indent style for reference-list entries. To
  achieve this in Sun Microsystem's OpenOffice Writer (available
  at openoffice.org), apply the numbered-list style, and then customize to delete the period and add brackets:

- Arrange the entries in the reference list in the order they occur in the text. In other words, the first bracketed citation in the body of the document should be [1].

GUIDELINES FOR IEEE REFERENCE-LIST ENTRIES: COMMON ELEMENTS

The following guidelines are the same across most resources such as books, articles, and other types.

- Begin the entry with the author or authors. Do not invert first name and last name. Convert the first and middle names to initials. For example, Hilda Gowans becomes H. Gowans; Patrick H. McMurrey becomes P.H. McMurrey (no space between initials). Unlike other documentation styles, IEEE does not use et al. for multiple coauthors; in IEEE, list all authors.
- When there are two or more authors, list them with commas and "and":

  [1] W.R. Ferrell and T.B. Sheridan. Supervisory control of remote manipulation. *IEEE Spectrum*, pages 81–88, October 1967.

  [2] J. Wright, F.R. Hartman, and B. Cooper, "Immersive Visualization for Mission Operations: Beyond Mars Pathfinder," *Proceedings of SpaceOps '98*, Tokyo, Japan, 2a006, (1998).

  [3] R.J. Wurtman, M.J. Baum, and J.T. Potts, *The Medical and Biological Effects of Light*. New York: New York Academy of Sciences; 1985.

- If no date is given, put **n.d.** in parentheses. If the item has not yet been published, end the entry with a comma followed by "in press." If the item has not been published, end the entry with a comma followed by "unpublished."
- After the article title, other identifying information can be included in brackets as in these examples: *[Letter to the editor]*, *[Special issue]*, or *[Abstract]*.

IEEE GUIDELINES FOR PRINT RESOURCES
ARTICLES

- For scholarly, research-oriented journals, include volume and issue numbers as well as page ranges and dates; use this format:

  [4] J.A. Veitch and R. Gifford, "Assessing Beliefs About Lighting Effects on Health, Performance, Mood, and Social Behavior." *Environment and Behavior*; vol. 28, no. 4, pp. 446–470, 1996.

[5] R.J. Wurtman, "The Effects of Light on the Human Body." *Scientific American*, vol. 233, no. 1, July, pp. 68–77, 1975.

- For nonscholarly, nonresearch-oriented periodicals, such as popular magazines and newspapers, supply the date and page range:

  [6] B.M. Thayer, "A Passive Solar University Center," *Solar Today*. March/April, pp. 34–36, 1996.

BOOKS

Guidelines for variations on author names are the same for books, articles, and other resources. See "Elements of IEEE Reference-List Entries."

- For books that have one or more editor names, use this format:

  [1] D.F. Beer, ed. *Writing and Speaking in the Technology Professions*, Greenwich, CT: Ablex, 1997.

  [2] S. Reber and L. Scarborough, Eds., *Toward a Psychology of Reading*, Hillsdale, NJ: Lawrence Erlbaum Associates, 1977.

- For the city of publication, provide only the city where the work was published. If the city might be mistaken for another or if the city might not be well known internationally, include the state or province name:

  [1] National Lighting Bureau, *Office Lighting and Productivity*. Washington, DC: National Lighting Bureau, 1988.

- If the book is a volume in a series, use this format:

  [1] M. Brill, *Using Office Design to Increase Productivity*. vol. 1. New York: Workplace Design and Productivity, 1985.

- To direct readers to a specific chapter or a specific page or pages of a book, use this format:

  [1] B.H. Evans, "The Nature of the Skies," *Daylight in Architecture*. New York: McGraw Hill, 1981, pp. 95–105.

- Notice that in references to the publisher, you can omit words and abbreviations such as *Publishers*, *Co.*, and *Inc.*, but not the words *Books* and *Press*.
- If the item has no author, move the title to the author position:

  [1] *Product Safety Label Handbook: Danger, Warning, Caution*. Westinghouse Electric Corporation, 1981.

- If the item is a translation, use this format.

  [1] J. Bertin, *Semiology of Graphics*, W.J. Berg, Trans., Madison, WI: Univ. of Wisconsin Press, 1983.

- If the author of the source is a corporation, agency, or organization, use that name of the publisher's names as the author:

  [4] Galileo Project. *Galileo Flight Operations Plan: Galileo Command Dictionary*. Technical Report PD 625–505, D-234, Jet Propulsion Laboratory, September 1989.

  [5] Purdue University. Development of optimal strategies for maintenance, rehabilitation and replacement of highway bridges. Lafayette, IN: Purdue University, 1993.

  [6] Bureau of the Census. *Statistical Abstract of the United States: 2007*, 127th ed., Washington, D.C. U.S. Gov. Printing Office, 2007.

- If the item is a second or subsequent edition, use this format:

  [1] L. Flower. *Problem-Solving Strategies for Writing*. 3rd ed. San Diego, CA: Harcourt, Brace, Jovanovich, 1984.

- If you cite an article in an edited book, use this format:

  [2] J. Doblin, "A structure for nontextual communications," in *Processing of Visible Language*, vol. 2, P.A. Kolers, M.E. Wrolstad, and H. Bouma, Eds. New York: Plenum, 1980.

  [3] A.M. Flynn, "Gnat robots (and how they will change robotics)," in *IEEE Micro Robots and Teleoperators Workshop*, Hyannis, MA: IEEE, November, 1987, pp. A88–42873–A88–42874.

  [4] B.L. McNaughton, "Neuronal mechanisms for spatial computation and information storage, in *Neural Connections, Mental Computation* (L. Nadel, L.A. Cooper, P. Culicover, and R.M. Harnish, Eds.). MIT Press: Cambridge, MA: 1989, pp. 181–193.

PAPERS PUBLISHED IN PROCEEDINGS

In your research, be sure to search for conference proceedings when you cite papers from proceedings:

  [5] N. Aylon, "Bridge management systems in the information era," in *1993 TAC Annual Conf. Proc.*, (Ottawa, Ontario, 1993, vol. 2, pp. B73–B90).

[6] A.M. Flynn, R.A. Brooks, W.M. Wells, and D.S. Barrett, "The world's largest one-inch cubic robot," in *Proc. IEEE Micro Electrical Mechanical Systems*, (Salt Lake City, Utah, February 1989, pp. 98–101).

TECHNICAL REPORTS AND GOVERNMENT DOCUMENTS

When you use technical reports, which are often published as government documents, include the document number:

[7] J.H. Connell, "A colony architecture for an artificial creature," MIT AI Lab Technical Report 1151, June 1989.

[8] J. Martel, "Analysis of the waste management practices at Bosnia and Kosovo base camps," US Army Corps of Engineers, Hanover, NH, Tech. Rep. ERDC/CRRELTR-03–6, April 2003.

THESES AND DISSERTATIONS

For masters and doctoral work, italicize the title and include the name of the institution:

[1] J.M. Maja, *Quantitative Sonar-Based Environment Learning for Mobile Robots*, M.S. Thesis, Massachusetts Institute of Technology, January, 1990.

[2] P. Grocoff, *Effects of Correlated Color Temperature on Perceived Visual Comfort*. Ph.D Thesis, University of Michigan, 1996.

INTERVIEWS AND OTHER UNPUBLISHED SOURCES

IEEE specifies that you formally cite only published works, soon-to-be published works, and unpublished materials available in libraries, depositories, or archives. Sources such as interviews are considered "non-recoverable" information; no formal entry for them in the references section is necessary. You can still indicate such sources informally in your text:

Red-winged and yellow-headed blackbirds were collected from wetland sites at each project. Three of the samples appear to be elevated. According to R. King of the Arizona Fish and Wildlife Service (personal communication), red-winged blackbirds collected near a lignite-fired power plant in Texas had a mean concentration of 33.1 ug Se/g dry weight in kidneys. Red-winged blackbirds from a control area had a mean selenium concentration of 7.7 ug/g dry weight in kidneys.

IEE GUIDELINES FOR ELECTRONIC SOURCES
ARTICLE BASED ON PRINT SOURCES

Use the following format for articles retrieved from online publications that are duplicates of print versions:

[1] S. Doughton, "Comet Dust Yields Surprises About Universe," in *The Seattle Times*, 15 December 2006. [Online] Available http://www. memagazine.org/Story.html?story_id=101342179&category=Engineer ing&ID=asme.

[2] "Sustainability: High Performance Buildings Deliver Increased Retail Sales." Seattle City Light. http://www.ci.seattle.wa.us/light/conserve/ sustainability/studies/cv5_ss.htm. Accessed March 19, 2001.

ARTICLES IN ONLINE JOURNALS

When an article occurs only online, include both the publication date shown on the document as well as the date you accessed that document. If no author is indicated, start with the title of the document:

[1] Seattle City Light, "Sustainability: High Performance Buildings Deliver Increased Retail Sales," http://www.ci.seattle.wa.us/light/ conserve/sustainability/studies/cv5_ss.htm. Accessed March 19, 2001.

[2] T. MacKay, "Seasonal Affected Disorder and Depression," http://www.sunalite.com/articles/sad4.html. Site last modified December 11, 1998; accessed June 6, 1998.

E-MAIL, ONLINE NEWSGROUPS, FORUMS, OR MAILING LISTS

As mentioned above, IEEE style restricts formal citations to published works, soon-to-be published works, and unpublished materials available in libraries, depositories, or archives. It excludes informal sources such as e-mail, newsgroups, forums, mailing lists, and blogs. You can still informally indicate these sources in your text:

According to J. H. Connell, the robot Herbert was programmed to head off in a random direction every 10 seconds and steal empty soda cans from employees' desks (e-mail communication).

12.1.3 IEEE MANUSCRIPT FORMAT

Guidelines for manuscript format vary from engineering field to engineering field and vary still more amongst academic programs in engineering. If you seek to publish in a professional engineering journal, search for guidelines in a recent issue of that journal, or contact the editor of that journal. For example, the American Society of Civil Engineers provides an author's guide at http://www.pubs. asce.org/authors/

If you are an engineering student, check with your instructors for specific manuscript requirements and have them confirm these minimal guidelines:

- Center the title of your document and your name at the top of the first page.
- Include an abstract below your name. Italicize the text of the abstract.
- Use headings throughout the document beginning with "Introduction."
- Use decimal-style numbering on the headings: for example, 2 Local Site Operator Control Station, 2.1 Interactive Perception, 2.2 Interactive Task Description.
- Place the references list at the end of the document on a new page.

For other details such as page numbering, titles, and format of the title page, see *IEEE Information for Authors* available at http://www.ieee. org/portal/cms_docs/pubs/transactions/auinfo03.pdf. Keep in mind that guidelines at *IEEE Information for Authors* are for people who intend to publish in professional engineering journals, not necessarily for students in academic programs in engineering.

12.1.4 SAMPLE PAPER: IEEE STYLE

Stephanie Beckett, aerospacing engineering major at the University of Texas at Austin, won the 2005 Braden Engineering-Communication Contest with her research paper entitled "The Danger of Oil Scarcity: A Two-Pronged Solution" shown in the following pages. She uses IEEE guidelines for manuscript formatting and documentation of sources that are presented in this handbook. Her in-text citations follow IEEE style, as does her list of references.

The Danger of Oil Scarcity:
A Two-Pronged Solution

Stephanie Beckett
Braden Engineering-Communication Contest
University of Texas at Austin
February 4, 2005

Petroleum scarcity has startling economical and political implications. On the economic front, Schwartz has suggested that a $10 increase in the cost of a barrel of oil may retard US economic growth by 0.5% per year [1:68]. Underlying this economic problem involving petroleum scarcity is an even more fundamental danger: if a decline in the supply of petroleum precipitates a rise in oil prices, oil-rich nations may gain an even more disproportionate degree of political clout in the world. Supply will fail to meet demand, and oil consumers struggling to alleviate the economic catastrophes associated with cost increases will be forced to accept any price offered by oil-rich nations.

> The information comes from source 1, page 68.

Many of these oil-rich nations are unfortunately unstable, as almost two-thirds of the world's oil supply is in the Middle East [2:88]. This dependency on the Middle East limits foreign policy options by forcing world leaders to appease and sometimes even actively support unjust autocrats in the region [3]. This hypocritical behavior certainly fuels terrorists' hatred of the US. Further, according to Lugar and Woolsey, consumers of oil contribute a great deal to the continuation of corrupt governments and perhaps even terrorism by direct infusions of money; over one trillion dollars will be funneled into the Middle East between 2000 and 2015.

> Notice that brackets are used, not parentheses.

Therefore, as Hans Mark, former Secretary of the Air Force, tersely put it, "The near-term problem with oil and gas lies in the political instability of the region where much of the resource is located" [4]. Moreover, the long-term solution to these problems involves (1) decreasing US dependency on foreign oil, (2) developing renewable energy resources, and (3) accepting the potential contribution of nuclear power to US energy needs.

> Overview of the document's contents in the introduction. You are not obligated to include page numbers in citation.

Decreasing the US Dependency on Middle Eastern Oil

Ensuring that the US is not subject to the whims of the Middle East during times of oil scarcity is a formidable problem because, according to Mark, fossil fuels currently provide 85% of the US's energy [4]. However, some energy sources, such as renewable and nuclear energy, are technically feasible alternatives to oil. Therefore, the strategy to decreasing the US's reliance on the Middle East should be two-pronged, involving the further research and implementation of both renewable and nuclear energy.

Japan and France have actively pursued a similar strategy in order to increase the security of their energy resources [5]. The strategy has proven effective in these two nations: within about twenty five years of implementing the program, each nation doubled the percentage of energy produced through domestic means [6;7].

The information comes from a combination of sources 6 and 7.

The Promise of Renewable Energy

Finding efficient renewable energy sources is highly desirable. Unfortunately, however, renewable energy technology is still in its infancy; every potential source of renewable energy seems to have some problem. While these problems have foreseeable solutions, depending entirely on renewable energy to replace oil would be a mistake.

For example, hydroelectricity is by far the most common renewable energy source in the US [4]; however, reservoirs and dams can cause problems with the environment and development [8:108]. Also, while wind power is the fastest-growing renewable energy source in the world, it is an immature industry: wind power only provided about 0.24% of worldwide electricity as of 2000 [9]. Further, wind power is subject to inconsistency due to seasonal changes [8]. Solar energy, while promising if produced by individual households, has limited potential as a centralized power generating system since the technique is so space-intensive [4]. Hydrogen fuel is most often created only after the burning of fossil fuel because the creation of pure hydrogen requires energy [10], making hydrogen power unsustainable. The use of biomass fuels may also be unsustainable [11:519]; at the very least, biomass fuel requires more research and development before it can become an economically viable alternative to petroleum [2:88].

Citations go inside the end punctuation.

However, these rather dismal analyses of the problems with renewable energy do not take into account all of the subtleties of the renewable-energy industry at large. For example, hydrogen fuel might become an efficient energy source if hydrogen is produced as the result of solar or wind power. However, the problems presented here indicate why renewable energy sources have not displaced oil despite oil's discouraging cost fluctuations: renewable energy is simply not yet a mature industry.

Because of its potential, the fledgling industry of renewable energy must be supported with increased incentives and funding for research and implementation. The US government is already helping renewable energy become a solution to energy problems; for example, according to Garman's testimony in the hearings on *Energy and Water Development Appropriations for Fiscal Year 2005*, the government appropriated about $1.23 billion to the US Department of Energy's Office of Efficiency and Renewable Energy in 2004 [12]. However, the US government should recognize renewable energy not only as an important environmental policy but also as a vital security interest necessary for ensuring the US's future economic prosperity.

Nuclear Power's Role in Limiting US Middle East Dependency

As detailed in the preceding, renewable energy is promising, but the technology is not yet mature. Nuclear power, on the other hand, has been in use for over fifty years and is a well-established technical alternative to petroleum energy production. In 2002, nuclear power supplied 20% of the US's electricity supply and 17% of the world's supply [13].

Because uranium is a plentiful element about as common as silver that has to be enriched only slightly to be used as a nuclear fuel [14] and because a small amount of uranium can produce a large amount of energy [6:76], nuclear energy cannot be monopolized by any particular region, which promotes the energy security of the entire world. However, despite nuclear power's importance in the energy arena, a number of problems decrease nuclear energy's attractiveness. These problems can be overcome with incentives, public education, and regulations.

The Massachusetts Institute of Technology report described the problems with nuclear power as fourfold [13]. First, nuclear power is expensive compared with petroleum power production when crude oil costs are low. This issue will become less important as the price of oil is driven up by scarcity. Further, government incentives are in place to favor nuclear power economically. For example, according to the Department of Energy's Office of Nuclear Energy, Science and Technology, the government is promoting the establishment of modern nuclear facilities by cost-sharing the construction of nuclear plants [15].

A second problem is that the public is greatly concerned with the safety of nuclear power plants. However, this problem is mainly one of popular perception. Europe and the US maintain safe nuclear plants [4]. To alleviate the fear of nuclear power, the government should inform the public of the safety statistics of nuclear facilities.

Third, the problem of waste disposal still plagues the nuclear program [13]. This issue is difficult to solve, but the foundation of a nuclear waste repository in the Yucca Mountains in Nevada is expected to confirm the feasibility of long-term geological disposal of nuclear waste. Meanwhile, the US should continue to study better locations and techniques for long-term waste storage: the MIT report, for example, suggests the exploration of deep borehole waste storage [13].

Finally, there is the serious concern that an increased use of nuclear energy will contribute to the proliferation of nuclear weapons [13]. This threat can only be averted if the developed world conforms to strict regulations concerning the enrichment of nuclear fuels and then insists that all developing nations follow the same guidelines on threat of being sanctioned and excluded from the general international community.

Thus, while the problems of nuclear energy are difficult, they are resolvable. Further, the potential for nuclear power to help the US gain energy independence is an important enough task to warrant facing such problems.

Conclusion

Because oil gives the Middle East a disproportionate influence in the world, it is vital that the US pursue a policy to decrease American dependence on Middle Eastern oil, particularly as the economic pressures of petroleum scarcity cause oil costs to rise. The two-pronged approach to decreasing the US's dependence on oil by increasing research, development, and implementation of renewable and nuclear energy will not completely eliminate the use of oil within the US overnight. Instead, a dedication to this strategy must be maintained over the long-term so that oil scarcity does not force the US to increase its dependency on an unstable region.

[1] N.D. Schwartz, "Oil—why prices will fall: because Iraq has been on the sidelines of the oil world for 20 years; soon it won't be," *Fortune*, vol. 147, no. 6, pp. 68+, 2003 March 31. Accessed January 30, 2005 from Expanded Academic ASAP database (A98880222).

References [This section would start a new page.]

[2] R.G. Lugar and R. J. Woolsey, "The new petroleum," *Foreign Affairs*, vol. 78 no.1, p. 88, 1999. Accessed January 29, 2005 from Expanded Academic ASAP database (A53545155).

[3] D. Plesch, "Ending oil dependency," *Observer*, http://observer. guardian.co.uk/waronterrorism/story/0,1373,564843,00.html. Accessed January 29, 2005.

[4] H. Mark, "The problem of energy," Lecture. University of Texas, Austin, TX, 2004.

[5] Organisation for Economic Co-operation and Development, *Proceedings of Business as Usual and Nuclear Power*, http://www.iea.org/dbtw-wpd/textbase/ nppdf/free/2000/busassual2000.pdf. Accessed January 29, 2005.

Abbreviations are used for authors' given names. Initials come first, before the last name.

[6] M. Mishiro. "Nuclear power development in Japan," in Organisation for Economic Co-operation and Development (Ed.), *Proceedings of Business as Usual and Nuclear Power*, pp. 75–79, 1999, http://www.iea.org/dbtw-wpd/textbase/nppdf/free/2000/ busassual2000.pdf. Accessed January 29, 2005.

[7] D. Maillard, "French nuclear industry and latest governmental decisions," in Organisation for Economic Co-operation and

Development (Ed.), *Proceedings of Business as Usual and Nuclear Power*, pp. 91–95, 1999, http://www.iea.org/dbtw-wpd/textbase/nppdf/free/2000/busassual2000.pdf. Accessed January 29, 2005.

[8] Energy Information Administration, *International Energy Outlook 2004*, http://www.eia.doe.gov/oiaf/ieo/. Accessed January 29, 2005.

[9] R.H. Williams, "Nuclear and Alternative Energy Supply Options for an Environmentally Constrained World," PowerPoint presentation at the Nuclear Control Institute's 20th Anniversary Conference, Washington, DC, 2000 April 9, http://www.nrel.gov/ncpv/thin_film/pdfs/nuclear_alt_energy_options_21st_cent_2001.pdf. Accessed January 29, 2005.

[10] The National Academies, "Hydrogen economy offers major opportunities but faces considerable hurdles," *The National Academies Press Release*, 2004, February 4, http://www.e85fuel.com/front_page/hydrogen_economy_offers.htm. Accessed January 29, 2005.

[11] T.W. Patzek, "Thermodynamics of the corn-ethanol biofuel cycle," *Critical Reviews in Plant Sciences*, vol. 23, no. 6, pp. 519–567, 2004.

[12] *Energy and Water Development Appropriations for Fiscal Year 2005: Hearings before the Subcommittee of the Committee on Appropriations*, Testimony of D.K. Garman, http://frwebgate.access.gpo.gov/cgi-bin/getdoc.cgi?dbname=2005_sapp_ene_1&docid=f:2910468.wais. Accessed January 30, 2005.

[13] Massachusetts Institute of Technology, *The Future of Nuclear Power*, 2003, http://web.mit.edu/nuclearpower/. Accessed January 29, 2005

[14] Los Alamos National Labs Chemistry Division, *Uranium*. 2004 January 5, http://pearl1.lanl.gov/periodic/elements/92.html. Accessed January 29, 2005.

[15] Department of Energy, Office of Nuclear Energy, Science, and Technology, *Nuclear Power 2010*, 2005 January 23, http://www.ne.doe.gov/nucpwr2010/NucPwr2010.html. Accessed January 30, 2005.

12.2 APA STYLE OF DOCUMENTATION

The APA documentation system uses parenthetical author names, dates of publication, and page numbers (called citations) which correspond to entries in the list of works at the end of the document. To see the source of the parenthetical citation, you go to the list of works section and find the author name (and date, if the document cites more than one work by that author). Including the date of the publication in the citation enables readers to put the discussion in the overall context of research in that field. See Figure 12.3 for an example.

Citations

Bibliographic
entries

Pyrolysis to Hydrogen and Carbon or Methanol

Steinberg and associates at Brookhaven National Laboratory have long considered processes based on high-temperature pyrolytic conversion of coal, biomass and other carbonaceous materials to hydrogen, carbon, methanol and light hydrocarbons.

In the Hydrocarb process™, Steinberg (1987a, 1989) describes a two-step process involving (1) the hydrogeneration of carbonaceous materials like coal and biomass to methane, followed by (2) thermal decomposition of the methane to hydrogen and a clean carbon-black fuel. For coal, a typical overall reaction would be:

$$CH_{0.8}O_{0.08} \rightarrow C + 0.32\ H_2 + 0.08\ H_2O$$

For biomass, the two-step reaction is:

$$CH_{1.44}\ O_{.66} = C + 0.06\ H_2 + 0.66\ H_2O$$

Preliminary pyrolysis experiments related to the Hydrocarb reactions are discussed in Steinberg et al. (1986), Steinberg (1986), and Steinberg (1987b). In later work (Steinberg, 1990), a is described process in which biomass and methane are converted to methanol plus carbon (Carnol). The overall stoichiometry is:

$$CH_{1.44}O_{0.66} + 0.30\ CH_4 = 0.64\ C + 0.66\ CH_3OH$$

As the names imply, Hydrocarb and Carnol processes emphasize the minimization of CO_2 and the production of elemental carbon.

REFERENCES

Beckersvordersan
in Sustainable
Hydrogen. On
Chapter 10, Ed
pp. 119–134.

Biohydrogen 200
www.ftns.wau

General Atomics
Gasification of
Study Technica

Gregoire-Padró,
Ed. C.E.

Gregoire-Padro a

Hall, D. O. (1999
pp. 147–156.

Modell, M.; Reid
Patent 4,113,4

Safrany, D.R. (19
Synthesis of in
to Vaporize Wa

Symposium Series:pp. 103–108.

Steinberg, M. (1986). The direct use of natural gas for conversion of carbonaceous raw materials to fuels and chemical feedstocks. *Int. J. Hydrogen Energy;* 11(11):pp. 715–720.

FIGURE 12.3

Excerpts: sample page with parenthetical citations and the corresponding reference page.

(Source: Milne, Elam, & Evans. (2001). Hydrogen from biomass: state of the art and research Challenges [Electronic version]. Retrieved December 22, 2006, from www.osti.gov/bridge/servlets/purl/792221-p8YtTN/native/792221.pdf.)

FIGURE 12.3 (*Continued*)

Steinberg, M.; Fallon, P.T. and Sundaram M. S. (1986). Flash pyrolysis of biomass with reactive and non-reactive gas. *Biomass* 9:pp. 293–315.

Steinberg, M. (1987a). Clean carbon and hydrogen fuels from coal and other carbonaceous raw materials.*BNL-39630*.

Steinberg, M. (1987b). The flash hydro-pyrolysis and methanolysis of coal with hydrogen and methane. *Int. J. Hydrogen Energy 12(4)*:pp. 251–266.

Steinberg, M. (1989). The conversion of carbonaceous materials to clean carbon and co-product gaseous fuel. Conference: 5. *European Conference On Biomass for Energy and Industry*, Lisbon, Portugal, 9–13, Oct 1989; BNL-42124; CONF-891034–1; DE9000156:p. 12.

Sung, S. (2001). Bio-hydrogen production from renewable organic wastes. *Proceedings of the 2001 U.S. DOE Hydrogen Program Review* (NREL/CP 570–30535).

The following guidelines are based on the *Publication Manual of the American Psychological Association*, 5th ed. (Washington: APA, 2001).

12.2.1 IN-TEXT CITATIONS: APA STYLE

As you can see in Figure 12.3, APA citations show the author's name and the year of publication. If the author's name has already been stated in regular text, only the year is cited. If the document uses more than one resource from the same author, lowercase letters are attached to the year of publication to differentiate those resources.

- For quotations, use the following format when the author's name occurs in the regular text:

  According to Wallman et al. (1996), "Products treated at 275°C proved acceptable as slurries" (p. 8).

- Use the following format when the author's name does not occur in the regular text:

  One researcher reports that "pumpable slurries from an MSW surrogate mixture of treated paper and plastic have shown heating values in the range of 13–15 MJ/kg" (Wallman et al., 1996, p. 9).

Page numbers are not required but help readers locate specific information in longer works.

- For paraphrases and summaries, use the format in the first of the following examples if the author's name occurs in the regular text; use the format in the second if the author's name does not occur in the regular text:

A technical note by Williams (1980) makes a case for efficient hydrogen production from coal using centrifuge separation of hydrogen from other gases following steam gasification at 1100–5000°C.

Hydrogen can be produced efficiently from coal using centrifuge separation of hydrogen from other gases following steam gasification at 1100–5000°C (Williams, 1980).

- When you have two authors, use the format in the first of the following examples if the authors' names occur in the regular text; use the format in the second if the authors' names do not occur in the regular text:

Pilot-scale experiments in the steam gasification of charred cellulosic waste material are discussed in Rabah and Eddighidy (1986).

Pilot-scale experiments have been conducted in the steam gasification of charred cellulosic waste material (Rabah & Eddighidy, 1986).

- When you have three to five authors, use the following formats for the first citation and subsequent citations, respectively:

Consideration of hydrogen from carbonaceous materials has a long history in the hydrogen literature. At the First World Hydrogen Energy Conference, Tsaros, Arora, and Burnham (1976) reported on three routes to hydrogen using sub-bituminous coal.

Consideration of hydrogen from carbonaceous materials has a long history in the hydrogen literature. At the First World Hydrogen Energy Conference, three routes to hydrogen using sub-bituminous coal were described (Tsaros, Arora, & Burnham, 1976).

Hydrogen yields of 93–96% of theoretical were predicted (Tsaros et al., 1976).

- When you have six or more authors, use the following format:

McDonald et al. (1981) proposed extracting protein from grass and lucern and using the residue for hydrogen production (among other fuels).

Extracting protein from grass and lucern and using the residue for hydrogen production (among other fuels) has also been proposed (McDonald, 1981).

- When the author is unknown, do one of two things: (1) find the corporate author (such as the name of a government or corporation organization) and use it as the author; or (2) cite the title of the work in the text and provide a shortened version of the title in the parenthetical reference:

  The U.S. Department of Energy (1997b) has shown that coals and other forms of solid carbonaceous fossil fuels could be oxidized to oxides of carbon at the anode of an electrochemical cell and hydrogen produced at the cathode.

  According to the editorial "Biomass—It's a Gas!" (2002), in the near- and mid-term, generating hydrogen from biomass may be the more practical and viable option because it is renewable and potentially carbon-neutral.

  In the near- and mid-term, generating hydrogen from biomass may be the more practical and viable option because it is renewable and potentially carbon-neutral ("Biomass—It's a Gas!" 2002).

- If the author is a corporation, organization, or agency, and the name is quite long, give the full name the first time you use it in a citation followed by the abbreviation in square brackets. In subsequent citations, use the abbreviation:

  Entry in reference list: Assembly of First Nations. (2000).

  First citation: (Assembly of First Nations [AFN] 2000)

  Subsequent citations: (AFN 2000)

- If you have authors with identical last name, include the authors' initials:

  A study by R.J. Jones indicates that . . .

- When you cite communications such as letters, memos, and e-mails, you do not need to include these items in your list of references. In the text of your document, provide the initial and last name of the person with whom you communicated, that person's title and the organization that individual is associated with (if available), and the date on which the communication took place:

  There has been very little experimental data supporting these proposed routes to hydrogen and methanol. No laboratory study of the integrated process has been published. Research is needed to validate the technoeconomic arguments of these methods (Milne, National Renewable Energy Laboratory, personal communication, November 19, 2001).

- When you use an electronic resource, obviously you cannot provide page numbers, but you can use the ¶ symbol or the abbreviation **para**:

  According to Milne, Elam and Evans (2001, ¶ 7), "The catalytic converter, using different steam reforming nickel catalysts and dolomite, was tested over a range of 660–830°C. Fresh catalyst at the highest temperature yielded 60% by volume of hydrogen."

- If you must acknowledge more than one source in a single paren-thetical citation, cite the sources in the order in which they appear in your list of references, and separate them with a semicolon:

  General Atomics is studying supercritical water partial oxidation of biomass for hydrogen (Spritzer, 2001; Johanson et al., 2001).

- If you must acknowledge two or more works by the same author in a single parenthetical citation, arrange the sources in chrono-logical order, adding *a, b, c,* and so on after the year:

  Thermodynamic calculations predict low carbon monoxide formation (Antal, 1994a, 1994b, 1995).

12.2.2 APA REFERENCE LIST

In APA style, the alphabetical list of works cited, which appears at the end of the document, is called "References." The following pres-ents guidelines for reference lists and model entries for common types of sources. If you require more information about formats not described here, see *Publication Manual of the American Psychological Association,* 5th ed. (Washington: APA, 2001).

GUIDELINES FOR THE APA REFERENCE LIST
As you can see in the model APA-style research paper on page 499, the following are standard format requirements for the reference list:

- Begin the reference list at the end of your document on a separate page. Center the title, *References,* but do not italicize it or put it within quotation marks.
- Number the page or pages of your reference list sequentially with the rest of the document.
- Double-space all entries in your reference list.

- Use a **hanging indent** for reference-list entries. Do not indent the first line of the reference list entry; however, do indent all subsequent lines five spaces (up to 1.25 cm, or 0.5 inch). To achieve this hanging indent, see Figure 12.4.

    Levenspiel, O. (1998). *Chemical reaction engineering*. New York: Wiley.

    Duncan, T.M. & Reimer, J.A. (1998). *Chemical engineering design and analysis: An introduction*. Cambridge, UK: Cambridge UP.

- However, APA permits paragraph-style indents.

    Duncan, T.M. & Reimer, J.A. (1998). *Chemical engineering design and analysis: An introduction*. Cambridge, UK: Cambridge UP.

- Arrange entries alphabetically by the authors' last names. If you have two or more works by the same author or authors, give the author name each time and list chronologically by year of publication:

    Zlokarnik, M. (1991). Dimensional analysis and scale-up in chemical engineering. New York: Springer.

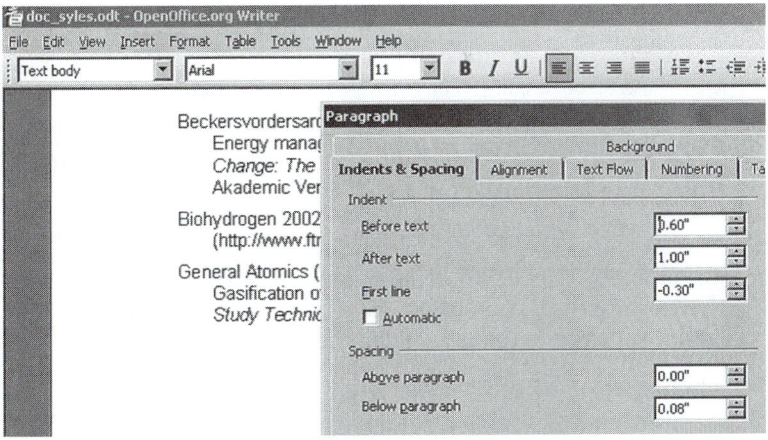

FIGURE 12.4

Customizing paragraph format in OpenOffice Writer to create APA-style bibliographic entries. (The hanging indent is created by the offset between the values for Before text and First line.)

Zlokarnik, M. (2006). *Scale-up in chemical engineering*. New York: Wiley.

- When the author is the sole author of one or more works and the first coauthor of other works, put the entries for the single-author works first. If the first coauthor name of two or more works by multiple authors is the same, alphabetize by the second author name.

GUIDELINES FOR APA REFERENCE-LIST ENTRIES: COMMON ELEMENTS

The following guidelines are the same across most resources such as books, articles, and other types:

- Format the author name (or names), last name first, using only initials for their given names. When there are two or more authors, use the ampersand (&). If a reference has more than six authors, after the sixth, use only **et al**. Do not write this abbreviation in italics.

Wang, T. (1984). *Computer rating system for bridge rating and fatigue life analysis*. Miami, FL: Florida International University, Department of Civil Engineering.

Romm, J. J. (1999) *Cool companies—how the best businesses boost profits and productivity by cutting greenhouse gas emissions*. Washington, DC: Island Press.

Hopwood, T. & V.G. Oka (1989). *Development of a priority ranking system for bridge rehabilitation or replacement*. Frankfort, KY: University of Kentucky, College of Engineering.

Krebs, D. & Blackman, R. (1988). *Psychology: A first encounter*. San Diego: Harcourt.

Griffin, R.W., Ebert, R.J., & Starke, F.A. (1999). *Business* (3rd Canadian ed.). Scarborough, ON: Prentice Hall.

- For publications of corporations, associations or government agencies where no individual author is named, use the name of organization as the author. Alphabetize according to the first significant word in the title; ignore initial articles—*A, An,* or *The*.

- Differentiate two or more works by the same author published in the same year by adding *a, b, c*, and so on to the year:

Zuk, W. (1991a). *An expert system as applied to bridges: testing phase*. Charlottesville, VA: Virginia Transportation Research Council.

Zuk, W. (1991b). Expert system for determining the disposition of older bridges. *Transportation Research Record, 1290*:145–148.

- If the author of the source is a corporation, agency, or organization, the publisher is often the same as the organization. In such instances, use *Author* to indicate the publisher's name:

  Ministry of Education and Training. (1990). Ministry of Education and Training style guide for editors and writers. Toronto: Author.

- If the reference has no author, start the entry with the title of the document:

  German for Travelers. (1986). Lausanne: Editions Berlitz.

- If no date is given, place **n.d.** in parentheses. For articles accepted for publication but not yet published, write **in press** in parentheses.

APA GUIDELINES FOR PRINT RESOURCES
ARTICLES

- This is the format to use if the article appears in a journal paginated by volume:

  Ismart, D. (1994). State management systems: Overview of ISTEA requirements and current implementation. *TRNews, 173*, 2–4.

- Include the issue number only if each issue of the periodical begins on page 1, and use the following format:

  Martin, R. (2002). The virtue matrix: Calculating the return on corporate responsibility. *Harvard Business Review, 80(3)*, 69–75.

  Muradov, N. Z. (1993). How to produce hydrogen from fossil fuels without CO_2 emission. *Int. J. Hydrogen Energy 18 (3)*, 211–215.

  Woodward, J. and Orr, M. (1998). Enzymatic Conversion of sucrose to hydrogen. *Biotechnology Progress, 14(6)*, 897–902.

- Use this format if the document cites multiple articles by the same author:

  Steinberg, M. (1987a). Clean carbon and hydrogen fuels from coal and other carbonaceous raw materials. *Biomass, 9*, 293–315.

  Steinberg, M. (1987b). The flash hydro-pyrolysis and methanolysis of coal with hydrogen and methane. *Int. J. Hydrogen Energy 12(4)*, 251–266.

- For articles that appear in magazines give the complete date (including the volume number if available):

  Stone, R. (2006, April). Inside Chernobyl. *National Geographic, 209*, 32–53.

  Marriott, C. (2005, July). IBC seismic requirements and HVAC systems: A short course. *Engineering System Solutions*.

- If the article appears in a newspaper, use the following format:

  Fountain, H. (2006). Observatory: Seals don't multitask. *New York Times*, F3.

- Use this format for a letter to the editor:

  McMurrey, J.A. (2006, June 6). Women, success and science. [Letter to the editor]. *Austin-American-Statesman*, A12.

- Use this format for a review:

  McMurrey, P.H. (2006, May 25). Slaves of invention. [Review of A hammer in their hands: A documentary history of technology and the African-American experience]. *The West Austin Dissenter*, 4.

BOOKS

- If the book has been developed by an editor, use the following format:

  Broughton, J. (Ed.). (1994). *Process utility systems*. Oxford, UK: Butterworth-Heinemann.

  Tucker, M.E. & Grim, J.A. (Eds.). (2002). *Worldviews and ecology: religion, philosophy and the environment*. Maryknoll, NY: Orbis Books.

- If you cite a specific article out of an edited book, use this format:

  Rewinski, S. (1991). Fuzzy algorithm in bridge and pavement management systems. In B. H. V. Topping (Ed.). *Artificial intelligence and civil engineering*, Edinburgh: Civil-Comp Press.

  David B. Johnson. (1996). Scalable support for transparent mobile host internetworking. In T. Imielinski & H. Korth (Eds.). *Mobile computing*. Norwell, NA: Kluwer Academic Publishers.

- If the book has been translated, use this format:

  Heidegger, M. (1977). *The Question concerning technology and other essays* (W. Lovitt, Trans.). New York: Harper & Row. (Original work published 1954).

- If the book is in an edition other than the first, use this format:

  Martin, M.W. & Schinzinger, R. (1993). *Ethics in engineering* (2nd ed.). New York: McGraw-Hill.

- If you cite a work that has multiple volumes, use this format:

  Coulson, J.M., Richardson, J.F., Backhurst, J.R., & Harker, J.H. *Chemical engineering* (Vols. 1–4). Oxford, UK: Butterworth-Heinemann.

- If you cite a dissertation abstract, use the following format:

  Karim, Y. (1999). Arab political dispute mediations (Doctoral dissertation, Wayne State University, 1999). *Dissertation Abstracts International, 61*, 350.

- This is the format to use if you cite an item from the published proceedings of a conference:

  Chorney, H. (1991). A regional approach to monetary and fiscal policy. In J.N. McCrorie & M.L. MacDonald (Eds.), *The constitutional future of the prairie and Atlantic regions of Canada* (pp. 107–121). Canadian Plains Research Center, University of Regina.

- Use this format if you cite technical reports and government documents:

  Solicitor General of Canada. (1995). *Annual report on the use of electronic surveillance*. Ottawa: Author.

APA GUIDELINES FOR ELECTRONIC SOURCES

The following guidelines for electronic sources are based on those found in the *Publication Manual of the American Psychological Association*, 5th ed. (Washington: APA, 2001). Updates can be found at http://www.apastyle.org/electref.html

To give credit to authors and to enable readers to find the source material, include author name, document titles, publication dates, retrieval dates, and exact web addresses (URLs). If you cannot determine the date in a document, use **n.d.** (no date) in the reference, and provide the date of retrieval. If a URL runs over a line, break it only after a slash or before a period, and make sure no hyphens are inadvertently added at line breaks. Do not place a final period at the end of the URL, and do not place the URL in angle brackets.

- If you access an article online and that article is a duplicate of a print version, use this format:

Strain, L.A. (2000). Seniors' centres: Who cares? [Electronic version].
Canadian Journal of Aging, 20: 471–491.

- If you access an article that occurs online only, use this format:

Sands, P. (2003, Fall). Pushing and pulling toward the middle. *Kairos,*
7(3). Retrieved May 2, 2003, from http://english.ttu.edu/kairos/7.3/
binder2.html?coverweb.html#gender

Smith, J. (2003, January 16). Journalism fails its sobriety test. *Salon.*
Retrieved January 19, 2003, from http://www.salon.com/news/
feature/2003/01/16/dui/index_np.html

- If the author of a document is not identified, begin the reference
  with the title of the document:

Erikson's development survey. (n.d.). Retrieved May 10, 2003, from
http://www.hcc.hawaii.edu/intranet/committees/FacDevCom/guidebk/tea
chtip/erikson

- To reference messages posted to archived online newsgroups,
  forums, or mailing lists, cite the author's name and the exact date
  of online posting. Follow this with the subject line of the posting
  and the address of the message group or forum beginning with
  Message posted to:

Nivalainen, M. (2002, December 17). The key and stupid web moments
of 2002 [Msg 3]. Message posted to Cybermind@listserv.aol.com

- If a document is contained within a large and complex web site
  (such as that for a university or a government agency), identify the
  host organization and the relevant program or department before
  giving the URL for the document itself. Precede the URL with a
  colon:

Ludlow, P. (Ed.). (1996). High noon on the electronic frontier:
Conceptual issues in cyberspace. Retrieved May 1, 2003, from
Georgetown University, Communication, Culture and Technology Web
site: http://semlab2.sbs.sunysb.edu/Users/pludlow/highnoon

- Do not include e-mails in the list of references; e-mails are a form
  of personal communication. Cite them in the text only, using this
  format:

. . .the design of the facility (Nelson Rafferty, personal communication,
March 4, 2003).

12.2.3 APA MANUSCRIPT FORMAT

To format a document using APA guidelines:

- Start the document with a separate title page.
- If an abstract is required, put it on page 2.
- Use headings, lists, tables, notices, figures as discussed in Chapter 2.
- Put the references list on a separate page at the end of the document.

For other details such as page numbering, titles, and format of the title page, see the following sample APA document.

12.2.4 SAMPLE PAPER: APA STYLE

Privacy Protection 1

Running head: Abbreviated title appears five spaces to the left of the page number, which is flush against the right margin. The paper's page numbering commences on the title page.

Running head: PRIVACY PROTECTION ON THE INTERNET

Privacy Protection on the Internet:
An Evaluation of Policies and Regulations
Erika Smith
Humanities IA03, Section 3
Professor Rockwell
March 25, 2002

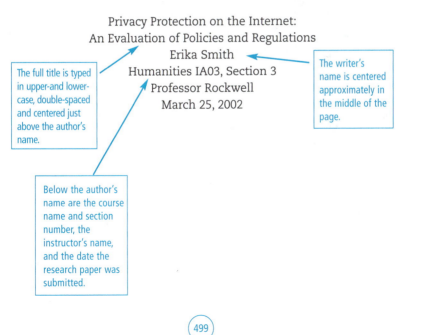

The full title is typed in upper-and lower-case, double-spaced and centered just above the author's name.

The writer's name is centered approximately in the middle of the page.

Below the author's name are the course name and section number, the instructor's name, and the date the research paper was submitted.

Privacy Protection 2

Privacy Protection on the Internet:

An Evaluation of Policies and Regulations

> Include full title, centered, at the beginning of the essay.

The Internet reaches nearly 50 million people worldwide, and this figure is growing at a rate of approximately 10% per month (Wang, Lee, & Wang, 1998, 63). It is a powerful medium, but its triumph as a global source of information will decline if proper privacy regulations are not enforced. At present, the issue of privacy protection remains

> Because the authors' names do not appear in the regular text, include them, and the date and page number, in parentheses. For the first citation of a work with three to five authors, cite all authors.

unsolved. According to the Internet Society (2002), this is a problem that some legislators and companies are not interested in addressing:

> The universal acceptance of the technology that includes e-mail and the World Wide Web has made this technology an appealing tool for many who believe it is a justification for a change in the rules and expectations of privacy. The CEO of Sun Microsystems was widely quoted as saying "You have zero privacy anyway. Get over it." (Internet Society, 2002, ¶ 3)

> Indent block quotations five spaces (about 1.25 cm, or 1/2 inch).

Committees such as the World Wide Web Consortium (W3C) and the Platform for Privacy Preferences (P3) are trying to enjoin the government, private industry, and users to resolve Internet problems. Three groups must work equally toward creating privacy: (a) governments must cooperatively create universal legislation to regu-

> Overview of the document is provided using the in-sentence list format.

late the Internet; (b) the computer industry must have self-regulation policies; and (c) Internet users must arm themselves against the invasion of privacy and contribute to the creation of legislation against the invasion of privacy.

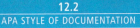
Need for Government Legislation

Many Internet users are apprehensive about giving any personal information over the Internet. A recent Louis Harris poll showed that out of 1000 users over the age of 18, more than half feared that seemingly harmless information would be linked to their e-mail address, then passed on to other parties (Engler, 1997). These concerns are not unfounded; cases related to the invasion of privacy are increasing. Such evidence includes the activities of junk e-mail marketing organizations and the activities of Web-based advertisements that track a user's usage history and preference through cookies. Concerns over programs constructed through security holes in Internet tools like JavaScript or ActiveX are also valid. ActiveX can obtain a person's credit information while JavaScript can access personal files (Wang et al., 1998). These and other malicious invasions of privacy demonstrate that there is inadequate legislation in place to prevent these offenses and that private industry cannot effectively regulate its own actions. Hence, consumers must work collectively with governments and the industry to improve privacy protection on the Internet.

Currently, committees are working to create a more unified, efficient system of privacy protection. W3C is one of the leading non-profit organizations specializing in proposing and enforcing standards on the Web (Wang et al., 1998). P3 is a concept devised by W3C. According to Tim Berners-Lee, the inventor of the Web and director of W3C, P3 is intended to create a base "on which technical, market, and regulatory solutions can inter-operate and build" (Engler, 1997, p. 81). It will enable users to control how much personal data they share with other websites by allowing each consumer to create a profile and register his or her own privacy practice. The profile

> Use headings so the organization is easy to follow.

> With quotations, a page number is required, but not with paraphrasing.

> When a work has three to five authors, in a subsequent citation that is the first in a paragraph give only the first name and **et al.**, plus the year.

> With quotations, a page number is required, but not with paraphrasing.

Privacy Protection 4

contains the personal details the user is willing
to reveal and the situations under which they
can be revealed (Engler, 1997). Internet consumers will be able to make informed choices

> For a quotation, a page
> number (preceded by **p.**)
> appears in parentheses.

about the use and disclosure of their private information
(Wang et al., 1998).

The P3 system also recommends that websites describe what
information is needed. This enables users to know what information
they must give to access a site. If a user is hesitant to give information, the site starts a "negotiation" phase where users can choose to
provide less data and choose a more limited option. W3C and P3
advocate an "informed consent" policy: users, not businesses, decide
how and when information is used (Engler, 1997). This is valuable
because it lets the user and the industry contribute equally to the
improvement of the Internet.

The government also has an important role in increasing the capabilities of the Internet. The U.S. Federal Trade Commission (FTC) has
held two workshops to determine if government regulation is needed
to create privacy standards. However, FTC has held off all decisions
until a report is given to the U.S. Congress (Engler, 1997). This demonstrates that processing of national legislation is often slow. Indeed,
governments must contribute to the regulation of the Internet since
they have the resources to enforce Internet regulations. However, if
countries combined resources to regulate the Web, the enforcement of
standards would occur more quickly and efficiently. Governments, private industry, and Internet customers must contribute equally to the
regulation of the Internet. If one group tries to regulate privacy individually, or dominates the two, privacy will not be adequately protected.

Governments from all nations must form
a universal legislation to prevent the invasion of privacy. The Internet is a worldwide medium; hence, it should be regulated
by all countries collectively. It is nearly
impossible for a single country to enforce
national legislation since

> Within a paragraph, you do
> not need to repeat the year in
> subsequent references to a
> source.

offenders may be from another jurisdiction. Thus, the violation of privacy is not deterred because it is difficult to enforce legislation. In the article "Consumer Privacy Concerns about Internet Marketing," Huaiqing Wang, Matthew K.O. Lee, and Chen Wang agree that

> The ability to conduct law enforcement against the violators of individual privacy is very limited. Even though many countries have enacted similar privacy protection legislation, the enforcement of such local legislation is difficult without the aid of international treaties and collaboration since the Internet has no national boundaries. (p. 70)

Without a structured universal system of legislation, users who fear the violation of privacy will be increasingly reluctant to use and/or contribute to the Internet. Clearly, the Internet will not reach its full potential as a medium without the co-operation of all governments.

Furthermore, international collaboration is necessary in order to prevent political agendas from hindering the right of privacy. In his article "Internet Malcontents of the World—Unite!" Wayne Madsen (1998) states,

> Privacy advocates have thought for many years that the real intention of back doors [controlled by the U.S. government] into cryptography is to crack down on political activists who are shielding their communications from the CIAs, NSAs, and FBIs of the world. (p. 28)

For example, Madsen criticizes U.S. politician David Aaron for adding "malcontents" (defined as persons dissatisfied with existing government, administrators, system, etc.) as new villains to the "list of international users of nonescrowed/nonrecoverable cryptography" (p. 28). Aaron's action demonstrates how a federal government can manipulate national legislation that regulates activities on the Internet. Universal legislation, however, would work to serve the interests of the people, not the motives of one political party. The private industry and users themselves should cooperate with international governments and take an active part in the creation of this legislation. Therefore, universal law regulating the Internet would improve the medium.

Privacy Protection 6

Need for Self-Regulation from the Computer Industry

Secondly, if the industry creates better self-regulation, both the Internet and the businesses that operate on it would benefit. Since the Internet is controlled mainly by businesses, it is logical that the industry would have the most innovative solutions to privacy issues. Many industries are introducing self-regulatory methods to ensure privacy; for example, some businesses have voluntarily written their own privacy codes (Wang et al., 1998). Unfortunately, problems may occur when businesses refuse to spend money on privacy devices or when the devices they implement are inadequate. Moreover, as Wang et al. state,

> Today's privacy enhancing technologies are not only primitive in nature, but also lacking the integrated environment under which most of the Internet consumers' privacy concerns can be dealt with. Such technologies are often cumbersome to use, unfriendly and require a degree of knowledge exceeding that of the common Internet user. (p. 70)

Universal legislation must enforce standards of privacy devices, and regulate the industry by making privacy protection a mandatory component of business on the Internet. Consequently, the Internet would improve its capability as a medium.

A positive aspect of industry self-regulation is that competition among companies increases the amount of technology available to society. This is important because if society has knowledge about technology, it can objectively judge the competence of government policies. Encryption lets one scramble data, so only individuals with a key can decode one's information. Privacy rights activists and computer industry groups have said that the Clinton administration regulations have fallen short (Kalin, 1997). For example, the U.S government has been promoting its own Data Encryption Standard (DES). DES enables the government to have "back door" methods of decryption. However, encryption experts have criticized DES, stating that the 20-year-old system is "obsolete and needs updating" (Engler, 1997, p. 82). Evidently, in order to protect privacy, governments, private industry, and users must work together to create a system that is updated, fair, and efficient. As a result, the Internet would be used more frequently and would expand.

Need for Users to Guard Their Own Privacy

Thirdly, one way to counter the invasion of privacy is for individuals to become informed before giving out information. Users should ask for adequate, relevant information as to why and for what purpose they are giving out personal information. Individuals should obtain information explaining their rights of redress, how they can remain anonymous, how to withhold personal information, and how to correct errors (Wang et al., 1998). The theory of supply and demand illustrates that if a consumer shows demand for a product (in this case, better privacy protection), then the market will meet the consumer's demand in order to make a profit. Hence, if users demand better protection, the industry will meet the demands of the consumer. Users must vocalize their needs and play an active role in the regulation of the Internet in order to have improved privacy protection. Thus, users must protect themselves while also contributing to the improvement of legislation and industry self-regulation. This would strengthen the Internet as a whole.

The relationship between government, industry, and users is illustrated when examining the current methods of restructuring the Web. Although W3C and P3 create some solutions to the privacy problem, there are many areas that need to be modified. W3C does integrate business and consumer needs concerning the regulation of privacy; however, it fails to recognize the limitations of the industry, or to see the importance of the government. The current situation with W3C divides government and industry. Furthermore, it does not allow consumers to be self-sufficient; for example, users still would rely on companies to provide them with options. W3C does call for the regulation of standards, but it does not propose methods of enforcement. The industry alone does not have the resources to enforce standards for the entire Web. It does not know how W3C would enforce binding contracts, even though its policies call for the implementation of agreements between business and user on the Internet (Engler, 1997). Consequently, there is no policy proposed to

Privacy Protection 8

make Internet standards a pre-requisite for business on the Internet. John Regale, a policy analyst with W3C, commented that "there are a lot of people interested in co-operation" (Engler, p. 82). This statement demonstrates that, currently, all participation and co-operation are voluntary and informal. There is no policy to define what will happen to companies not willing to implement privacy standards. Clearly, the resources and powers of governments, updated industrial technology, and the user's demands and self-protection are all equally important factors when considering the regulation of Internet privacy protection.

In the meantime, consumers can consult Web sources that will help them to secure more privacy. For example, browsers like Internet Explorer 6 will allow you to view machine-readable privacy policies at businesses where you shop on-line simply by pressing the Tools button on your browser (Microsoft, 2001).

In conclusion, privacy protection on the Internet is a complex problem. Many areas must be examined when reforming the Web. Firstly, all governments must contribute to the creation of an international standard for privacy protection. Also, the industry must contribute to regulation. Finally, users must protect themselves and contribute to the reformation of the Internet. All three groups must contribute in order to prevent political or financial motive from driving regulation policies. W3C and P3 are important organizations because they provide some solutions to privacy protection; however, there are areas that are still not covered under these proposed policies. The Internet is already a powerful source of information that is constantly expanding. However, it will not reach its potential strength if consumers and governments are afraid to use it. Consequently, companies will not invest in a medium that people are reluctant to use, and the amount of business supporting the Internet will decline. The Internet holds special value because it displays intellectual property. Humankind in its entirety will suffer if the growth of this intellectual medium is hindered.

Privacy Protection 9

References ← References begin on a new page with this heading.

Engler, C.E. (1997, August). Trading in on some loss of personal privacy. IEEE *Spectrum, 34,* 81–82.

Internet Society. (2002). Internet Privacy. Retrieved March 3, 2003, from http://www.isoc.org/ internet/issues/privacy/

Kalin, S. (1997, January 3). More grumbling about encryption export reform. IDG News Service Online. Retrieved March 3, 2003, from http://www.pcworld.com/noews/daily/data/ 0197/97010311290.html

References are listed alphabetically by author's last name, and entries are double-spaced. In each entry, the second and following lines are indented.

Madsen, W. (1998, June). Internet malcontents of the world—Unite! *Communications of the ACM, 41,* 27–28.

Microsoft Corporation. (2001). Safe Internet: Microsoft Privacy & Security Fundamentals. Retrieved March 4, 2003, from http://www. microsoft.com/ Privacy/SafeInternet/topics/browsing.htm

Wang H., Lee, M.K.O., & Wang, C. (1998, March). Consumer privacy concerns about Internet marketing. *Communications of the ACM, 41,* 63–70.

12.3 CSE STYLE OF DOCUMENTATION

"CSE" stands for Council of Science Editors, which before 2000 was called the Council of Biology Editors (CBE). Full guidelines to the CSE style of documentation can be found in *Scientific Style and Format: The CSE Manual for Authors, Editors, and Publishers,* 7th ed. (Reston, VA: Council of Science Editors, 2006). While originally the *CBE Manual* focused on biology and medicine, the most recent edition covers all scientific disciplines.

The *CSE Manual* presents two styles of documentation:

1. The **name-year** system is commonly used in the biological and earth sciences. As in the APA system, sources are cited in the text using parenthetical name-year references, which corresponds to an alphabetized author list of sources at the end of the document.

2. The **citation-sequence** system is most often used in the applied sciences, such as chemistry, computer science, mathematics, physics, and health. It is the style outlined in the following subsections.

12.3.1 IN-TEXT CITATIONS: CSE STYLE

Use the following guidelines for the CSE citation-sequence system to indicate the source of borrowed information in the text of an engineering document:

- Use superscript numbers to indicate that you have used borrowed information at those points in the document, starting at 1 and continuing throughout, except when you repeat a source.
- When you repeat a source, use the number you originally assigned that source.
- Although superscript numbers are preferred, you may use numbers in parentheses.
- Place the number right after the reference in the text and before any punctuation, like this[1].
- If you mention the authority's name, place the number after the name.
- If a single reference points to more than one source, list the source numbers in a series—for example, 1, 4, 7. Use a hyphen to show more than two inclusive source numbers—for example,[8–10].

Here is an excerpt that uses CSE-style textual citations:

McCarthy and Masson[1] wrote a book that not only touched on a subject not much examined before—animals' emotions—but became a popular nonfiction work as well. It continues work that Charles Darwin had begun, and like Savage-Rumbaugh and Lewin's[2] work on Kanzi, the chimpanzee who understands a good deal of spoken English, it expands our notions of what constitutes animal intelligence.

12.3.2 CSE REFERENCE LIST

Use the following guidelines for the CSE citation-sequence system to format the references list:

- Begin your reference list on a separate page with the centered title "References" or "Cited References" at the top.
- Single-space entries, leaving a space between them. Use a flush-left style for entries.

- List references according to the number used in your text and in that order. The references for the preceding excerpt would look this way:

  1. McCarthy S, Masson JM. When elephants weep: the emotional lives of animals. New York: Delacorte; 1995.

  2. Savage-Rumbaugh ES, Lewin R. The ape at the brink of the human mind. New York: Wiley; 1994, 299 p.

- Invert author names and use initials, without a space between the first and middle initials or a comma between an author's last name and initials or first name.
- Separate names of multiple authors with commas, and do not add *and* before the last author's name
- Titles of works should not be underlined or placed in quotation marks. Use the punctuation and capitalization style for titles of books and articles that you see in the following examples.
- For books, give the city of publication and, if necessary, the abbreviation for the state or country, in brackets. Omit common words in publisher names such as *Inc.* and *and Sons*, but retain the words *University* and *Press*. Use the punctuation and capitalization style for publication details that you see in the following examples.
- Use standard abbreviations for journal names unless the name is only one word. Use the spacing, punctuation, and capitalization style for journal names, volumes, issues, dates, and pages that you see in the following examples.
- For a chapter or other part of a book and for journal articles, indicate the beginning and ending pages. Use the punctuation and abbreviation style for page numbering that you see in the following examples.
- This is the format to use for books with one author:

  1. Hawking SW. The universe in a nutshell. New York: Bantam; 2001.

- Use this format for books with more than one author:

  2. McCarthy S, Masson JM. When elephants weep: the emotional lives of animals. New York: Delacorte; 1995.

- Use this one for books developed by an editor:

  3. Bowling AT, Ruvinsky A, editors. The genetics of the horse. New York: Oxford University Press; 2000.

- To cite a chapter from an edited book, use this format:

  4. Polanyi JC. The transition state. In: Zewail AH, editor. The chemical bond: structure and dynamics. Boston: Academic Press; 1992.

- For books that are in subsequent editions, use this one:

  5. Lyon MF, Searle AG, editors. Genetic variants and strains of the laboratory mouse. 2nd ed. Oxford (UK): Oxford University Press; 1989.

- This is the format to use for articles in a journal paginated by volume:

  6. Reimann N, Barnitzeke S, Nolte I, Bullerdick J. Working with canine chromosomes: current recommendations for karyotype description. J Hered. 1999;90(1):31–34.

- For articles in a journal paginated by issue, use this format:

  7. Lee TG, Liu W, Polanyi JC. Photochemistry of advanced molecules. Surf Sci. 1999;426:173.

- For articles in a journal with discontinuous pages, use this format:

  8. Crews D, Gartska WR. The ecological physiology of the garter snake. Sci Am. 1981;245:158–64,166–8.

- If you borrow information from a scholarly project or reference database, use this format:

  9. "Mass Flow Rate." *Science and Engineering Encyclopaedia* [Internet]. Version 2.3. Warwick (England): Dirac Delta Consultants, Ltd. [date unknown; cited 2007 May 25]. Available from http://www.diracdelta.co.uk/science/source/m/a/mass%20flow%20rate/source.html.

In this example "date unknown" refers to the fact that the web page in which the article is shown has no date of publication.

- This is the format to use if you borrow information from a professional or personal web site:

  10. Bridge Engineering Home Page [Internet]. Sunnyvale, CA: SC Solutions; c2006 [cited 2007 May 25]. Available from: http://www.scsolutions.com/bridge.html

- If you borrow information from an online book, use this format:

  11. Lienhard, John IV, Lienhard, John V. A Heat Transfer Textbook [Internet]. 3rd ed. Cambridge, MA: Phlogiston Press; c2006 [cited

2007 May 25]. Available from: http://web.mit.edu/lienhard/
www/ahttv124.pdf

- If the information comes from an article in online scholarly journal, use this format:

  12. Jorgensen, Niels. Putting it All in the Trunk, Incremental Software Development in the FreeBSD Open Source Project. Information Systems Journal [Internet] 2001 cited 2007 May 25]; 11: 321–336. Available from: http://webhotel.ruc.dk/nielsj//research/papers/freebsd.pdf.

- If the information comes from an article in an online magazine, use this format:

  13. Seattle City Light, "Sustainability: High Performance Buildings Deliver Increased Retail Sales." [updated 2006 Jun 6; cited 2007 May 25]. Available from: http://www.ci.seattle.wa.us/light/conserve/ sustainability/studies/cv5_ss.htm.

- If the information comes from an article in an online newspaper or newsletter, use this format:

  14. Lee, G. Fred, Jones-Lee, Anne. 2007 January 24. Summary of Water Quality Issues in the San Joaquin River and Stockton Deep Water Ship Channel. Stormwater Runoff Water Quality Newsletter [Internet]. [updated 2007 Jan 24; cited 2007 May 25]. Available from: http://www.members.aol.com/annejlee/swnews101.pdf.

- If you use information from an e-mail or from other material not formally published on the Internet, do not include it as an item in the reference; instead, put the source information in parentheses in the related text:

  . . . the 600-ton moveable mount is easily steered with a one-horsepower electric motor (2007 e-mail from J. Lowe to me, unreferenced, see "Acknowledgments").

In this example, "Acknowledgments" is a section of the document establishing that the writer of the e-mail granted permission to the writer to cite his e-mail.

- If you cite other types of online information, use labels such as the following to indicate type of information the source provides:

  15. Searle, Ronald. Magazine cover. Cartoonbank.com [Internet]. [updated 27 May 1972; cited 2007 May 25]. Available from: http://www.cartoonbank.com.

16. City of Seattle, WA. Map. Yahoo! Maps [Internet]. [updated 2002 Mar 11; cited 2007 May 25]. Available from: http://maps.yahoo.com/yahoo.

12.3.3 CSE MANUSCRIPT FORMAT

To format a document using CSE guidelines:

- Start the document with a separate title page.
- Begin the next page (numbered page 2) with the abstract of the document at the top just below the centered heading ABSTRACT.
- Immediately following the abstract, begin the introduction after the centered heading INTRODUCTION.
- Throughout the text, use italicized, left-aligned headings that use sentence-style capitalization.
- Immediately following the body of the document, begin the conclusion under the centered heading CONCLUSION.
- On a new page following the conclusion, place the references under the centered heading CITED REFERENCES.

For other details such as page numbering, titles, and format of the title page, see *The CSE Manual for Authors, Editors, and Publishers*, 7th ed. (Reston, VA: Council of Science Editors, 2006).

12.3.4 SAMPLE PAPER: CSE STYLE

The Danger of Oil Scarcity: A Two-Pronged Solution

Stephanie Beckett
Braden Engineering-Communication Contest
University of Texas at Austin
February 4, 2005

Petroleum scarcity has startling economical and political implications. On the economic front, Lugar and Woolsey have suggested that a $10 increase in the cost of a barrel of oil may retard US economic growth by 0.5% per year[1]. Underlying this economic problem involving petroleum scarcity is an even more fundamental danger: if a decline in the supply of petroleum precipitates a rise in oil prices, oil-rich nations may gain an even more disproportionate degree of political clout in the world. Supply will fail to meet demand, and oil consumers struggling to alleviate the economic catastrophes associated with cost increases will be forced to accept any price offered by oil-rich nations.

Many of these oil-rich nations are unfortunately unstable, as almost two-thirds of the world's oil supply is in the Middle East[1]. This dependency on the Middle East limits foreign policy options by forcing world leaders to appease and sometimes even actively support unjust autocrats in the region[2]. This hypocritical behavior certainly fuels terrorists' hatred of the US. Further, according to Lugar and Woolsey, consumers of oil contribute a great deal to the continuation of corrupt governments and perhaps even terrorism by direct infusions of money; over one trillion dollars will be funneled into the Middle East between 2000 and 2015.

Therefore, as Hans Mark, former Secretary of the Air Force, tersely put it, "The near term problem with oil and gas lies in the political instability of the region where much of the resource is located" [3]. Moreover, the long-term solution to these problems involves (1) decreasing US dependency on foreign oil, (2) developing renewable energy resources, and (3) accepting the potential contribution of nuclear power to US energy needs.

Decreasing the US Dependency on Middle Eastern Oil

Ensuring that the US is not subject to the whims of the Middle East during times of oil scarcity is a formidable problem because, according to Mark, fossil fuels currently provide 85% of the US's energy[3]. However, some energy sources, such as renewable and nuclear energy, are technically feasible alternatives to oil. Therefore, the strategy to decreasing the US's reliance on the Middle East should be two-pronged, involving the further research and implementation of both renewable and nuclear energy.

Japan and France have actively pursued a similar strategy in order to increase the security of their energy resources[4]. The strategy has proven effective in these two nations: within about twenty five years of implementing the program, each nation doubled the percentage of energy produced through domestic means[5,6].

[Full text of this report can be found on pages 482 to 487.]

Cited References

[1] Schwartz, ND. Oil—why prices will fall: because Iraq has been on the sidelines of the oil world for 20 years; soon it won't be. Fortune. 2003 March 31; 147(6):68+.

[2] Plesch, D. Ending oil dependency. Observer [Internet]. [updated 2001 Oct 2001; cited 2005 Jan 29]. Available from: http://observer.guardian.co.uk/waronterrorism/story/0,1373,564843,00.html.

[3] Mark, H. The problem of energy. Lecture. University of Texas, Austin, TX; 2004.

[4] Organisation for Economic Co-operation and Development. Proceedings of Business as Usual and Nuclear Power [Internet]. [updated 1999 Oct 15; cited 2005 Jan 29]. Available from: http://www.iea.org/dbtw-wpd/textbase/nppdf/free/2000/busassual2000.pdf.

[5] Mishiro, M. Nuclear power development in Japan. In Proceedings of Business as Usual and Nuclear Power [Internet]. [updated 1999 Oct 15; cited 2005 Jan 29]. In: Proceedings of Business as Usual and Nuclear Power; 1999, Oct 14–15; Paris, France: OECD Publications; c2000. p. 75–79. Available from: http://www.iea.org/dbtw-wpd/textbase/nppdf/free/2000/busassual2000.pdf.

[6] Maillard, D. French nuclear industry and latest governmental decisions [Internet]. [updated 1999 Oct 15; cited 2005 Jan 29]. In: Proceedings of Business as Usual and Nuclear Power; 1999, Oct 14–15; Paris, France: OECD Publications; c2000. p. 91–95. Available from: http://www.iea.org/dbtw-wpd/textbase/nppdf/free/2000/busassual2000.pdf.

12.4 MLA STYLE OF DOCUMENTATION

The MLA system uses parenthetical author names and page numbers (called citations) in text which correspond to entries in a section at the end of the document called "Works Cited." To see the source of the parenthetical citation, you go to the works-cited section and find the corresponding author name.

Documentation guidelines within this section are consistent with the MLA style described in *MLA Handbook for Writers of Research Papers*, 6th ed. (New York: MLA, 2003).

12.4.1 IN-TEXT CITATIONS: MLA STYLE

As mentioned previously, parenthetical citations containing author names and (optionally) page numbers occur in the text of a document that uses MLA documentation style.

- If you do not mention the author name in the text, cite both the author name and the page number in parentheses:

  One commentator notes that America is "divided between affluence and poverty, between slums and suburbs" (Schrag 118).

- If you mention the author name in the text, you do not need to repeat it in the parenthetical citation:

  Peter Schrag observes that America is "divided between affluence and poverty, between slums and suburbs" (118).

- If you use more than one work by the same author and you mention only the author's name in the text, cite the title of the work in abbreviated form in the parenthetical reference.

  - If you mention both the author name and one of that author's works in the text, cite the page number in parentheses:

  In *Lament for a Nation*, George Grant claims that "modern civilization makes all local cultures anachronistic" (54).

  - If you mention only the author name in the text, cite the title of the work and the page number in parentheses:

  George Grant claims that "modern civilization makes all local cultures anachronistic" (*Lament* 54).

  - If you mention neither the author name nor the title of the work in the text, cite all three in parentheses:

  Some propose that "modern civilization makes all local cultures anachronistic" (Grant, *Lament* 54).

- If the works-cited entry begins with an editor or translator's name, use that name in the parenthetical citation (but without the *ed.* or *trans* after the name).
- If you mention the title of the work and the author name in the body text, just use the page number in the parenthetical citation.
- You can further abbreviate MLA in-text citations using the list of common abbreviations list in *MLA Handbook for Writers of Research Papers*.

- Place MLA in-text citations as close to the borrowed material as possible without interfering with readability, preferably at the end of the sentence, or else at a pause (comma, semicolon).
- Except for long quotations, place the punctuation mark after the parenthetical citation.
- When both the author name and page number appear in the citation, do not insert punctuation between the two.
- When the information you are borrowing spans several pages, provide the page range using these formats: 245–46 and 1206–1316.
- As in the above examples, italicize or underline book titles, and use quotation marks for article titles.
- If a work has two or three authors, use the following format:

  According to Clarkson and McCall, even late in the decade of the Quiet Revolution, "Trudeau saw the constitutional question as only one facet of his general mandate for the Justice Department" (258).

  Even late in the decade of the Quiet Revolution, "Trudeau saw the constitutional question as only one facet of his general mandate for the Justice Department" (Clarkson and McCall 258).

  With three authors, use a serial comma in the reference: (Wynkin, Blynkin, and Nodd viii).

- If the work you are citing has more than three authors, use this format:

  One position is that "in cultures whose religion, unlike Christianity, offers no promise of an afterlife, a name that will live on after one's death serves as the closest substitute for immortality" (Abrams et al. 3).

- If you borrow information from a corporate author (a company, agency, or institution credited with authorship of a work), use these formats:

  The Toyota brochure states that "every Toyota built in Canada has a recyclable content of at least 85%—and meets or exceeds today's most stringent emission standards" (6).

  At least 85% recyclable content is used in every Canadian-built Toyota, which meets or exceeds today's most stringent emission standards (Toyota 6).

- If you cannot find the author name, use the entire title in the text or a short version of the title in the parenthetical citation:

  German authorities had drafted legislation to "ban the sale and importation of 'dangerous dogs,' including but not limited to pit bulls, Staffordshire bull terriers and American Staffordshire terriers" ("Breed Ban" 8).

- If you use works by authors with the same last name, add the full first name in the text, or add the first initial in a parenthetical citation:

  When considering mixed-race issues in Canada, Lawrence Hill contends "the terms 'black' and 'white' ultimately acquire meaning only in opposition to each other" (208).

  Some have claimed that "the terms 'black' and 'white' ultimately acquire meaning only in opposition to each other" (L. Hill 208).

- If you borrow information from a multivolume work, include the volume number in the parenthetical citation, use the following format (excluding *vol.* or *volume*):

  Abram et al. state that "the period of more than four hundred years that followed the Norman Conquest presents a much more diversified picture than the Old English period" (1: 5).

- If you cite an author quoted in a work by another writer, begin the citation in the parenthetical reference with the abbreviation *qtd. in* (for 'quoted in"):

  To Woody Allen, the successful monolog is a matter of attitude: "I can only surmise that you have to give the material a fair shake at the time and you have to deliver it with confidence" (qtd. in Lax 134).

- If your text combines information borrowed from more than one source, use a semicolon between the citations:

  An understanding of the business cycle is fundamental to successful investing (Gardner 69; Lasch 125).

- If the work has no page numbers, use paragraph numbers to direct readers to a specific text location (*par.* or *pars.* for "paragraph" and "paragraphs," respectively):

  (Barcza, par. 7)

- When the electronic source has an author and fixed page numbers, provide both (but not the page numbers of a printout of a document on the Web) using the same format as described previously.
- The electronic source has an author but no page number, use some other available means of numbering, such as paragraphs or sections, specify them by using the abbreviations *par., pars., sec., screen*, or *screens*.

Fackrell asserts the accommodation for animals is adequate, "We have lodgings for up to 12 dogs at a time in our indoor/outdoor runs" (par. 9).

- If the electronic source has no author, use the complete title in the text and a shortened form of the title in the parenthetical citation:

According to the Web page sponsored by Children Now, an American organization that provides support for children and families, "52% of girls and 53% of boys say there are enough good role models for girls in television, although more girls (44%) than boys (36%) say there are too few" ("Reflections in Media").

12.4.2 MLA WORKS-CITED LIST

In MLA style, the list of works cited appears at the end of the document and is called "Works Cited." The following presents guidelines for setting up a works-cited list and example entries for common types of sources cited.

GUIDELINES FOR THE MLA WORKS-CITED LIST

- Place the list of works cited on a new page at the end of the document, and entitle it "Works Cited."
- Include in the works-cited list only those sources from which you quoted, paraphrased, or summarized information.
- Continue the page numbering of the text through the works-cited list.
- Double-space both within and between entries.
- Do not indent the first line of each source entry, but do indent any additional lines five spaces (about 1.25 cm or 0.5 inch).
- Alphabetize works-cited list entries using the last name of the author, using the letter-by-letter alphabetizing system. If a source has no author or editor, alphabetize according to the first word of its title, ignoring initial articles such as *A, An,* or *The.*

MLA GUIDELINES FOR PRINT RESOURCES
ARTICLES

- If you use information from an article in a weekly or monthly magazine, punctuate the article title with quotation marks, italicize the name of the magazine, omit volume and issue numbers, and otherwise use the format shown in the following examples:

  Begley, Sharon. "The Schizophrenic Mind." *Newsweek* 11 Mar. 2002: 44–51.

  Ehrlich, Paul. "The Ecological Impact of Nuclear War." *Mother Earth* Sept. 1981: 142–50.

- If you borrow information from an article in a journal paginated by volume, use this format:

  Wright, Julian. "International Telecommunications, Settlement Rates, and the FCC," *Journal of Regulatory Economics* 15 (1999): 2267–291.

- If you borrow information from an article in a journal paginated by issue, use this format:

  Douthitt, Bill. "Voyage to Saturn." *National Geographic* 210:6 (2006): 38–57.

- If you borrow information from a daily newspaper, use this format:

  McGhee, Tom. "Companies Turning to GPS to Track Their Employees." Austin American-Statesman 31 Dec. 2006: H2.

- If the newspaper article is unsigned, use this format:

  "Top 10 Percent Rule, Deregulation on Lawmakers' Plates." *Austin American-Statesman* 31 Dec. 2006: G3.

- If the newspaper article is an editorial, use this format:

  "Privatization Lessons Learned." Editorial. *Austin American-Statesman* 29 Dec. 2006: A14.

- If the article is a letter to the editor, use this format:

  Kennedy, Paul. Letter. *Harper's* Sept. 2002: 4.

- If you borrow information from a review, start with the reviewer's name and the title of the review, if one is provided, and otherwise use the following format:

  Ed Nawotka. "Austin Literati Recommend Their Favorites." Rev. of *Challenger Park*, by Stephen Harrigan. *Austin American-Statesman* 31 Dec. 2006: J6.

BOOKS

- For a book by one author, use the following format:

  Ferguson, Eugene S. *Engineering and the Mind's Eye*. Cambridge, MA: MIT Press, 1994.

- For a book by two or more authors, provide the authors' names in the same order as they appear on the title page:

  Murray, David W., and Bernard F. Buxton. *Experiments in the Machine Interpretation of Visual Motion*. Cambridge, MA: MIT Press, 1990.

- For books by three authors, use this format:

  Evans, David S., Andrei Hagiu, and Richard Schmalensee. *Invisible Engines: How Software Platforms Drive Innovation and Transform Industries*. Cambridge, MA: MIT Press, 2006.

- For books by more than three authors, use the abbreviation *et al.* (for "and others" in Latin):

  Sharples, Mike, et al. *Computers and Thought: A Practical Introduction to Artificial Intelligence*. Cambridge, MA: MIT Press, 1989.

- If the book has been developed by an editor, use this format:

  Kinnear, Kenneth E., ed. *Advances in Genetic Programming*. Cambridge, MA: MIT Press, 1994.

- If the book is a translation, use this format:

  Habermas, Jürgen. *Between Facts and Norms: Contributions to a Discourse Theory of Law and Democracy*. Trans. William Rehg. Cambridge, MA: MIT Press, 1996.

- If the book has a corporate author (a company, institution, association, or agency credited with authorship), use this format:

  PriceWaterhouseCoopers Inc. *Technology Forecast: 2000*. Menlo Park: PriceWaterhouseCoopers Technology Center, 2000.

- If you cannot find the author of the work, begin the entry with the title of the work, alphabetizing according to the first main word in the title (ignoring initial articles (*A*, *An*, or *The*):

  An Engineer's Perspective on the Bible. San Francisco, CA: Anonymous Press, 2000.

- If you borrow information from two or more works by the same author, provide the name of the author in the first entry only, and, in succeeding entries, type three hyphens followed by a period in place of the author's name:

  Petroski, Henry. *Success through Failure: The Paradox of Design*. Princeton, NJ: Princeton UP, 2006.

  ---. *To Engineer Is Human: The Role of Failure in Successful Design*. London, UK: Vintage, 1992.

- If the work is in a subsequent editing, use this format:

  Strunk, William, Jr., and E.B. White. *The Elements of Style*. 3rd ed. New York: Macmillan, 1979.

- If the work consists of multiple volumes, use this format:

  Chang, S.K., ed. *EnHandbook of Software Engineering and Knowledge Engineering*. 2 vols. Hackensack, NJ: World Scientific Publishing, 2002.

- If you use only one of the volumes, use this format:

  Hanna, Awad S. *Concrete Formwork Systems*. Vol. 2. Boca Raton, FL: CRC Press, 1998.

- If you borrow from an encyclopedia, dictionary, or other reference work and you cannot find the author name, use this format:

  "waveguide." *The Columbia Encyclopedia*, 6th ed. New York: Columbia University Press, 2004.

- If you borrow from a previously published article that you find in a collection, use this format:

  Shurmer, Mark. "Standardization: A New Challenge for the Intellectual Property System." *Innovation and the Intellectual Property System*. Eds. Andrew Webster. Heidelberg: Verlag, 1996. Rpt. in *Vierteljahrshefte Zur Wirtschaftsforschung* 4 (1996): 482–93.

  Jackson, Matthew O. "The Stability and Efficiency of Economic and Social Networks." *Advances in Economic Design*. Eds. Semih Koray and

Murat R. Sertel. Springer-Verlag: Heidelberg, 2003. Rpt. in *Networks and Groups: Models of Strategic Formation*. Eds. Bashkar Dutta and Matthew O. Jackson. Heidelberg: Springer-Verlag, 2005.

MLA GUIDELINES FOR ELECTRONIC SOURCES

In the following examples, notice that most have two dates: the first is the publication date of the electronic source (which is not always available); the second is the date the researcher (for example, you) accessed the source.

- If you borrow information from a scholarly project or reference database, use this format:

  "Charles George Douglas Roberts." *The Electronic Text Centre*. Dir. Alan Burk. 1996. U of New Brunswick Libraries. 5 Mar. 2002 <http://www.lib.unb.ca/Texts/research.htm>.

  Frost, Robert. "Mowing." *A Boy's Will*. New York: Henry Holt, 1915. *Project Bartleby Archive*. Ed. Steven van Leeuwen. Dec. 1995. Columbia U. 6 Mar. 2002 <http://www.bartleby.com/117/19.html>.

- If you borrow information from an entire online scholarly project, use this format:

  Early Modern English Dictionaries Database. Ed. Ian Lancashire. 1999. U of Toronto. 7 Mar. 2002 <http://www.chass.utoronto.ca/english/emed/emedd.html>.

- If you borrow information from a professional or personal web site, use this format:

  Lancashire, Ian. Home page. 28 Jan. 2002. 8 Mar. 2002 <http://www.chass.utoronto.ca/~ian/>.

  Epic Records. 2001. Sony Music Inc. 18 Mar. 2002 <http://www.epicrecords.com>.

- If you borrow information from an online book, use this format:

  Austen, Jane. *Pride and Prejudice*. 1813. Ed. Henry Churchyard. 1996. 12 Apr. 2002 <http://www.pemberley.com/janeinfo/pridprej.html>.

- If you borrow information from an online book in scholarly project or reference database, use this format:

  Dickens, Charles. *A Tale of Two Cities*. 1859. *An Online Library of Literature*. Ed. Peter Galbavy. 29 June 1999. 17 Mar. 2001 <http://www.literature.org/authors/dickens-charles/two-cities/>.

- If the information comes from an article in an online scholarly journal, use this format:

  Herman, David. "Sciences of the Text." *Postmodern Culture* 11.3 (May 2001): 29 pars. 16 Mar. 2002 <http://www.iath.virginia.edu/pmc/text-only/issue.501/11.3herman.txt>.

- If the information comes from an article in an online magazine, use this format:

  Nyham, Brendan. "Spinsanity." *Salon* 5 Mar. 2002. 7 Mar. 2002 <http://www.salon.com/politics/col/spinsanity/2002/03/05/dissent/index.html>.

- If the information comes from an article in an online newspaper, use this format:

  Webber, Terry. "Bank of Canada Stand Pat." *globeandmail.com* 5 Mar. 2002. 6 Mar 2002 <http://globeandmail.com/servlet/RTGAMArticleHTMLTemplate/D, B/20020305/wbankcan?hub=homeBN&tf=tgam%252Frealtime%252Ffullstory.html&cf=tgam/realtime/config-neutral&vg=BigAdVariable Generator&slug=wbankcan&date=20020305&archive=RTGAM&site=Front&ad_page_name=breakingnews>.

- If you use information from an e-mail, use this format:

  Chamberlain, Tim. "Re: Credibility in Magazines." E-mail to the author. 12 Nov. 2001.

- If you cite other types of online information, use labels such as the following to indicate type of information the source provides:

  Searle, Ronald. Magazine cover. 27 May 1972. *Cartoonbank.com*. 16 Mar. 2002 <http://www.cartoonbank.com>.

  City of Victoria, BC. Map. *Yahoo! Maps*. Yahoo: 2000. 11 Mar. 2002 <http://maps.yahoo.com/yahoo>.

MLA GUIDELINES FOR OTHER SOURCES
GOVERNMENT PUBLICATIONS

- If you do not know the author of the work, use the government agency as the author:

  Ontario Human Rights Commission. *Human Rights: Employment Application Forms and Interview*. Toronto: Ontario Human Rights Commission, 1991.

- If the author is known, it is optional whether you start with the author's name or the name of the agency; if the latter, give the author's name after the title of the work, preceded by *By*.
- If you borrow information from a published dissertation use this format:

  Haas, Arthur G. *Metternich, Reorganization and Nationality, 1813–1818.* Diss. U of Chicago, 1963. Knoxville: U of Tennessee P, 1964.

- If the dissertation is unpublished, put the title in quotation marks:

  Mercer, Todd. "Perspective, Point of View, and Perception: James Joyce and Fredric Jameson." Diss. U of Victoria, 1987.

- If you get information from the published proceedings of a conference, use this format:

  Cassidy, Frank, ed. "Reaching Just Settlements." *Proc. of Land Claims in British Columbia Conf.*, Feb. 21–22, 1990. Lantzville and Halifax: Oolichan and Inst. for Research on Public Policy, 1991.

- If you use information from a personal letter you have received, use this format:

  Joslin, Simon. Letter to the author. 23 Oct. 2003.

- If you borrow information from a lecture or public address, provide the name of the organization sponsoring the lecture or address, the location, and the date it was given:

  Hill, Larry. "Navigating the Void and Developing a Sense of Identity." Traill College, Trent University, Peterborough. 30 Jan. 2002.

- If you borrow information from an interview, provide the title of the interview (or *Interview*, if there is no title), publication or broadcast information, and other information formatted as shown in the following:

  Bellow, Saul. "Treading on the Toes of Brahmans." *Endangered Species.* Cambridge, MA: De Capo, 2001: 1–60.

  Smith, Michael. Interview. *Morningside*. CBC-AM Radio, Toronto. 10 Oct. 1993.

- If you use information from an interview you have conducted, use this format:

  Henry, Martha. Personal Interview. 19 Apr. 1998.

- If you cite a map or chart, follow the same format as you would for a book with an unknown author:

  Great Britain/Scotland. Map. Paris: Michelin, 2001/2002.

12.4.3 MLA MANUSCRIPT FORMAT

To format a document using MLA guidelines:

- Start the document with a separate title page.
- If an outline is required, put it on a separate page, numbered lowercase roman numeral i.
- Start the body of the document on a new page, numbered arabic 1.
- Put the works-cited list on a separate page at the end of the document.

For other details such as page numbering, titles, and format of the title page, see *MLA Handbook for Writers of Research Papers*, 6th ed. (New York: MLA, 2003).

12.4.4 SAMPLE PAPER: MLA STYLE

The Danger of Oil Scarcity:
A Two-Pronged Solution

Stephanie Beckett
Braden Engineering-Communication Contest
University of Texas at Austin
February 4, 2005

Petroleum scarcity has startling economical and political implications. On the economic front, Nelson Schwartz has suggested that a $10 increase in the cost of a barrel of oil may retard US economic growth by 0.5% per year (68). Underlying this economic problem involving petroleum scarcity is an even more fundamental danger: if a decline in the supply of petroleum precipitates a rise in oil prices, oil-rich nations may gain an even more disproportionate degree of political clout in the world. Supply will fail to meet demand, and oil consumers

> Nelson Schwartz's name is cited in the text; thus, only the page number (68) is needed.

struggling to alleviate the economic catastrophes associated with cost increases will be forced to accept any price offered by oil-rich nations.

Many of these oil-rich nations are unfortunately unstable, as almost two-thirds of the world's oil supply is in the Middle East (Lugar and Woolsey 88). This dependency on the Middle East limits foreign policy options by forcing world leaders to appease and sometimes even actively support unjust autocrats in the region (Plesch). This hypocritical behavior certainly fuels terrorists' hatred of the US. Further, according to Lugar and Woolsey, consumers of oil contribute a great deal to the continuation of corrupt govern-ments and perhaps even terrorism by direct infusions of money; over one trillion dollars will be funneled into the Middle East between 2000 and 2015 (88).

> In the first instance, Lugar and Wollsey's names are not mentioned in the text; therefore both their names and the page number must be cited in parentheses.

Therefore, as Hans Mark, former Secretary of the Air Force, tersely put it, "The near term problem with oil and gas lies in the political instability of the region where much of the resource is located." Moreover, the long-term solution to these problems involves (1) decreasing US dependency on foreign oil, (2) developing renewable energy resources, and (3) accepting the potential contribution of nuclear power to US energy needs.

> Overview of the main sections of this report is provided in this in-sentence list.

Decreasing the US Dependency on Middle Eastern Oil

Ensuring that the US is not subject to the whims of the Middle East during times of oil scarcity is a formidable problem because, according to Mark, fossil fuels currently provide 85% of the US's energy (Mark). However, some energy sources, such as renewable and nuclear energy, are technically feasible alternatives to oil. Therefore, the strategy to decreasing the US's reliance on the Middle East should be two-pronged, involving the further research and imple-mentation of both renewable and nuclear energy.

Japan and France have actively pursued a similar strategy in order to increase the security of their energy resources (Organisation 14). The strategy has proven effective in these two nations: within about twenty five years of implementing the program, each nation doubled the percentage of energy produced through domestic means (Mishiro 75; Maillard 91).

> When an author name is not available, use the organization name, or some identifiable part of it.
>
> Use a colon to punctuate a citation in which two sources must be indicated.

Works Cited

> Full text of this report can be found on pages 482 to 487.

Lugar, Richard G. and R. James Woolsey. "The New Petroleum." *Foreign Affairs* 78.1 (Jan./Feb. 1999): 88. Jan. 29, 2005. Expanded Academic ASAP database (A53545155).

> MLA courses are listed alphabetically.

Maillard, Dominique. "French Nuclear Industry and Latest Governmental Decisions." *Proceedings of Business as Usual and Nuclear Power*, Oct. 14–15 1999. Paris, France: OECD Publications, 2000. <http://www.iea.org/dbtw-wpd/textbase/nppdf/free/2000/busassual2000.pdf>.

Mark, Hans. "The Problem of Energy." Lecture. University of Texas: Austin, TX, 2004.

Mishiro, Masaaki. "Nuclear Power Development in Japan." *Proceedings of Business as Usual and Nuclear Power*, Oct. 14–15 1999. Paris, France: OECD Publications, 2000. <http://www.iea.org/dbtw-wpd/textbase/nppdf/free/2000/busassual2000.pdf>.

Organisation for Economic Co-operation and Development. *Proceedings of Business as Usual and Nuclear Power*, Oct. 14–15 1999. Paris, France: OECD Publications, 2000. <http://www.iea.org/dbtw-wpd/textbase/nppdf/free/2000/busassual2000.pdf>.

Plesch, Dan. "Ending Oil Dependency." *Observer* Oct. 2001. <http://observer.guardian.co.uk/waronterrorism/story/0,1373,564843,00.html>.

Schwartz, Nelson D. "Oil—Why prices Will Fall: Because Iraq Has Been on the Sidelines of the Oil World for 20 Years; Soon It Won't Be." *Fortune* 147.6 (Mar. 31 2003): 68+. Jan. 30, 2005. Expanded Academic ASAP database (A98880222).

13

PROFESSIONAL COMMUNICATIONS AND RÉSUMÉS

13

Professional Communications and Résumés

13   Professional Communications and Résumés

This chapter provides you with strategies, formats, and examples that you can use to create résumés and write application letters for employment searches. It also provides contents, organization, formats, and strategies for common business communication situations including memos and e-mail. This chapter concludes with preparation and delivery strategies for oral presentations.

13.1 EMPLOYMENT COMMUNICATIONS

Professional employment applications usually require a résumé, application letter (often referred to as a cover letter), or both. Check the job announcement or check with the employer to find out which is required.

13.1.1 RÉSUMÉS

The **résumé** is a detailed record of your work experience, education, training, and other relevant material that demonstrates your qualifications for employment. Years ago, the résumé was a much more stable thing than it is today. Then, you updated it whenever you changed jobs and perhaps whenever you completed a major project. In these times, however, people update their résumés constantly—preparing for the next job change, the next lay-off, the next downsizing, or the next technology advance. Instead of one life-long career, people now have three, four, or maybe even five careers during their lifetimes.

Despite all that, the design of the résumé, while quite diverse, is rather stable and predictable. You can use the templates for résumés that most word-processing software provides, or you can design your own. Your goal is to create a résumé that shows your best credentials, that provides strong detail, that can be scanned rapidly, and that looks well designed and professional.

HEADING

Start with the top of the résumé. It can contain any combination of the following, as long the result does not look overcrowded:

Name	Mailing address	Phone numbers
E-mail address	Web site	Occupational title
Objective	Highlights	Position sought

If you are still in college, it is customary to list two mailing addresses—your college address and your home address. If you have earned the right to claim a professional title, such as Professional Engineer, you can put that in the résumé heading as well. Other titles not based on professional licensing or certification can be placed in the heading—for example, accounts manager, wafer fab technician, or business analyst. Although not a common practice, the heading can include the name of the position you seek.

Goals and objectives statements within headings are the subject of some debate. Some would argue that an objective statement such as the following limits your opportunities:

Questionable: Research and development engineering in the nanotechnology industry

However, if you tailor your résumé for each employment position you seek, what's the risk? Others disparage objective statements like the following as meaningless fluff:

Questionable: Seeking a challenging, rewarding career in the exciting nanotechnology field, with plenty of room to grow and advance.

Such statements say next to nothing. If anything, they indicate enthusiasm, albeit in a rather bogus way. Disparaged also are the objective statements that add self-congratulatory self-description:

Questionable: Bright, eager, talented, hard-working young professional seeking a challenging, rewarding career in the exciting nanotechnology field, with plenty of room to grow and advance.

Still, objectives statements such as these are commonplace. Those who disparage statements like the preceding would prefer simple descriptive statements like those shown in Figure 13.1.

The highlights section—variously called "Professional Objective, "Engineering Profile," "Summary," "Highlights, and "Qualifications"—provides a quick glimpse of your qualifications. If the potential employer is too rushed to read your résumé carefully, the highlights section gives you at least a chance to be noticed. As with objectives sections, but even more so, highlights sections can be marred by self-congratulation.

Sometimes, these sections are formatted as regular paragraphs (Figure 13.2), especially when the writer has so much information that the résumé could become overcrowded:

Objective: To obtain a position as a Principal Design/Development Engineer.

Career focus: Electrical engineer.

Objective: Research and development position in the nanotechnology industry.

Objective: Graduate of major university engineering program seeking research and development position in the nanotechnology industry.

Objective: Experienced mechanical engineer seeking a professional position in chemical engineering.

Objective: Seeking a senior engineer position in a research and development or product development enterprise in which to use my mechanical engineering experience.

FIGURE 13.1
Examples of objective sections. Objective and goal sections are a requirement in résumés.

Summary of Qualifications:

Superior ProEngineer Skills, Proficient with Cimatron, AutoCad, CNC machine programming (mill and lathe), PLC programming, basic networking and CAD administration, strong machine shop and industrial automation experience, basic knowledge of plastic injection molding, C++, Visual Basic, Pascal, Windows XP, Windows NT, UNIX, and all Microsoft Office applications. Trained in SPC and QS 9000. Excellent communication and interpersonal skills.

FIGURE 13.2
Highlights section. While it enables potential employers to get a quick overview of your background and credentials, the highlights summary is not a requirement of résumés.

In Figure 13.2, you can see that the writer engages in a mild bit of self-praise ("Excellent communication and interpersonal skills"). Using bulleted lists for highlights summaries is much more readable and quickly scannable, as you can see in Figure 13.3.

As the reformatting of the qualifications paragraph in Figure 13.3 shows, bullets take up much more room—which can be a good thing if you are at the beginning of your career and don't have much information to put on your résumé. Other variations on the highlights section, shown in Figures 13.4 and 13.5, focus on achievements and competencies.

Summary of Qualifications

- Superior ProEngineer skills.
- Proficient with Cimatron, AutoCad, CNC machine programming (mill and lathe), PLC programming, basic networking and CAD administration.
- Strong machine shop and industrial automation experience.
- Basic knowledge of plastic injection molding, C++, Visual Basic.Net, Pascal, Windows XP, Windows NT, UNIX, and all Microsoft Office applications.
- Trained in SPC and QS 9000.

Excellent communication and interpersonal skills.

FIGURE 13.3
Highlights section—bulleted format. Bullets take up more résumé space but are more immediately scannable.

MAJOR ACHIEVEMENTS

- Saved the GrayBear Corporation $50,000 a year by designing and implementing an improved quality-assurance system.
- Led the successful effort to increase the acceptance level of finished goods from 96% to 99%.
- Enabled the corporation to achieve ISO 9000 for the production line by rewriting quality-assurance procedures.
- Solved major quality-assurance problems that temporarily halted production in 2005.
- Earned a 20% salary increase as a reward for my leadership and innovation.

FIGURE 13.4
Achievements section—a variation on the highlights summary section at the top of the résumé.

BODY

The body of the résumé presents the details on all your relevant work experience, projects, accomplishments, education, training, and other relevant information. Two organizational approaches are commonly used for the body of résumés:

Performance approach: Also called the "chronological" or "reverse chronological" approach, this approach divides the content into work experience and education and training, in whichever order is most effective. Military experience can be presented as another, separate section or folded into the experience and training sections. Information in these sections is sequenced according to the company or the educational institution and within these

CORE COMPETENCIES

Project life-cycle management	Laboratory turbine testing	Product development management
Team training and mentoring	Hardware design and integration	Quality assurance standards
Engineering documentation	Engineering Change Order	Regulatory compliance

FIGURE 13.5
Skills, competencies section—another variation on the highlights summary section at the top of the résumé.

subsections by the job, position, degree, or training course. The résumé example in Figure 13.8 uses the performance approach.

Functional approach: Also called the "thematic" approach, this approach presents separate sections for each major area of your background—for example, process design, administration, quality assurance, project management. Each of these subsections presents all activities related to that function: positions held, achievements, projects, education, training, awards, certifications, licenses. See Figure 13.6 for an example of the functional approach. In the functional approach, you do include a brief section on your education and training. The details of your studies and training should already have been presented in the body sections.

Gas Plant

- Project management and design review for grass-roots cryogenic expander gas plant on the Texas Gulf Coast using a Houston-based engineering firm.
- Revision of existing absorption-type gas plants, including compressor modifications, new compressor installation designs, amine treating, dehydration and fractionation.
- Modification and troubleshooting old existing gas-processing plants including refrigeration systems. This type activity required day-to-day operations knowledge and close work with plant personnel.
- Installation of three 3500 kw generators in an existing plant in East Texas, using electrical engineering contractor for electrical design and different contractor for civil and mechanical design.

Production

- Modification of existing facilities and design of grass-roots facilities in production facilities, as well as project management in all areas of gas production, oil production, saltwater gathering, and disposal systems.
- Design and installation of all production facilities beginning at the wellhead, including gathering systems, separation and metering, pumping, compressing, and storage facilities.

Risk Management

- Participation in initial round and second round of HAZOP activities on several major plants defined as PSM facilities.
- Utilization of a knowledge-based system for reviewing all mechanical and safety systems within the operating plants. For all new facilities, conducted HAZOP review before operator training and startup.

FIGURE 13.6
Example from a résumé using the functional approach.

Whichever organization approach you use, supply as much specific detail as you can: company names, dollar amounts, numerical percentages, formal product names, version and release numbers, beginning and ending dates, and so on. Numbers and capital letters, which these details necessitate, cause the reader to slow down, take notice, be impressed, and possibly remember details about you.

Whichever approach you take, you have several formatting choices for the actual detail you present: paragraph format or vertical-list format. The paragraph format enables you to present much more information, but that information is less readily scannable. The bulleted-list format is much more scannable but takes up more résumé space. See Figure 13.7 for some examples.

Whichever format you use, keep the following strategies in mind:

- Include as much specific detail as you can: numerical data, formal product and organization names, version and release numbers, model numbers, dates.

1993 – Present
Chief Supervising Engineer. The GrayBear Company, Gun Barrel City, TX
Provide design and consultation for industry, private development, and private utilities. Designed a major wastewater treatm
Corp. Developed
Jacinto Timber C
when necessary. J
neers. Developed
treatment process
ganese removal,
water. Have provi
four occasions.

1993 – Present
Chief Supervising Engineer. The GrayBear Company, Gun Barrel City, TX

- Provide design and consultation for industry, private development, and private utilities.
- Designed a major wastewater treatment plant for Gun Barrel Chemical Corp. Developed industrial site design for San Jacinto Timber Co.
- Manage office staff up to fifteen.
- Joint-venture with other engineers.
- Have developed numerous water and wastewater treatment processes including gas stripping, manganese removal, high-strength industrial wastewater.
- Have provided expert witness testimony on four occasions.

FIGURE 13.7
Two formatting approaches for résumé detail.

- Drop "I" from sentences. Instead of writing "I performed quality assurance audits, which includes analyzing procedures and making recommendations for improvements," write "Performed quality assurance audits, which includes analyzing procedures and making recommendations for improvements."
- Present information reverse-chronological order. In an education section, list the last degree you got first then list the one previous to that and so on.

CONCLUSION

You can choose from a variety of possibilities for the final section of the résumé—or none at all:

- *Memberships in professional organizations.* Show that you are a professional by joining the essential organizations in your field and listing them on your résumé.
- *Professional licenses and certifications, with registration numbers.* This information, however, may be important enough to put in the heading of your résumé.
- *Security clearances.* This information may also be better placed in the heading.
- *Personal interests and nonwork activities.* Writers hesitate to put this kind of information in their résumés, thinking it has nothing to do with career or employment. That's true, but information of this sort humanizes you for potential employers, giving them something to talk to about at those odd moments such as waiting for the elevator.
- *Publications.* A list of publications—formal or informal—is another way to demonstrate your professionalism.
- *References.* Listing references' mailing and e-mail addresses, phone numbers, and company and position names can take up a lot of résumé space. Most writers prefer "References available upon request" at the bottom of the résumé.
- *Date of preparation.* On some résumés, you'll notice a date at one of the bottom corners. Some believe that indicating the date you wrote the résumé is a bad idea; potential employers might see a relatively old date and quickly toss the résumé without looking at it at all. Just as likely, however, potential employers might like your qualifications, wonder what else you've been doing since that date, and contact you.

As mentioned at the beginning of this list, it is a valid option to end a résumé with the body section and omit a résumé conclusion altogether. However, if you have trouble filling up the page, using the conclusion is a good option.

13.1.2 APPLICATION LETTERS, COVER LETTERS, AND RELATED MESSAGES

Typically, a résumé is accompanied by an application letter, also known as a cover letter. To some, a cover letter is quite a different thing from the application letter. To some, a **cover letter** can mean a simple letter that states the job sought and the fact that a résumé is

---

(371) 131-6216 Home
(371) 131-9138 Office
(371) 759-2459 Cell

jhmmpe@graybear.com
6397 Kerbey Lane
Port O'Conor, TX 77982

J. Hughes-McMurrey, P.E.

Objective   To obtain a position in nuclear-power-plant safety research and regulation.

Experience

2000–present   Mid-Coastal Utilities   Port O'Connor, TX

Director—Nuclear Engineering

- Provide project management and leadership to 3 groups that included 45 engineering professionals.
- Provide design and outage support to all engineering, maintenance and operations.
- Resolve several long-standing problems with mechanical and electrical systems.
- Responsible for fuels budget and technical support and analyses for all fuel activities.
- Provide training in problem solving and root cause evaluation.
- Received company's highest employee rating.
- Interview, hire, and train employees; reward and discipline employees; appraise performance; plan, assign, and direct work; address complaints and resolve problems.
- Responsibility for developing and implementing a local area network (LAN) with 3 IBM UNIX servers and 45 workstations.
- Received two design patents from the United States Patent Office. Also received several International patents for the same designs.

FIGURE 13.8
Complete résumé, using the performance (reverse-chronological) approach.

FIGURE 13.8 (*Continued*)

	Supervisor and Senior Engineer—Site Engineering	
	• Was responsible for full system and design support for the 8 mechanical systems.	
	• Was responsible for developing and implementing a probabilistic safety analysis (PSA) that was actively used for decision making and outage management.	
1996–2000	White & Rivers Consulting Engineers	New Orleans, LA
	Engineer III—Systems Engineering	
	• Project Engineer and Manager for power plant safety valve testing and qualification.	
	• Provided engineering support for design and construction of power plant systems.	
	• Developed fire protection designs and specifications for several power plants.	
1992–1996	USM National Engineering	Gulfport, MI
	Senior Engineer—Code Verification	
	• Provided technical support to the regulatory agencies for a General Electric test facility.	
	• Provided code development and verification of several large computer programs.	
1982–1986	Mobile Engineering Corporation	Mobile, AL
	Engineer—Safety Analysis	
	• Provided design, safety analyses and shielding assessments for a power plant.	
	• Provided criticality and thermal hydraulic analyses for 7 fuel rack designs.	
Education	University of California, Berkeley	Berkeley, CA
	• Master of Science, Nuclear Engineering	
	• Graduate Student Council	
	University of Texas at Austin	Austin TX
	• Bachelor of Science, Mechanical Engineering	
	Bachelor of Science, Industrial Engineering	
Certifications, Registrations	Professional Engineer in Alabama, Mississippi, Louisiana, Texas Mechanical	
	CCNA—Certified Cisco Network Associate	
	MCP—Microsoft Certified Professional	
	Proficient in the use of many computer programs including RELAP4, RELAP5, RETRAN, GOTHIC, COMPARE, CONTEMPT, KYPIPE and ADLPIPE.	
Other	Interest in community and church activities, running, genealogy, writing poetry, carpentry, and computers.	

attached. This type of cover letter does no promotion of the writer at all. The **application letter** promotes you, showing how your background and credentials make you a good candidate for the position you seek. Make sure you know which type is expected. It would be a shame to miss an opportunity to make a case for yourself as a good candidate for a job, which is the purpose of an application letter.

The primary purpose of the application letter is to get you into an interview. While it's certainly possible to go straight from application letter to employment, the interview is usually the gateway where multiple candidates are considered. In any case, the job of the application letter is to show that you are right for the job. Think of the application letter as a person actively pointing out relevant information on the résumé to a potential employer.

BODY

As with the résumé, you can take one of two approaches (or a combination) to the application letter.

- *Education–experience approach.* Similar to the performance approach for résumés, this approach summarizes relevant education and training information in one section and relevant work experience in another. Begin with whichever section is your strongest.
- *Functional approach.* Like the functional approach for résumés, this approach summarizes information in each relevant area of your experience, education, and training in separate sections. For example, you might have a paragraph on everything you've done relating to project management, another paragraph on everything relating to quality assurance, and so on.

The functional approach can be used to particular advantage; you can address each major job requirement in a separate paragraph of the application letter. Typically, job announcements cite three to five major requirements. In the body of your application letter, you can have a paragraph presenting your experience, education, and training for each of those requirements.

INTRODUCTION

That first paragraph of an application letter is critical. You've got to "hook" your reader—the potential employer—into reading the rest of the letter. The standard elements of an introduction include the following:

- Indicate the purpose.
- Indicate the official title of the position you are applying for, including identifying numbers if available.
- Indicate your source of information—how you found out about the job.
- Provide a bit of information that will motivate the reader to keep reading. For example:
  - State your best qualification.
  - Mention someone within the organization who knows you and can act as a reference (or someone external to the organization who can do so).
  - Mention some detail about the organization, showing that you've done your homework.
  - State the most important requirement of the job and show that you match that requirement.
  - Express your enthusiasm about the work or the organization.

Obviously, you cannot do all of these motivational things in the introduction. The introduction must be brief: no more than three lines.

CONCLUSION

In the conclusion to an application letter:

- Provide information on how to contact you.
- Mention that the résumé is attached, if you have not done so in the body of the letter.
- Encourage the potential employer to get in touch with you and arrange for an interview
- Express your interest or enthusiasm for the job.

See Figure 13.9 for a complete example of an application letter.

Be aware of the other employment-seeking messages related to the application letter, memo, or e-mail:

- *Follow-up message.* If you've not heard from the potential employer, write a brief message in which you state that you sent an application letter on a specific date (10 days to two weeks ago), that you are still interested in the position, and that you are just checking

1000 Sequoia View Apts.
Phoenix, AZ 85003
(602) 555-4523
phmcmurrey@arizona.edu

March 25, 2009

Mr. John Wilson
Personnel Director
Wilson & Company, Engineers & Architects
3507 Desert View Suite 1001
Albuquerque, NM 87125

Dear Mr. Wilson:

In albuquerque.craigslist.org (posting ID 266613498) May 3, I discovered your posting for a Mechanical Engineer possessing substantial experience with municipal and industrial buildings and with AutoCAD Architectural Desktop 2005. My strong background in these areas makes me a good candidate for the position you have open.

The past two years at Hughes, Gano Consultants, have given me substantial experience with AutoCAD Architectural Desktop 2005. During this time, I have performed numerous pipe stress analyses and design for new coal-fired, simple-cycle and combined-cycle power plants, cogeneration facilities, and other generation facilities.

Your craigslist post also mentions LEED. At Hughes, Gano, I have used this green-building rating system on three occasions to assess the environmental sustainability of industrial building designs.

In addition to this practical experience, I hold a Bachelor of Science degree in Mechanical Engineering from the University of Arizona. While there, I took leadership roles as president of the student chapter of Engineers Without Borders and spearheaded a mission to a remote area of Sonora, Mexico, to construct facilities for adequate sanitation and safe drinking water.

I would welcome the opportunity to interview with you. I will be in the Albuquerque area during the week of April 12th and would be available to speak with you at that time. In the next week to ten days I will contact you to answer any questions you may have.

Thank you for your consideration.

Sincerely,

Jane H. McMurrey
Jane H. McMurrey

Enclosure

FIGURE 13.9
Complete example of an application letter.

to see whether your letter was received. You can also repeat your contact information and availability times.

- *Acceptance message*. If you've been offered the position and accept it, write a brief message in which you state that you accept the position, refer to the position number if there is one, indicate when you will be ready to start work, and provide or request any other relevant information.

- *Decline message*. If you've been offered the position and must decline it, write a brief message in which you state that you cannot accept the position (referring to its number if available) with or without explaining your reasons. Thank the employer for the offer. This bit of professionalism may serve you well in your career; professional communities can be rather small, and word gets around. Don't burn any bridges.

13.2 TYPES OF BUSINESS MESSAGES

The following subsections discuss strategies for writing the typical kinds of messages that are sent within and out of organizations. The term **message** is meant to encompass formal business letters, memoranda, and e-mail. True, some of the situations in these examples could be handled face to face without written correspondence. However, the communication strategies are the same. This discussion of common types of business messages relies on several important ideas.

- *Rhetorical pathways*. Business messages can be written from management to employee, from employee to management, from peer to peer (employees at the same organizational level), from government-agency official to citizen, from citizen to government-agency official, from customer or client to a nongovernmental organization (such as a company providing products or services), or from that nongovernmental organization to a customer or client.
  In a business communication, it's not enough merely to consider the audience. You must take into consideration who you are as the message sender in relation to the recipient.

- *Communication objectives*. Every business message has a **primary objective**. For example, you must let your manager know you will work overtime this weekend; you must refuse a direct order from your manager; you want to ask fellow employees to follow a process that will make your work more efficient; you want to congratulate someone on a promotion.

However, business messages typically carry with them one or more **secondary objectives**. If you confirm that you will work overtime this weekend, you may also want your manager to know that you will be making some sacrifices and that scheduling and staffing may need to be reviewed to reduce the need for overtime. If you must refuse to carry out a direct order, the secondary objective might be to protect the organization, your manager, and yourself from the damage caused by carrying out the direct order.

- *Communication risks.* With every business message, you must consider the communication risks. There are two types: inevitable (unavoidable) and inadvertent. If you must refuse to carry out the directive of your manager, there are obvious, unavoidable risks—such as getting fired. If you must ask fellow employees to follow a process that will make your work easier, there may be inadvertent risks—such as sounding officious, managerial, or pushy.

Consider the rhetorical pathways, primary and secondary communication objectives, and communication risks in the following types of messages.

13.2.1 REQUEST MESSAGES

A **request** is a message that seeks cooperation from people at your same level in the organization. It can also seek agreement or approval from people above you in the organization. Addressing your fellow employees, you can tactfully request that they enter scheduling data online so that you know what your future workload will be. Addressing your manager, you can request temporary contract help for a period when you will be overloaded with work.

The difference between requests to peers (those at your same level in the organization) and requests to managers (those at a higher level) is not all that great. In neither case can the language you use be demanding; you cannot use imperatives. (Of course, if your repeated requests have gone unheeded, then stronger language may be necessary.)

Consider this example request from a writer to her peers:

> Hey, everyone. In the online tracking system, will you please enter the dates you expect to deliver your documents to me along with page counts? I need this info in order to estimate my workload and to see if there are any periods where I need to bring in a temp. Otherwise, I'll be late in completing my work on your documents and make us miss the release date. (I just don't think I can handle that guilt!)

Notice what is going on in this message:

- *Establish some cordiality.* The "Hey, everyone" establishes a collegial, friendly tone. The goofy humor at the end of this message is also an element of cordiality. The writer is being sarcastic about the guilt and assumes that people in the organization know her well enough to understand that final statement.
- *Make the request.* At some point, the message must ask people to do something or to stop doing something. If you do not find this message pushy or demanding, it may be because of the brief bit of cordiality that precedes it. Imagine the difference if this message began with the actual request.
- *Provide some rationale.* In a request, you are obligated to explain why you need people to do something or to stop doing something. Perhaps managers can get away without stating this element, but peer-to-peer and peer-to-management requests cannot. People are much more likely to cooperate with your request if they know why they need to do what you are asking them to do.
- *Problems, consequences.* Another good strategy in a well-crafted request is to explain the problem that necessitates the request and the consequences if the request is not heeded. In the preceding example, the writer cannot project future workloads and thus may not be able to keep up, which would in turn cause the entire organization to be late with its part of the project.

Consider a related request addressed to a manager:

> Mark, I've been checking the schedule for the next 9 months and am seeing a big logjam starting in June and lasting about 6 weeks. The page count per day jumps to 350; as you know, my average is 150. No amount of overtime on my part could cover that workload, and I'll end up being the bad guy who keeps us from making our overall deliverable date. Therefore, I am requesting your approval to hire a temp for a two-month period, starting a week ahead.

As you can see, the components of this message addressed to management are much the same as those in message addressed to peers:

- *Cordiality.* Here, addressing the manager by his first name—which is quite common in many organizations in the United States—establishes a bit of cordiality. So does the phrase "bad guy."
- *Actual request.* Here, the request is to hire a temporary contract employee to help during the overload period. Notice that the request is not stated until details, rationale, and consequences have been stated—a strategy known as the **indirect approach**. Imagine the difference if this message had begun with the request.
- *Rationale.* The reason for the request is that the individual cannot handle the workload for an upcoming period, and as a result, the overall project due date will be missed.
- *Details.* This request goes a bit further than the request addressed to peers. It provides details to show how bad the problem is—in this case, statistical detail. This individual needs to work a bit harder on the manager; after all, hiring a temp is a budgetary issue, something the manager has to justify to her or his peers.
- *Problem, consequences.* As implied above, problem, rationale, consequences work as a sequence, almost like a domino effect. The individual could go even further: "And if we miss our deadline, that'll make us look bad, and make you look bad in particular, Mark." But Mark knows that!

You cannot demand or order managers or supervisors to do anything—instead, you request.

13.2.2 APPROVAL MESSAGES

An **approval** is simply an agreement to a request. There are two obvious types of approvals: manager-to-employee approvals and peer-to-peer approvals. In the preceding example, the manager is likely to approve the employee's request to hire a temporary; the budget hit is a far lesser evil than missing the project deadline. Consider the following approval of the request to hire a contract worker:

> Becky, I'm glad you have anticipated this workload problem in June. And I'm glad we finally got people to put in their scheduling info.

Otherwise, you'd have no idea that this tidal wave was coming until it was too late. Go ahead and start the paperwork; I'll sign.

In this approval, the manager congratulates the writer for being proactive, looking ahead and anticipating problems. But what if the approval was tinged with regret?

Becky, I'll go ahead and sign the paperwork for the temp this time, but let's get together and discuss how you are doing this review work. Maybe together we can figure out ways to speed up your work. Or perhaps you could train someone else in our department who is less busy during those periods? I know we are a bit too close to June to try these options, but be thinking about these ideas during the June cycle.

In the preceding approval, the manager expresses some concern that employee is not working intelligently or efficiently. The manager also implies that despite all her anticipation, she may have waited too late to implement less costly solutions. Consider a less-cordial version of the preceding approval-with-regrets:

Becky, I will approve your request for a temp this time, but in the future I need you to reexamine your process to see if you can speed it up. I also need you to look around our department and our area to see if there are people whom you could train to help out in these overload periods.

If you were Becky, would you feel like crawling in a hole and hiding?

Peer-to-peer approvals are also common types of messages. A good example is a situation in which one peer has sign-off authority over the work of other peers. Some organizations give editors sign-off authority on documents that ship with products. The editor is a peer to the writers whose work he signs off on. Their work can't go out the door without his signature!

If you approve the work or request of another employee at your same level, you don't run many communication risks. Certainly, you could inadvertently sound begrudging of your approval. Avoid that. Also, the tone of your approval could inadvertently sound officious, as if your ego had become enlarged and you now think of yourself as having managerial status. Avoid that too! A peer-to-peer approval communication can:

1. Begin by stating the subject and purpose of the communication.
2. Combined with the preceding or in a separate sentence, state the approval.

3. Make any relevant comments about the request or the approval.
4. Perhaps congratulate the fellow employee on work well done.
5. Make a statement about the next step in the organizational process (for example, an action that either the requester or the approver will take).

13.2.3 DIRECTIVE MESSAGES

A **directive**, on the other hand, is an order, a communication in which you direct people to do something. You're the boss now. However, it's not so easy giving out orders, telling people what to do. Would you want to work for a boss with these communication skills—or lack of?

KEEP THE PRINTER ROOM NEAT!

STOP USING THE FAX MACHINE FOR PERSONAL BUSINESS!

PLAN TO ATTEND THE FRIDAY AFTERNOON MEETING!

As written, these directives sound as though they are being shouted, barked out by some power-mad would-be dictator. In contemporary business culture, this style of directive will not work.

Consider how each of the foregoing orders can be modulated for less hierarchical business contexts.

KEEP THE PRINTER ROOM NEAT!	People are having problems finding their printouts and complaining about the general mess in the printer room. Please get your printouts promptly. Put other printouts behind the appropriate tab. Put unneeded paper in the recycling bin.
STOP USING THE FAX MACHINE FOR PERSONAL BUSINESS!	Please do not use the fax machine for personal business. It wastes this resource and ties up this resource for legitimate business needs.
PLAN TO ATTEND THE FRIDAY AFTERNOON MEETING!	Please plan to attend the Friday 2:00 p.m. meeting on how the company is being reorganized.

Notice what is being added to each of these directives. In the printer-room directive, the problem is explained as well as the solution. Notice too the use of the word "people"; this turns the problem into a group issue, not just a pet peeve of the supervisor. The fax-machine directive works much the same way: here, readers learn that the problem is the use of the fax machine for personal business. Maybe some had no idea that it was a waste of business resources. And finally consider the meeting directive. It now includes the reason for the meeting. True, recipients of this directive do not have a choice; it would not matter what the topic might be—they must attend. However, letting them know about the purpose of the meeting is a courtesy, and it might get people to start thinking about the meeting and be more alert and responsive.

Thus, if you generalize on these three examples, you can see that these elements are added to directives to make them less dictatorial:

- Explain the problem—maybe people don't know it exists or how bad it is. Do a bit of legitimate guilt-tripping. Explain the consequences of the problem (for example, in terms of expense).
- Explain how to solve the problem. People may know the problem exists but not know how to solve it.
- For other situations not involving problems and solutions, just provide a reason for the directive.
- Most importantly, ensure that the problem is seen as a problem for the group as a whole—not just your pet peeve as the manager. Make it clear how the solution will be a solution that benefits all.

These four elements by no means exhaust the possibilities for ways to modulate directives. For example, how would you craft a directive for all employees to attend the Friday afternoon ice cream social?

But consider now this example of a directive that has been modulated in different ways for less hierarchical business contexts:

Original message: KEEP THE PRINTER ROOM NEAT!

People are having problems finding their printouts and complaining about the general mess. Please pick up your printouts promptly. Put other printouts in the appropriate tab. Put unneeded paper in the recycling bin.	Quietly managerial in tone, this one does state the problem, explains how to solve it, and makes it a group problem.
Hey, people! The department we share the printer room with is complaining about us. Imagine that! They think we are a bunch of slobs. We better clean up our act. Be sure and get your printouts right away . . .	Styled more like a request. Humor is injected, the assumption being that people know the manager well enough to understand his sense of humor. The use of the word "we" firmly establishes the problem as a group problem.

These efforts to soften directives might seem like wishy-washy reluctance to take charge and be the boss—an effort to be the good guy, to be everybody's friend. It all depends on you, your relation to those whom you lead, the expectations of those whom you lead, the business culture in which you operate, and the style of leadership in that culture.

13.2.4 CONFIRMATION AND AGREEMENT MESSAGES

Confirmation (agreement) messages, as their name suggests, acknowledge or agree to something. They are unlike approval messages, because the writer is confirming the receipt of a request message and agreeing to take the requested action. When you respond to a directive from management, you normally agree. Ordinarily, you should send back a message confirming your receipt of the directive and agreeing to carry it out. For example,

Mark, I received your e-mail about investigating that new Adobe meeting software called Connect. It sounds like an exciting but simple way to have meetings with people at other corporate sites. I'll get on it right away and have a review for you by next Wednesday.

When the directive is easy and requires no pain, writing the confirmation message is easy. In the preceding example, notice that the message:

1. Identifies the previous message and its topic.
2. Expresses some interest in doing the task, stating some knowledge about it.
3. Agrees to do the task.
4. Specifies a date by which the task will be finished.

However, some directives may create problems for you—for example, because your manager requests overtime work this weekend, you'll miss your daughter's championship volleyball game. Your confirmation message needs to make management aware of these things—diplomatically, of course!

> Mark, I have received your orders for the department to come in this weekend. I completely understand and support the need to finish the current project on time. You know me—I'm not a complainer, not a whiner, but I will be missing Jane's championship volleyball game on Saturday, which I have promised her that I would attend. Maybe I can get Phoebe to record it. See you Saturday a.m.!

Who knows? Maybe Mark will let this employee off long enough to attend the volleyball game. In any case, Mark needs to know about the sacrifices that he is demanding of his department. Notice that this example message:

1. Refers to the previous message and its subject matter but does not get into details.
2. Acknowledges the reason for the directive.
3. Seeks to avoid sounding negative, then points out the problem that the directive creates.
4. Seeks a solution for the problem.
5. Agrees to the directive.

A confirmation or agreement message to another employee at your same level is not much different from one to management. If anything, such a message need not be so cautious about career-limiting statement. Perhaps it need not express enthusiasm and can be shorter and more informal. And if the agreement entails some hardship, perhaps the writer can be blunt and less diplomatic about it.

What if you simply cannot comply with a directive or a request? That's the challenge of the refusal message.

13.2.5 REFUSAL MESSAGES

A **refusal message** turns down a request. As with requests and approvals, the types of refusal messages, again, depend on the rhetorical pathway. As a manager, you can turn down a request from an employee. As a peer, you can turn down a request from another peer. As an employee, you may have to refuse a directive from a manager.

Unfortunately, you may have to turn down an order or request from management—in which case things get very tricky! You risk your status in the organization if not your job there as well. At the same time, your management or the organization as a whole might be at risk if the directive were carried out. Management may be unaware of the problems associated with the directive. Worse still, the directive may violate certain ethical principles. Situations like these present a parallel set of risks—one set of risks for the organization (and the manager) and another set of risks for the writer who must construct the refusal.

Consider this situation: you have received a directive to approve products for shipment to customers even though you know that those products have a high probability of defects, based on your samplings. Essentially, you are being asked to look the other way and not do your job. Is management willing to risk customer complaints, lawsuits, revenue losses, and firings? Obviously, situations like these may be better *not* captured in written documents, but for the sake of this exploration of communication risks and strategies, consider this approach. An upward refusal might:

1. Begin by stating the subject and purpose of the communication—but not the actual refusal. For example, "I'm writing in response to your directive to approve. . ." or "I am writing to express some concerns about your directive to approve. . ."
2. Using the indirect approach, summarize your understanding of the directive—perhaps it's all a big mistake and management never intended you to go against your best judgment.
3. Still keeping with the indirect approach, explain the actions you have taken, the data you have collected, the observations you have made, and the conclusions you have drawn. This is the

evidence supporting your refusal; you present it first, before stating the actual refusal.

4. State your refusal along with your rationale for that refusal. You can carry the indirect approach to the limits by stating the reasons for your refusal first then the actual refusal.

5. Conclude with an offer to work things out, to find a way to resolve this problem. Don't just seem to stand back in righteous indignation and moral superiority.

Obviously, in a communication of this sort, the risks are high. You could lose your job; you could make your management look bad; problems could arise if you don't take the shortcut ordered by management; problems could arise if customers complain, seek costly reimbursement, sue, or cease doing business with your company. Your business-communication skills can take you only so far—your business sense must do the rest.

As with confirmation and agreement messages, a refusal message to someone at your level in an organization is likely to be much the same as the same kind of message to management. However, because a fellow employee cannot have such control over your position as your manager does, you may not need to work quite so hard to be cautious and diplomatic. You can be blunt and direct.

13.2.6 COMPLAINT AND CLAIM MESSAGES

Both **complaint** and **claim letters** obviously contain complaints. Claim letters differ because they also contain a claim—a request for compensation, or "adjustment." Messages like these can be external to organization as well as internal.

Organization-external complaints and claims. Addressing themselves to a commercial organization, customers may complain about products or services, requesting a solution to the problem—for example, a billing error that the billing company's employees refuse to correct. Figure 13.10 shows an example of a routine from customer claim letter.

Notice that this message:

1. Begins calmly by indicating the purpose of the message, making clear that a problem exists and even attempting to establish some good will.

1416 Crestwood Dr. ← heading and date
Hero, TX 78541
November 12, 2008

Customer Service ← insider address
Microsoft Corp.
One Microsoft Way
Redmond, WA 98052

Dear Customer Service representative: ← salutation

I am writing to you concerning the Microsoft Wireless Optical Desktop 1.0 that I purchased on July 1st, 2008 for $59.95 plus tax. As a long-time customer of Microsoft products and as a technologically savvy person, I am disappointed with the problems that I have had with this product. ← introduction

On July 1st, 2008, I purchased the Microsoft Wireless Optical Desktop 1.0 from Fry's Electronics for $364.90, which included tax. I brought the set home and set it up with my Acer laptop. It functioned beautifully until last week, specifically on November 8th, 2008. I began having problems with the mouse communicating with the receiver. I replaced the batteries with fresh ones, but with no improvement. I ran through the entire troubleshooting guide on your website, www.microsoft.com. I even tested it by placing the set on a different computer. Nothing changed. I also had intermittent, irreproducible problems with the keyboard. ← body

I realize there is a 90-day warranty for this product and that that date has passed. However, 4 months is too short of a lifespan for a desktop set costing almost $370. Contrary to my previous experiences with Microsoft products, this product has not lived up to Microsoft's standards. From my research online, I have discovered that this product has been discontinued, which furthers my suspicions about the product.

Therefore, I am requesting either the new-model equivalent as a replacement or a full refund of the purchase price plus tax, for a total of $364.90. If you require that I ship the item back to you, I also request that cost be reimbursed as well. I have enclosed a photocopy of my receipt.

If you have any further questions, please call me at 512–259–0000. I look forward to having this issue resolved quickly and painlessly so I can continue to recommend Microsoft products. ← conclusion

Regards, ← complimentary close
Ariel Beacher ← signature

Ariel Beacher ← signature block
Encl: photocopied receipt ← indication of an enclosed item

FIGURE 13.10
Example complaint (claim) letter, which uses the block-letter format.

2. Describes specific details about the problem—dates, locations, costs, product name—in a strictly objective manner. This is the evidence for the case that the writer is building.
3. Explains the writer's efforts to solve the problem on her own and her research on this particular product.
4. Requests a reimbursement or similar product.
5. Ends with an encouragement to the recipient to honor her request and a veiled suggestion that she could take her business elsewhere and might even say less than positive things about the recipient's company to her friends.

Outside government organizations, citizens may complain about services of those government agencies—for example, a factory polluting the local environment. Notice how the discussion is managed in the complaint letter shown in Figure 13.11. The letter:

1. Begins by calmly stating its purpose, to discuss a problem, but not diving into the details of the problem.
2. Discusses the objective facts of the matter: citizen complaints, sources of the problem, government agency inaction.
3. Presents reasons why the facility is out of compliance and then calls for the agency to shut the facility down.
4. Ends by providing contact information, encouraging the recipient to take action, and hinting at more drastic action if the problem is not resolved.

Organization-internal complaints and claims. Within an organization, individual employees or departments can complain about each other, requesting that they stop doing something or start doing something. They can even request some sort of compensation. Imagine system-test engineers having to complain about software engineers not scheduling product tests. The test engineers don't know when to expect work or how to staff.

On behalf of System Test, I'm writing you about problems we are having with scheduling system tests on your deliverables.

Developers are not consistently indicating when they will bring their deliverables for system test. Without a reasonably accurate schedule, system-test engineers cannot staff properly and thus cannot guarantee on-schedule tests. Obviously, if we miss our deadlines, the product will not ship on time.

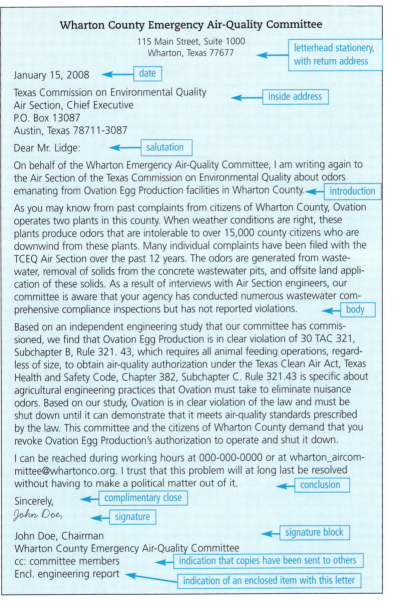

Wharton County Emergency Air-Quality Committee

115 Main Street, Suite 1000
Wharton, Texas 77677

→ letterhead stationery, with return address

January 15, 2008 ← date

Texas Commission on Environmental Quality ← inside address
Air Section, Chief Executive
P.O. Box 13087
Austin, Texas 78711-3087

Dear Mr. Lidge: ← salutation

On behalf of the Wharton Emergency Air-Quality Committee, I am writing again to the Air Section of the Texas Commission on Environmental Quality about odors emanating from Ovation Egg Production facilities in Wharton County. ← introduction

As you may know from past complaints from citizens of Wharton County, Ovation operates two plants in this county. When weather conditions are right, these plants produce odors that are intolerable to over 15,000 county citizens who are downwind from these plants. Many individual complaints have been filed with the TCEQ Air Section over the past 12 years. The odors are generated from wastewater, removal of solids from the concrete wastewater pits, and offsite land application of these solids. As a result of interviews with Air Section engineers, our committee is aware that your agency has conducted numerous wastewater comprehensive compliance inspections but has not reported violations. ← body

Based on an independent engineering study that our committee has commissioned, we find that Ovation Egg Production is in clear violation of 30 TAC 321, Subchapter B, Rule 321. 43, which requires all animal feeding operations, regardless of size, to obtain air-quality authorization under the Texas Clean Air Act, Texas Health and Safety Code, Chapter 382, Subchapter C. Rule 321.43 is specific about agricultural engineering practices that Ovation must take to eliminate nuisance odors. Based on our study, Ovation is in clear violation of the law and must be shut down until it can demonstrate that it meets air-quality standards prescribed by the law. This committee and the citizens of Wharton County demand that you revoke Ovation Egg Production's authorization to operate and shut it down.

I can be reached during working hours at 000-000-0000 or at wharton_aircommittee@whartonco.org. I trust that this problem will at long last be resolved without having to make a political matter out of it. ← conclusion

Sincerely, ← complimentary close

John Doe, ← signature

John Doe, Chairman ← signature block
Wharton County Emergency Air-Quality Committee
cc: committee members ← indication that copies have been sent to others
Encl. engineering report ← indication of an enclosed item with this letter

FIGURE 13.11
Complaint letter addressed to a government agency.

We realize that you cannot be 100% certain about your deliverable dates, but your best guess is better than nothing at all. We also realize that an automated database approach to system-test scheduling would be the best solution to this problem, but until then please remember to enter your system-test dates as early in the cycle as you can. Thanks!

Because this is a "lateral" communication, between peers, the writer cannot just angrily demand cooperation. In doing so, the communication risk might be even less cooperation! Notice the communication strategies used in the preceding example. The message:

1. States that there is a problem (in other words, complains), but does not dive into details.
2. Describes the problem, its immediate as well as ultimate consequences.
3. Acknowledges that there is a legitimate reason for the problem (implicitly, that it's not a result of incompetence or negligence).
4. Cites a solution to the problem (although unavailable at the moment).
5. Makes the request.

Imagine having to complain about a fellow employee and office-mate who spends way too much time on personal phone calls or about an employee whose work is consistently defective. Obviously, there are two communication risks in situations like these: the subsequent ill-will of the employee you complain about and the perception of you as a complainer and tattletale—someone who "rats" on others, perhaps for personal gain. Here's an example that attempts to avoid these risks:

Mark, I regret having tell you about this problem, which I hope we can keep confidential.

For some time now, I and others have been having to correct numerous coding errors in Don's work. We've tried to work with him, but to no avail. The rest of us have to take time out of our day, sometimes out of our nights, to correct his errors in time for system builds.

We would all appreciate it if you could find some way to intervene. I certainly don't want to jeopardize Don's employment here. Perhaps some additional training, reassignment, or just talking to him would make the difference?

Notice the communication strategies used here. The communication:

1. Begins by indicating the purpose (to report a problem) and expressing reluctance, but without getting into details.
2. States the problem.
3. Describes employees' efforts to resolve the problem, without having to go to management.
4. Explains the consequences of the problem.
5. Asks the manager to take some action, expressing sympathy for the problem person and suggesting some possibilities.

13.2.7 ADJUSTMENT MESSAGES

Replies to complaint and claim messages are typically called **adjustment messages**. Obviously, an adjustment message can either grant the request or refuse to grant the request. Either way, there are some communication risks:

- If you grant the requested adjustment, the main risk is that you may inadvertently make your organization look incompetent, careless, or indifferent.
- If you cannot grant the requested adjustment, the communication risk is that you may lose some business, create an enemy, or both. These refusals test your diplomacy and tact as a writer.

Here are some suggestions for either type of adjustment message:

1. Begin with a reference to the date of the original message of complaint and to the purpose of your message. If you deny the request, don't state the refusal right away unless you can do so tactfully.
2. Express your concern over the writer's troubles and your appreciation that she or he has written you.
3. If you deny the request, explain the reasons why the request cannot be granted in as cordial and noncombative manner as possible. If you grant the request, don't sound as if you are doing so in a begrudging way; assert your organization's concern for quality and its customers.
4. If you deny the request, try to offer some partial or substitute compensation or offer some friendly advice (to take the sting out of the denial).

557

5. Conclude the message cordially, perhaps expressing confidence that you and the writer will continue doing business.

13.2.8 TACTICAL MESSAGES

Like it or not, you may need to engage in tactical communications within your organization. **Tactical messages** are communications about success, accomplishments, failure, or mistakes. For example, if something went wrong with the project, you might need to clarify that it was not your fault. You might need to engage in this self-defense either as an individual or as a representative of a group (for example, as a manager of a department). Conversely, you might need to communicate your successes or those of your group—in other words, "toot your horn." Self-promotion of this sort is a little distasteful to most people. Few people are comfortable with what feels like jumping up and down, shouting "look at me! look what I did!" Actually, however, legitimate self-promotion is possible; you can maintain your integrity and the respect of your coworkers, while still achieving some important recognition for yourself or your group.

Consider the simple monthly or quarterly status report. It reports your accomplishments, your ongoing work, your problems. And after all, most supervisors must do a yearly performance appraisal on you. What are they going to use if they lack status information that you could have provided? Here is a tactical message we'd all be happy to write:

> Mark, this is to let you know that we have completed the design project—on time and under budget. Going in, we certainly did not think this would be possible. Also, I'm happy to report that reviewers in Manufacturing, in Quality Assurance, and in Distribution are all quite pleased with the design—no rework at all.

As you can see in this example, the writer takes a low-key approach and simply describes what has transpired.

But what to do when things go wrong and it's *not* your fault? Your objective is to make it clear to the right people that you were not at fault and to seek a solution to the problem. There are several risks, however; your message can appear to be shameless finger-pointing and blame-throwing. In addition, your message can alienate not only those who were at fault but many others in the organization as well; your message can appear to be an effort at shameless self-promotion at the expense of others.

> Mark, we missed that deadline because the scheduler never changed the dates. My team had no idea that the deadline was a week earlier than originally scheduled. Mistakes happen, I guess. I certainly don't want to create any problems for Mel, though, and would just as soon keep this matter to ourselves. Could we get together to figure out how to fix the damage?

Notice what's going on here. There is no apparent defensiveness— no "it's not my fault!" The cause of the problem is stated in a flat, quiet way. The name of the scheduler is not stated right away. The next two sentences are an effort to protect the scheduler, rather than to condemn him. The last sentence seeks to look past the problem, the cause of the problem, and the person responsible for the problem. Instead, it moves on to something positive and productive: let's fix the problem.

What to do when things go wrong and it is your fault? What if you are Mel, the scheduler, and you failed to enter those changed due dates, causing some of your coworkers to miss deadlines? Your objective is to take responsibility for the problem and find a solution. There are some obvious risks in this message: you must avoid sounding irresponsible and incompetent; you must avoid sounding defensive when you explain why you were at fault; you must avoid a whining tone; you must avoid sounding excessively contrite.

> Mark, as you probably already know, somehow I screwed up and failed to enter those changed schedule dates and caused the design team to miss a deadline. I'm still trying to figure how this happened. I may need to review my process with you and a few others to figure out where this exposure is. In any case, I sincerely regret this and have met with the design team to figure out a solution.

This message attempts to sound honest and forthright: hey, I messed up, I don't know how, I'm trying to find out so it won't happen again, and I'm trying to clean up the mess. This message does not attempt to shift the blame, at least on the surface: however, the suggestion to have a meeting and the word "exposure" subtly suggest the problem might be more of an organizational one—not just poor incompetent Mel. To review the sequence, this message:

1. Admits the error right away (or it could state the subject, "I'm writing you about the deadline that we all missed").
2. Describes the error.
3. Seeks to explain why the error occurred, not shifting the blame but making sure not to sound incompetent or indifferent.
4. Promises to find a way to avoid the error in the future.
5. Promises to clean up the mess caused by the error.
6. Expresses regrets; apologizes.

13.2.9 GOODWILL MESSAGES

Some people might call it "schmoozing" or even "sucking up," but as with self-promotion there is a legitimate role for goodwill communications up, down, and across the organizational hierarchy. What's wrong with sincerely congratulating your boss on getting an award? What's wrong with sending condolences to your supervisor on the death of her father? Goodwill messages within an organization remind us that we are human beings not workplace automatons and that we do, in spite of everything, create and live in our own communities in the workplace.

13.3 BUSINESS-LETTER FORMAT

Traditionally, there have been several business-letter formats to choose from, but the block style, shown in Figures 13.10 and 13.11, is far and away the most commonly used. In the block style:

- Align all paragraphs flush against the left margin.
- Do not indent the first lines of paragraphs.
- Insert a blank line between paragraphs and other letter elements (such as the heading, inside address, and signature block).
- Unless you are using company stationery, place your address (the heading) at the top of the letter, flush left.
- If you are using company stationery, no return address is necessary. Include the date, flush left.
- Place the inside address—the address of the person to whom you are writing—flush against the left margin.
- Place the salutation (the word or phrase you use to greet the person to whom you are writing) flush against the left margin.

If you can't get the name of the recipient, use the company's name (*Dear Miracle Air Conditioning*), a department name (*Dear Human Resources Department*), or to use a person's title (*Dear Human Resources Manager*). If you are writing to a woman whose marital status or title you do not know, use *Ms.*

- Punctuate the salutation with a colon.
- The signature block includes the complimentary close (for example, *Sincerely yours,*), the signature, the typed name, and end notations:

  o Place all of the elements of the signature block on the left margin.

  o Capitalize only the first letter of the complimentary close.

  o For your signature, leave four blank lines between the complimentary close and your typed name.

- The end notations include abbreviations and indicators of enclosures, typists, and others who have been sent a copy of the letter.

  o To indicate enclosed material, use *Encl.* and, optionally, a descriptive phrase for what is enclosed (*Encl. résumé*).

  o To indicate that you have copied someone else on the letter, use *cc* followed by a colon and the name of the person.

13.4 MEMO FORMAT

Memos, short for memorandums, are internal company or organization communications. Memos are written for the same purposes as those discussed in "Types of Business Messages."

When producing written communications within any company or organization, make sure to identify and follow the particular format and style preferences of that company or organization. In most cases, memos should be brief—if possible, a single page.

Memos must be written and designed so they can be read and understood quickly. Thus, a memo's content, organization, and features should help a reader scan and skim the document for essential information. Most memos follow a fairly standard format and include features shown in Figure 13.12, such as headings, bullets, and short paragraphs that make it quickly scannable.

Date:	Fri, 6 June 2009
To:	Patrick Hughes, Project Manager
From:	Jane A. McMurrey, Chief Civil Engineer
Subject:	Downtown construction update

Patrick, the following is an update on downtown construction work:

SD42 Downtown Sioux Falls

The work downtown is complete except for sealing the joints. This work would be quick, but traffic should be alert to workers and equipment when they are present.

18th Street bridge over I229

Modifications are nearing completion on the structure over I229:

- The steps on the south half of the structure for pedestrian movements are in place.
- Bridge rail work along the sidewalk on the north half of the structure has begun and should be completed next week.
- The sidewalk on the NE corner has been removed and replaced to provide a flatter slope onto the bridge sidewalk.

Traffic should be alert for construction equipment.

Interstate 29 and Madison Street in Sioux Falls—Minnehaha County

Industrial Builders, Inc., continues building footings for the new interstate bridges over Madison Street. They have also started construction on the south-bound bridge over the Ellis and Eastern Railroad.

Runge Enterprises is continuing the grading on the new embankment for the southbound interstate lanes. Grading has started on Madison Street west of I-29.

Metro Construction has started installing storm sewers on Madison Street east of I-29 and will continue installation through next week.

FIGURE 13.12
Example memo. Headings, bullets, and short paragraphs make this memo quickly scannable.

13.5 E-MAIL STRATEGIES AND FORMAT

Here are some strategies for e-mail:

- Write a specific subject line that clearly identifies the subject and purpose of the e-mail message, but without making the subject line overly long (and thus not entirely viewable in an e-mail in-box).
- State your business right away in the first sentence.
- Segment paragraphs according to content and function.
- Keep paragraphs short.

- Put the essential information in the first several paragraphs to ensure that busy recipients read what you want them to read.
- Indicate specifically what you want the recipient to do.
- Proofread your e-mail message very carefully. People often compose e-mail in a hurry and send it without proofing. Typos are embarrassing of course, but far worse is omitting words such as *no, never, not.*
- Double-check the e-mail addresses you are sending to, especially when you are replying to someone else's e-mail. Lots of people have been embarrassed by sending out replies to others who were not supposed to see those replies.
- Store both your sent and your received e-mail messages in folders with descriptive names (such as *clients, department, projects, research*).

13.6 ORAL PRESENTATIONS

From time to time in your college career and your work career, you'll find yourself with the opportunity (or requirement) to make an oral presentation. You may need to present information to clients, coworkers, management, government officials, and even citizens. Typically, such presentations are brief, under 10 minutes. But you might need a lot more time if you must present extensive design plans or research results.

13.6.1 PREPARATION

Whichever is the case, do some careful planning and preparation well ahead of time. Oral presenters use a variety of methods to prepare and deliver their material, including:

- Traditional note-cards
- Detailed outline
- Double-spaced script using a large font

Others may use some combination of these methods. Each of these methods has its own weaknesses. You must fill in the transitions yourself as you proceed from note-card to note-card. You must know your material well to use a detailed outline effectively. Reading a script too closely can result in a dull, mechanical monotone.

Two methods that are not likely to work well at all are the memorization method and the off-the-cuff method. Memorization is tedious work; a memorized delivery is likely to break down in front of a crowd and can easily result in a dull, mechanical monotone. The off-the-cuff—or extemporaneous—method is a disaster waiting to happen unless you have plenty of experience with oral presentations and you know your material very well.

If you lack experience with oral presentations or if you seriously worry about nervousness, the script method of preparation (shown in Figure 13.13) may be your best choice. First, it ensures that you gather all your information; it enables you to rehearse your presentation, making sure of the timing and practicing the difficult parts. After you have rehearsed with your script a few times, you can glance away from it more and more. If you get an attack of the nerves or forget something, the script is still there to save you. You can combine the outline method with the script method by placing the outline items at the appropriate locations in the script (also illustrated in Figure 13.13). Doing so will enable you to find where you are if you have to refer to your script during the actual oral presentation.

13.6.2 VISUALS

As you prepare for an oral presentation, develop the visuals, which these days are likely to be slides in presentation software such as Microsoft PowerPoint (but remember that OpenOffice offers presentation software for free). Here are some ideas for visuals:

- *Title screen*. This should present your name, title, and organization; the title of your presentation; and other such identifying information.
- *Outline*. This is the sequence of points, topics, and subtopics you will be covering.
- *Diagrams or photos of key objects*. If you are discussing some mechanism in detail, include illustrations so that your audience can have two simultaneous sources of information.
- *Flowcharts of key processes and charts of organizational structures*. If you are discussing the steps in a process, or the phases in a historical event, create flowcharts to provide your audience with two simultaneous sources of information.

- *Data presented in tables, graphs, or charts.* If you have numerical data of any complexity at all, show it in tables, graphs, or charts rather than orally present it. But don't forget to explain the key points or trends in that data.
- *Key points.* If you have key points such as conclusions or recommendations to make, present them visually as a series of bullets.
- *Conclusion.* Consider showing a final slide with your name, contact information, and perhaps sources for further information.

Whichever method you use to present your visuals—presentation software, transparencies, flipcharts—don't forget to explain or at least point to those visuals. If you show a table, point to the essential data and start with some phrase like "As you can see in this table. . . ." With all types of visual presentation, don't cram too much detail on any single visual.

In recent years, there has been a backlash against presentations using presentation software—specifically, against PowerPoint-based presentations (only because it is the most widely used). The complaints are essentially as follows:

- Oral presenters cram so much information on individual slides that audiences find the slides unusable.
- Slides are often ugly, unsightly, garish conglomerations of bad formatting, with too many colors, too much animation, and too many slides.
- Oral presenters mistakenly believe that the presentation software will actually do the oral presentation; all the oral presenter has to do is click the Next button.

13.6.3 DELIVERY

As for the actual delivery of an oral presentation, here are some suggestions:

- *Make plenty of eye contact with your audience.* Avoid reading from your script and scarcely ever looking up at your audience.
- *Take your time: speak at a slightly slower pace than you would ordinarily.* When people are nervous or excited, they tend to

speak more rapidly. Your listeners will not get much out of your oral presentation if you speak too fast.

- *Practice avoiding nervous or thought-gathering verbal mannerisms such as "uh," "like," and "you know."* Instead of saying "uh" while you think of what next to say, don't say anything at all. Brief moments of silence are okay; they give your listeners time to think, reflect, absorb, and even rest.

- *Watch out also for nervous physical movement.* Busy, frantic hand gestures are distracting.

- *Plan, practice, and start with a good introduction.* Say who you are, whom you represent, what you are going to talk about, and what topics you are going to cover. In the introduction to your oral presentation, include only just enough background to enable listeners to understand and get interested in what follows. In fact, spend a lot of time rehearsing the introduction. Inexperienced oral presenters most often have trouble getting started. (See Figure 13.13 for labeled examples of an introduction.)

- *Use verbal headings.* Just as you use headings in printed documents, you should use verbal headings in oral presentations. These verbal headings should echo the same words you used in the overview in the introduction. Verbal headings tell listeners, "Okay, we've covered topic 2, and now we're moving on to topic 3, and here's how they are related." Verbal headings tell your listeners that your oral presentation is well organized and enable your listeners to follow. (See Figure 13.13 for examples of verbal headings.)

- *Carefully plan and end with a conclusion.* It's easy to ruin an otherwise great oral presentation by trailing off into a mumble because you cannot think of a logical, natural, graceful way to finish. Consider the following possibilities:

  - Summarize your main points.
  - End with one last interesting fact.
  - End with an interesting anecdote.
  - Loop back to something you mentioned earlier, especially something in the introduction if possible.
  - Offer to answer questions.

Figure 13.13 shows some excerpts from an oral report.

Hello and thank you for having me here this afternoon. My name is Christie Thomassen; I represent Alternative Fuels Research Consultants. As you know, your management has invited me here today to update you on (1) hydrogen as an alternative fuel for transportation, (2) the California Hydrogen Highway Network Action Plan, and (3) the price that the public may have to pay for this technology. With the current instability in the Middle East, interest in this technology has skyrocketed, and development has quickened.

Speaker introduces himself,

states topic and purpose,

provides overview of his presentation, and

builds some interest.

Let's begin by considering the hydrogen fuel cell. A hydrogen fuel cell is like the engine of a car. . . .

Speaker indicates first topic (verbal heading).

FIGURE 13.13A

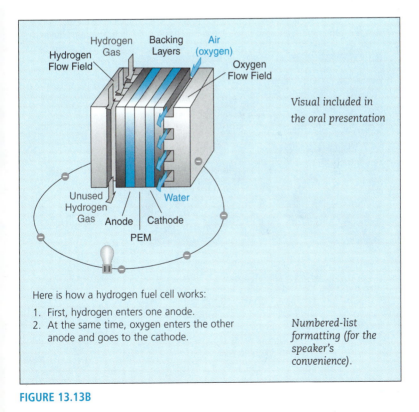

Visual included in the oral presentation

Here is how a hydrogen fuel cell works:

1. First, hydrogen enters one anode.
2. At the same time, oxygen enters the other anode and goes to the cathode.

Numbered-list formatting (for the speaker's convenience).

FIGURE 13.13B

FIGURE 13.13B *(Continued)*

3. Then, a platinum catalyst splits the hydrogen into positive ions and negative electrons.
4. The electrons produce electricity by traveling along an external circuit, while the positive ions go to the cathode.
5. Finally, the electrons, positive ions, and oxygen all come together in the cathode to produce water, which exits the fuel cell as waste product.

Using hydrogen fuel cells in vehicles offers numerous advantages

Another verbal heading, introducing the next major topic.

Let's turn our attention now to the most ambitious plan for transportation based on hydrogen fuel cells.

Another verbal heading, announcing next major topic.

Slide in presentation software

Governor Schwarzenegger's Hydrogen Highway Network Action Plan is an initiative to create a rapid transition to what he called a "hydrogen transportation economy." Already, 40 fuel cell vehicles have been displayed in demonstrations, and, for 2007 demonstrations, California is working on fuel cell hybrid buses and light-duty vehicles. Also in the plan for next year, California

FIGURE 13.13C

FIGURE 13.13C (*Continued*)

plans to have internal combustion hydrogen hybrid vehicles available for purchase. Most ambitiously, the California plan envisions "tens of thousands" of hydrogen fuel cell vehicles, not hybrids, available to the public by the year 2010. To implement this vision, the plan calls for hydrogen-fueling stations every 20 miles along California's major highways, at a cost of $75–200 million.

Script of this oral presentation

FIGURE 13.13C
Excerpts from the script for an oral presentation with outline elements. Notice the important elements in the introduction as well as the verbal headings that lead listeners from one section to the next.

INDEX

CREDITS

Image Credits

Chapter 1

Page 11: LiftPort Group/Nyein Aung;

Chapter 2

Page 51: NASA Goddard Space Flight Center/Peter Rossoni; page 72: (both figures) Courtesy of United States Air Force.; page 73: United States Air Force.; page 74: NASA Glenn Research Center; page 77: NASA/JPL-Caltech. Found at: http://mars.jpl.nasa.gov/missions/present/globalsurveyor.html.; page 77: NASA/JPL-Caltech. Found at: http://mars.jpl.nasa.gov/classroom/pdfs/MSIP-MarsActivities.pdf.

Text Credits

Chapter 1

Page 24: Northwestern University. Found at: http://www.iti.northwestern.edu/publications/technical_reports/tr11.html.; page 25: NASA. Found at: www.jpl.nasa.gov/news/press_kits/merlandings.pdf (p. 44).; page 25: Source: http://whatis.techtarget.com/definition/0,,sid9_gci816722,00.html.; page 29: Washington State Office of Superintendent of Public Instruction. Found at: energyperformance.pdf.; page 30: Adapted from Yuri Artsutanov, *To the Cosmos by Electric Train*. Young Person's Pravda (July 31, 1960) Trans. Joan Barth Urban and Roger G. Gilbertson. Found at: http://www.liftport.com/files/Artsutanov_Pravda_SE.pdf.; page 40 : Mechanical Engineering Magazine. Found at: http://www.memagazine.org/contents/current/webonly/wex30905.html.; page 41: Source: http://www.greatachievements.org/?id=3713 & http://www.greatachievements. org/?id=3711; page 42: Source: http://www.greatachievements.org/?id=3713; page 44: Mechanical Engineering Magazine. Found at: http://www.memagazine.org/contents/current/webonly/wex30905.html.

Chapter 2

Page 54: Adapted from the Superconducting Quantum Interference Devices (SQUIDs) Research Group, Bristol University Physics Department. Found at: http://homepages.nildram.co.uk/~phekda/richdawe/squid/.; page 70: NASA."Apollo 16 Traverse Comparison with Apollo 15." Found at: http://www.hq.nasa.gov/alsj/a16/A16_PressKit.pdf. page 35.

Chapter 4

Page 171: Michio Kaku, *Hyperspace : A Scientific Odyssey Through Parallel Universes, Time Warps, and the 10th Dimension*. Copyright © Bantam Doubleday Dell Publishing Group, Inc.

Chapter 5

Page 202: Alice Munro. *Friend of My Youth*. Copyright © Alfred A. Knopf (1990).; page 202: Source: http://www.greatachievements.org/?id=3717.; page 203: Al Fasoldt. Copyright © 1985, The Syracuse Newspapers. Found at; http://aroundcny.com/technofile/texts/compupoet85.html.; page 203: Source: http://www.greatachievements.org/?id=3717.; page 204: Copyright © Dresser, Inc. Found at; http://en.wikipedia.org/wiki/Edward_Teller.

Chapter 10

Page 370: *A Second Act for Illinois Coal?* Perspectives: Research and Creative Activities at SIU, Spring 2005, Southern Illinois University Carbondale. Found at: http://www.siu.edu/perspect/05_sp/coal1.html.; page 370: U.S. Department of Energy. *Gasification Technology R & D*. Found at: http://www.fossil.energy.gov/programs/powersystems/gasification.

Chapter 11

Page 406: Copyright © EERI. Found at: http://www.eeri.org/lfe/pdf/USA_ca_landers_
big_bear_1992_eeri_preliminary_report.pdf.; page 407: U.S. Department of Energy. Found
at: http://www.nrel.gov/docs/fy04osti/36273.pdf.; page 407: Copyright © EERI. Found at:
http://www.eeri.org/lfe/pdf/USA_ca_landers_big_bear_1992_eeri_preliminary_report.
pdf.; page 408: U.S. Department of Energy. Found at: http://www.nrel.gov/docs/fy04osti/
36273.pdf.; page 409: Copyright © EERI. Found at: http://www.eeri.org/lfe/pdf/USA_ca_
landers_big_bear_1992_eeri_preliminary_report.pdf.; page 428: Adapted with permission from *A Proposal to Implement a Monitoring and Control System into Virginia Tech's 2005
Solar House.* Michael Christopher, Ken Henderson, Dan Mennitt, Josh McConnell of the
Solar Decathlon Team, Mechanical Engineering Department, Virginia Tech. Found at:
http://www.writing.eng.vt.edu/design/sample_proposal.pdf (2003). Accessed February
23, 2007; page 429: Adapted with permission from *A Proposal to Implement a Monitoring
and Control System into Virginia Tech's 2005 Solar House.* Michael Christopher, Ken
Henderson, Dan Mennitt, Josh McConnell of the Solar Decathlon Team, Mechanical
Engineering Department, Virginia Tech. Found at: http://www.writing.eng.vt.edu/
design/sample_proposal.pdf (2003). Accessed February 23, 2007.; page 433: Adapted
with permission from *A Proposal to Implement a Monitoring and Control System into Virginia
Tech's 2005 Solar House.* Michael Christopher, Ken Henderson, Dan Mennitt, Josh
McConnell of the Solar Decathlon Team, Mechanical Engineering Department, Virginia
Tech. Found at: http://www.writing.eng.vt.edu/design/sample_proposal.pdf (2003).
Accessed February 23, 2007; page 438: U.S. Environmental Protection Agency and TAMS
Consultants, Inc., Hudson River PCBs Reassessment RI/FS Phase 3 Report: Feasibility
Study (December 2000).; page 448: Energetics, Inc., and the US Department of Energy.;
page 449: Energetics, Inc., and the US Department of Energy.; page 450: Energetics, Inc.,
and the US Department of Energy.; page 455: *Initial Results of the Imager for Mars
Pathfinder Windsock Experiment.* R. Sullivan Greeley, et al. Space Sciences, Cornell
University; Department of Geology, Arizona State University; Jet Propulsion Laboratory,
Lunar and Planetary Laboratory, University of Arizona. Found at: http://mars.jpl.
nasa.gov/MPF/science/lpsc98/1901.pdf.; page 456: Douglas A. Skowronek, Stephen E.
Ranft, and A. Scott Cothron. *An Evaluation of Dallas Area Hov Lanes, Year 2002.* Texas
Transportation Institute, Texas A & M University and Texas Department of
Transportation Research and Technology Implementation Office. Found at:
http://tti.tamu.edu/documents/4961-6.pdf.; page 457: Douglas A. Skowronek, Stephen E.
Ranft, and A. Scott Cothron. *An Evaluation of Dallas Area Hov Lanes, Year 2002.* Texas
Transportation Institute, Texas A & M University and Texas Department of
Transportation. Research and Technology Implementation Office. Found at:
http://tti.tamu.edu/documents/4961-6.pdf.; page 460: S. Tolba, A.H. El-Baz, and A.A.
El-Harby, *Face Recognition: A Literature Review.* International Journal of Signal Processing
vol. 2, no. 2, 2005.

Chapter 12

Page 474: P.G. Backes, J. Beahan, M.K. Long, Robert D. Steele, Bruce Bon, and Wayne
Zimmerman. *A Prototype Ground-Remote Telerobot Control System.* Found at: http://trs-
new.jpl.nasa.gov/dspace/bitstream/2014/35262/1/93-0842.pdf.; page 482: Used by permis-
sion of Stephanie Beckett.; page 488: Milne, Elam, & Evans. (2001). *Hydrogen from biomass:
State of the Art and Research Challenges* Found at: www.osti.gov/bridge/servlets/purl/792221-
p8YtTN/native/792221.pdf. Retrieved December 22, 2006.; page 503: Huaiquing Wang,
Matthew K.O. Lee, and Chen Wang. *Consumer Privacy Concerns about Internet Marketing.*
Communications of the ACM. Volume 41, Issue 3 (March 1998) pg. 63–70.; page 503:
Internet Malcontents of the World—Unite! Communications of the ACM archive. Wayne
Madsen. Volume 41, Issue 6 (June 1998) pages: 27–28 ISSN:0001-0782.

Chapter 13

Page 568: Excerpt adapted by permission from Chris Thomas, Technical Communication
student at Austin Community College.